T0136235

Birds of the Lower Colorado River Valley

BIRDS OF THE LOWER COLORADO RIVER VALLEY

Kenneth V. Rosenberg *Robert D. Ohmart*

William C. Hunter *Bertin W. Anderson*

THE UNIVERSITY OF ARIZONA PRESS   TUCSON

The University of Arizona Press

Copyright © 1991
The Arizona Board of Regents
All Rights Reserved

This book was set in 10/12 Linotype CRT Trump.
⊗ This book is printed on acid-free, archival-quality paper.
Manufactured in the United States of America.

95 94 93 92 91 90   5 4 3 2 1

Library of Congress Cataloging-in-Publication Data

Birds of the lower Colorado River Valley / Kenneth V. Rosenberg . . .
   [et al.].
      p.   cm.
   Includes bibliographical references and index.
   ISBN 0-8165-1174-8 (alk. paper)
      1. Birds—Colorado River Valley (Colo.–Mex.)   2. Birds—Arizona.
3. Birds—California.   4. Riparian ecology—Colorado River Valley
(Colo.–Mex.)   5. Riparian ecology—Arizona.   6. Riparian ecology—
California.   I. Rosenberg, Kenneth V., 1951–   .
QL683.C63B57   1990
598.29791–dc20   90-11120
                                                                      CIP

British Library Cataloguing in Publication data are available.

Contents

Figures and Tables

FIGURES

TABLES

Foreword

Some 15 years ago, I was spending much pleasurable time gathering notes on Arizona birds from other observers for use in writing a new annotated checklist. This had grown to a sizable task because an expanded body of active and talented bird observers were sending in so many records.

Then my work took on an added dimension. A great influx of bird records began to arrive from the lower Colorado River Valley, taxing my ability to keep up with them. I wondered about them, and discovered their source before long—Dr. Robert Ohmart of Arizona State University had dispatched a small band of eager young ornithologists to make a detailed survey of the valley. This survey was designed to furnish material to aid in planning the best future land and water use along the Colorado River from Davis Dam to the border with Mexico. During the next 12 years, these young men and women accumulated an absolutely astounding amount of information about the birds of the valley, running the gamut from distributional status to social behavior and food habits.

The data gathering was carefully planned and executed. Dr. Bertin Anderson was the training, supervisory, and synthesizing element for the field personnel. The nitty-gritty work fell to the field crew, which changed from time to time during the study. One who really dug in and was a spark plug from the start was Ken Rosenberg, the senior author of this book. Later, William C. (Chuck) Hunter, another of the authors, joined up and made an exceptional contribution. This is not to detract from those not sharing authorship, as their enthusiasm added greatly to the synergism that culminated in this volume. Their names appear throughout, and they are to be applauded along with the authors.

Fieldwork, even swamping out transect lines, is a pleasant enough occupation along that stretch of the Colorado River under consideration, at least in the winter months. But when the weather turns hot, as it does in April, lasting into October—that's different. The valley heat then is truly enervating and is accompanied by extremely annoying mosquitoes and "no see 'ems." The unpleasant exudate of saltcedars gets on everything, and stinking sticky mud must be traversed. One's perspective and enthusiasm are not easy to keep under such conditions. The lower Colorado River Valley group cheerfully endured all this.

I can speak with some wisdom here, as for over 16 years beginning in 1942, I was often in the field along the lower Colorado when I worked on federal wildlife refuges. I would note that Havasu Lake Refuge is now named Havasu Refuge, following a considerable diminution in acreage when the lake proper was deleted about 1960. Imperial Refuge lost its area from Martinez Lake south about the same time.

During the time I was manager of these refuges (after 1954 I was also manager of the Cabeza Prieta and Kofa Mountain Wildlife Refuges), my duties allowed me to make incidental bird observations and keep records of them. The authors asked if I would put my records at their disposal, to which I acquiesced and for which I am grateful, as it is a way for my data to be used and published, which I doubt would have taken place otherwise.

As is detailed in this book, great changes have taken place in the lower Colorado River Valley at the hands of man since the giant Hoover Dam was completed in 1936, a bare five years before I began looking at the birds along the river. The authors have ably described these momentous alterations. In my own mind, I think of the drowning of the mesquites that have stood dead in the Topock Marsh for 50 years, a testament to the durability of mesquite wood. Of course they are destined to disappear. The forest of cottonwood tops, which for nearly 30 years protruded from the surface of Lake Havasu providing sites for cormorant nests, gradually disappeared as the holding roots decayed and the once giant living trees drifted to the shores and coves. I was there when almost the entire flow of the Colorado River spread out over Topock Marsh before the Bureau of Reclamation channelized the river between Topock and Davis Dam. I saw this same flow run through what is now Martinez Lake before the channel was stabilized in its present location by deposited sands.

Once there was solitude the birder could enjoy. I recall boating the entire distance from Needles to the Bill Williams Delta without seeing another boatman except for Bob Bolam, the squatter who started a boat rental at Needles Landing (now called Havasu Landing or simply "Havasu Lake"). Lake Havasu City was then still in the future. Compare this to the present day, with thousands of people living by Lake Havasu, with speedboats and water sports enthusiasts usurping the desert quiet and crowding out birdlife.

Well, I could reminisce at considerable length and perhaps at the reader's expense, but I will close this Foreword with congratulations to the authors for turning out an excellent piece of work that will shine on any shelf of regional bird accounts. This is a historic book that will stand as an authentic, accurate, detailed, and interesting contribution to the natural history of John Wesley Powell's romantic river.

GALE MONSON
Tucson, Arizona

Preface

In June 1976, as Rosenberg prepared to move from Ithaca, New York, to Blythe, California, he was forewarned with tales of bleak, desolate landscapes, extreme heat, physical discomfort, and isolation. The only dissenting voice was that of Kenn Kaufman, then a Southwest regional editor for *American Birds*. Kenn spoke of an exciting ornithological frontier and the great potential for discovery in the lower Colorado River Valley. When Rosenberg arrived in Blythe it was 118°F, and the intrepid crew of biologists already assembled by Ohmart and Anderson were in the midst of conducting their monthly bird censuses. The very next day, another of many discoveries was made—a breeding population of Indigo Buntings at Topock Marsh near Needles. From that time on, and for years to come, Kaufman's predictions proved correct on countless occasions, with the regional list having grown by more than 75 species since 1975. Because of this work, we slowly pieced together a modern view of this poorly understood region, resulting eventually in the creation of this book.

The primary charge of the many field biologists working on the "Lower Colorado River Project" was to unravel the mysteries of ecological relationships among desert riparian wildlife and their dependencies on native vegetation and habitats. Ohmart was instrumental in encouraging the federal agencies to move forward with these studies, which started in 1972, and he oversaw all administrative and laboratory work. Anderson was responsible for organizing and supervising the large-scale field effort that generated most of the ecological data used in this and all other publications resulting from the Lower Colorado River Project.

The constant discovery and documentation of new regional records, out-of-season stragglers, and rare migrants were recognized early by Ohmart as fringe benefits that made working in this hostile environment all the more rewarding. However, it was Rosenberg who first developed a system for recording patterns not easily discerned during standard day-to-day field effort. Rosenberg's system was maintained diligently through to the study's official end in 1984. Hunter joined the project team in 1981, primarily to study status changes among bird species since the turn of the century, but he also periodically updated species accounts beyond the official end of the project. Until now, much of our information on distribution patterns of species, status changes, migration dates, and local natural history has remained unpublished. In this book, in addition to summarizing this information, we attempt to fit it within the broader scientific context, to maximize the benefit to the professional biologist as well as the hobbyist birder.

Thus, in 1976 we began a detailed file on all lower Colorado River birds, which later formed the primary source for the individual species accounts in this book. Virtually all of the 50-plus biologists who worked on the Lower Colorado River Project contributed in some way to this file. Among the earlier project biologists, Alton Higgins was quick to recognize the need to direct our efforts towards a new regional bird book. Higgins' encyclopedic recall of Arizona birdlore and his enthusiasm for building on the works of Grinnell and of Phillips, Marshall, and Monson were an inspiration from the start.

As records accumulated, they were periodically summarized for the regional reports in *American Birds*, with the enthusiastic support of Kenn Kaufman and Janet Witzeman in Arizona, and Guy McCaskie in southern California. From 1976 to 1978, this effort was directed by Rosenberg, with invaluable help from Alton Higgins and Paul Mack. In August 1978, Bret Whitney assumed the primary responsibility of compiling the records file, and his energetic pursuit of birds throughout the valley contributed greatly to our overall efforts. The role of coordinator was then passed on to a succession of dedicated field ornithologists from Sharon Goldwasser to Richard Martin, Mark Kasprzyk, Janet Jackson, Dave Krueper, and finally to Hunter, who joined Rosenberg, Ohmart, and Anderson in the completion of this book.

The many other biologists, working between 1975 and 1984, censused birds nearly every day to form our primary data base. Most made the extra effort to record and document their significant bird sightings to ensure continuity of our coverage. For this cooperation we thank Lawrence Abbott, Dave Baker, Jim Bays, John Bean, Tim Brush, Larry Clark, Ken Clough, Dan Cohan, Jeff and Kathleen (Conine) Drake, Robert Dummer, Bruce Durtsche, Bruce Edinger, Ron and Carolyn Engel-Wilson, Erik Ferry, Lee Grimm, Cathy Harris, Elizabeth Hayes, Ron Haywood, Valerie Hink, Bill Howe, Mary Hunnicutt, Alexa Jaffurs, Anne Kasprzyk, Linda LaClaire, Mike Lange, Andy Laurenzi, Diane Laush, Julie (Duff) Martin, Julie Meents, Dion Powl, Larry Pyc, Camille Romano, Gary Rosenberg, John Shipley, Dewey Shrout, Doug Wells, Ed Wickersham, Helen Wood, and Brian Woodbridge.

In addition, Tim Brush, Bill Butler, Kathleen (Conine) Drake, Robert Dummer, Ron Engel-Wilson, Deborah Finch, Alton Higgins, and Kathy Voss-Roberts, along with Hunter and Rosenberg, produced masters' theses in the Department of Zoology at Arizona State University, which focused on particular aspects of the birdlife along the lower Colorado River. Many of these same people continued their interest in Colorado River birds long after their official work was over, adding even further to our knowledge.

Besides these professional ornithologists, several local residents and many visiting birders contributed individual records, impressions, and companionship in the field. Mrs. Ione Arnold, a long-time resident of Blythe, not only provided the bulk of our information on hummingbirds in the Colorado River Valley, but as a teacher she performed an invaluable service to both the valley's children and its birdlife. One of Ione's "students" was Sue Clark, an owner of the Clark Ranch north of Blythe. Sue's unending hospitality, charm, and dedication to protecting the riverine environment and healing injured wildlife were an inspiration to all of us who had the pleasure of visiting her ranch. Most notable, perhaps, was Sue's contribution as an amateur ornithologist. She added numerous significant records to our files and provided a continuous picture of the changing birdlife on the Clark Ranch.

At the southern end of the valley, where our own fieldwork was limited, Gwen and Dick Robinson and Katherine and Ralph Irwin added many important records and a modern perspective of the birds around Yuma. At the opposite end of the valley, Chuck Lawson and Vince Mowbray ventured down from Las Vegas to scour the small portion of our study area in Nevada. Most significant was their "discovery" that Davis Dam was a magnet for wintering waterbirds.

On 20 August 1977, an important ornithological meeting, organized by Arizona raptor expert Rich Glinski, took place at the Bill Williams Delta. There, under the majestic cottonwoods, Rosenberg met with Gale Monson, Guy McCaskie, and Jon Dunn (co-author of *The Birds of Southern California*). This meeting suffused a sense of history into our present fieldwork and opened new lines of communication to other birders in both states. Among the many individuals who contributed their expertise and their friendship, we especially wish to acknowledge Steve Cardiff, David Gaines, Kimball Garrett, Kenn Kaufman, Steve Laymon, Paul Lehman, Bob McKernan, Van Remsen, Dave Stajskel, John Sterling, Doug Stotz, Scott and Linda Terrill, and Bob and Janet Witzeman. Many others added significant records while visiting the valley or taking part in Christmas Bird Counts. Among the regular visitors were Alvin and Sandy Anderson, Charles Babbitt, Bob Bradley, Troy Corman, Bix Demaree, Rich Ferguson, Alan Gast, Mary Halterman, Chuck Kangus, Reed Tollefson, and Mary Whitfield.

Finally, our effort to document the avifauna of the lower Colorado River Valley was greatly facilitated by the many federal and local government biologists and land managers with whom we frequently interacted. Most notable were the managers of the Colorado River's three

U.S. Fish and Wildlife Service national wildlife refuges who provided assistance and information over the years: Bill Belrends, Ty Berry, George Constantino, Dan Dinkler, Gerald Duncan, Terry Fears, Sean Furniss, Jim Good, Dan Ledford, Wes Martin, Will Nidecker, Les Peterson, and Greg Wolf. Other federal personnel contributing to the knowledge of the valley's birdlife included Dave Busch, Bill Butler, Herb Guenther, Ed Lundberg, Jim Rorabaugh, Phil Sharpe, and Mike Walker of the U.S. Bureau of Reclamation, and Susanna Henry of the U.S. Bureau of Land Management. State field biologists figuring prominently in our work with Colorado River birds included Gordon Gould, Ron Powell, and Melody Serena of the California Department of Fish and Game, and Todd Soderquist and Richard Todd of the Arizona Department of Game and Fish. In addition to government biologists, John and Louise Disano were instrumental in the success of the revegetation efforts described in Chapter 4.

Once the bulk of the fieldwork was completed, the task of putting together a book had still barely begun. This work was formally begun in 1980 at Arizona State University's Department of Zoology and the Center for Environmental Studies. The first species accounts were prepared as part of our reports to the U.S. Bureau of Reclamation, the principal funder of this study. In fact, the growth of the book was tied invariably to the many other scientific reports resulting from the Lower Colorado River Project. Many of the same people acknowledged in those reports helped, at least indirectly, with our present efforts. In particular, we must recognize the expertise of Ed Minch, Bill Warner, and Jeanne Weilgus, who identified thousands of insect and plant parts in hundreds of bird stomachs, resulting in one of the largest such information files ever compiled. Similarly, Melodie Carr, Stan Cunningham, and Jim Fiedler assisted in analyzing the contents of the many duck gizzards. Invaluable were Diana Gabaldon, Jack Gildar, John Murnane, Linda (Richardson) Parsons, Greg Pollock, Jake Rice, and Kurt Webb who succeeded in computerizing and analyzing the massive data sets collected in the field. Jake Rice, in particular, designed many of our most innovative analyses and was a constant source of ideas and advice. Bea Anderson, Penny Dunlop, Jane Durham, Susan Penner, and Cindy D. Zisner read and edited most of the previously published Colorado River Project material referenced in this book. Developing and editing the present work fell solely on Cindy D. Zisner, who masterfully converted our scribbled thoughts into typed text and oversaw the production of the entire manuscript.

Additionally, we would like very much to express our gratitude to the U.S. Bureau of Reclamation. The Center for Environmental Studies at Arizona State University, under the direction of Duncan T. Patten, provided invaluable logistical and financial support throughout the production of this book. Also, Barbara Beatty and other staff at the University of Arizona Press are thanked for patiently guiding us along toward what seemed an unreachable goal, at times, in the publishing of this book.

For thoughtful reviews of the entire manuscript, we are indebted to Gale Monson and two anonymous referees for The University of Arizona Press. Gale Monson also generously shared his copious field notes from years of work, as well as his impressions and insights on historical changes in this region. We also thank Jon Dunn for reviewing all the species accounts and adding a needed perspective on significant records from the California side of the river. Other helpful comments and discussions were provided by John Bates, Bill Eddleman, Peg Gallagher, Shannon Hackett, Tom Jones, Van Remsen, Gary Rosenberg, and Dave Stajskal. Mark Swan prepared the final bar graphs and some of the figures.

KENNETH V. ROSENBERG
ROBERT D. OHMART
WILLIAM C. HUNTER
BERTIN W. ANDERSON

Birds of the Lower Colorado River Valley

1 Introduction

The Colorado River has played a major role in shaping the physical and cultural history of a large portion of the western North American continent. This great river has been likened to the Nile as a source of life and intrigue in an arid landscape (Reisner 1986). Like the Amazon, the Colorado River carries the waters of melting snows, heavily laden with sediments, in a pulse of flooding and retreating. The wildness of the Colorado River has been celebrated since the first explorers gazed upon its vast canyons and attempted to navigate its treacherous rapids.

Today, however, the Colorado River symbolizes a profound human influence on the environment in the struggle to dominate and convert natural resources for human needs. The modern river now inspires politicians and lawyers, engineers and recreationists. Its recent history is a realization of the engineers' dream to "reclaim" the West by harnessing the river's power and irrigating desert lands with its waters. Politics of water use is a fundamental aspect of life in the Southwest today and is the determinant of the Colorado River's future. Much has been written on these subjects, and the present situation is well described by Phillip Fradkin (1981) in *A River No More* and by Marc Reisner (1986) in *Cadillac Desert*.

Our book tells of the Colorado River from a very different perspective. We describe the birdlife that has evolved in the river's lower valley(s), which has changed rapidly with the influence of modern human inhabitants. For such a small geographical area (roughly 1,800 km^2), the lower Colorado Valley attracts a remarkable variety of birds—400 species have been recorded with certainty through 1989. This high diversity is influenced by the proximity of wetlands and desert and the recent development of deep reservoirs and agriculture.

The remoteness of this valley and its uninviting climate have limited the attention given to it by early naturalists and modern ornithologists. Until now, no single work has compiled and summarized the available knowledge on lower Colorado River birds. In 1878, Elliot Coues published part one of his "Birds of the Colorado Valley," an eloquent, thorough account of contemporary American birdlore that briefly touched on observations he made along the Colorado River from 1864 to 1866. Joseph Grinnell's "An Account of the Mammals and Birds of the Lower

Colorado Valley" (1914) was a detailed work that serves as our primary source on the historic status of many bird species. More recently, both Arizona and southern California have been treated in regional works (Phillips et al. 1964; Monson and Phillips 1981; Garrett and Dunn 1981). However, these accounts regard the Colorado River as a relatively little-known frontier of minor significance to their region. As such, each has picked and chosen among specific Colorado River Valley records, reporting only those that can be attributed to their respective states. Similarly, the chronicle of Colorado River bird occurrences since 1948 has been divided among three regions by the National Audubon Society's journal, *American Birds*: Southwest (Arizona), Southern Pacific Coast (California), and Mountain West (southern Nevada). Needless to say, many potential patterns of bird distribution have been obscured, and the birdlife of the valley has remained poorly described.

We attempt to offer more than a mere listing of species and their status. Chapter 2 outlines the history of human use in the valley and the resultant changes in vegetation and birdlife. Chapter 3 summarizes our 12 years of research on the ecology of bird species presently inhabiting the floodplain. Chapter 4 applies the data generated from this research to our efforts to reestablish native riparian vegetation, so vital to an important subset of the avifauna. Also included in Chapter 4 are examples of the types of active conservation that will be necessary to reverse the spiral of habitat degradation in this region.

Chapter 5 offers perspectives on the seasonal patterns of bird occurrence, as well as on geographic variation exhibited by individual species. Understanding these patterns will lead to a greater appreciation of the dynamic nature of this region's birdlife. Migration, in particular, contributes greatly to the ever-changing bird species composition of the region.

For the ornithologist or birdwatcher who considers a visit to the lower Colorado River Valley, Chapter 6 describes the major, accessible bird-finding localities in each section of the valley. Brief accounts on arrival and departure of bird species throughout the year are also provided in Chapter 6.

The heart of any regional bird book lies in the species accounts. We attempt to summarize all that is known about each species in the valley in Chapter 7. New information is provided on the habitat affinities, ecological requirements, and food habits of many resident species. This information will increase the opportunities for managing and maintaining healthy populations of resident species in the future. In addition, trends in distribution and seasonal occurrence are described for transients and rare species in light of patterns found in adjacent regions and the Southwest in general. This summary, like all other regional accounts, will be incomplete and somewhat outdated before it is even published, as new patterns are discovered every year. Only through the interest that this book generates and with the diligent recording of future events will our understanding of this region's natural history ever near completion.

A GEOGRAPHICAL PERSPECTIVE

The Colorado River, as presently named on most maps, originates in Rocky Mountain National Park, Colorado, amid snow-capped mountains. It eventually merges with the Green River in Canyonlands National Park, Utah. Interestingly, the Green River is the longer of the two major tributaries and, technically, should have been considered the master stream of the Colorado River. However, because of the first of many political controversies regarding control of the Colorado River, it was the shorter Grand River that came to bear the name of the river we recognize today as the Colorado (Fradkin 1981). The Colorado River travels some 3,000 km and drops over 4,000 m in elevation before it (theoretically) empties into the Gulf of California in Mexico. Between the mountain peaks and the sea, the Colorado's journey has been a famous one. It has left behind a wonder of sculptured gorges and canyons through Colorado, Utah, and the unsurpassed Grand Canyon of Arizona. Below the Grand Canyon the river turns south toward the Gulf of California and forms the present-day boundary of Arizona with Nevada, California, and Baja California. This lower stretch of the Colorado passes through a level and rather broad valley, historically reaching its mouth in a vast delta of alluvial silt beds, marshes, and forest.

Today the Colorado River encounters a series of major man-made obstacles. In Arizona, the river first encounters Glen Canyon Dam on the Utah–Arizona border, then Hoover (Boulder) Dam below the Grand Canyon, and then a series of lesser dams to the south. Virtually every drop of Colorado River water is allocated and used by the seven states in the United States (Colorado, Wyoming, Utah, Nevada, New Mexico, Arizona, and California) and two states in Mexico (Baja California Norte and Sonora) that constitute its watershed basin (Fig. 1). The controlled flow now barely reaches the Mexican border below Yuma, Arizona, and scarcely a drop has reached the delta in three decades. However, very high flows from 1983 to 1986 allowed the river waters to reach the Gulf of California temporarily.

Although this region has been markedly altered by the placement of dams, knowledge of the natural events that shaped the floodplain is essential in understanding present-day plant and animal life. Joseph Grinnell (1914) appreciated this fact and described in detail the geography of the lower Colorado River Valley. Presently, the valley is defined as a north-south stretch of roughly 322 km from Davis Dam, near Bull-

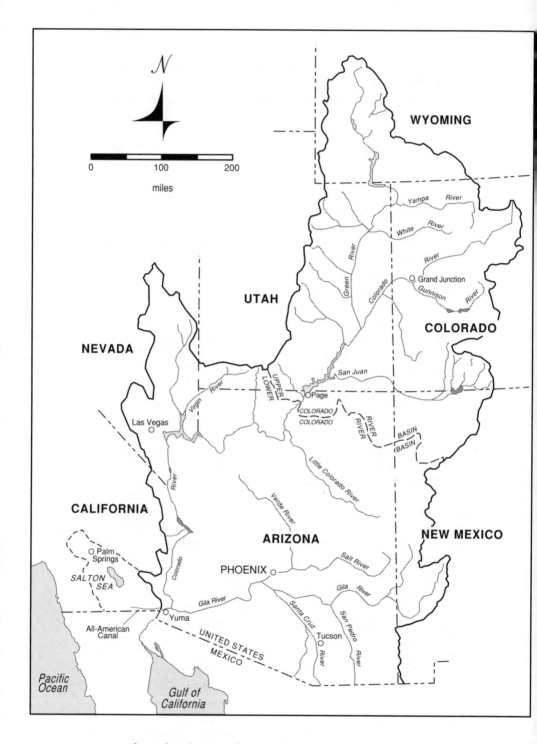

FIGURE 1. The Colorado River drainage basin.

head City, Arizona, to the Mexican boundary at San Luis, Arizona. Two noteworthy characteristics of the river were largely responsible for floodplain formation. One was the unusually large load of sediments carried by the river, which contributed both to the erosive actions of the current and to the deposition of large expanses of alluvial soil (Sykes 1937). The other characteristic of the river was its enormous fluctuation in water levels, with an annual period of flooding between 15 May and 1 July. Grinnell (1914) cited the variation in flow as 100–3,000 m^3/sec, with the peak determined largely by the annual snowmelt in the mountains. The only additional major flows into the Colorado River below Davis Dam are provided by the Bill Williams and Gila rivers, both tributaries entering from the east. The Gila River drains much of the higher portions of Arizona and southwestern New Mexico, but most water currently entering the Colorado River is irrigation return from the Wellton-Mohawk Valley.

The historic channel of the lower Colorado River, like all unmanaged rivers, was constantly on the move, except where constrained by bedrock. The river created several broad alluvial valleys by constantly eroding the bank along the outside of each meander arc and depositing new and enriched soils on the inside bank. The effect of variation in flood stages from year to year created a series of terraced "bottoms," the first bottom being leveled and replenished annually by inundation. The second and therefore higher terraces were inundated only intermittently, allowing a slower cycle of building and destruction, which consequently formed a more stable bank.

These alluvial valleys, separated by narrow bedrock formations, are marked today by human settlements and agricultural lands. The northernmost valley is the Mohave, which extends from just below Davis Dam to the head of Topock Gorge; it supports the towns of Bullhead City, Arizona, and Needles, California. The next valley south is the Chemehuevi, which once supported a thriving population of Native Americans. This valley now lies completely under Lake Havasu. At Parker, Arizona, the floodplain opens up again on the Arizona side and stretches south to Ehrenberg, Arizona, and Blythe, California. At Blythe the floodplain shifts to the California side of the river and extends to the town of Palo Verde. The only other broad valley is in the Yuma area, from the vicinity of the Gila River confluence south into Mexico.

The bedrock portions of the river, with relatively straight and fixed channels, are as well-marked today as they were to the early explorers, because they are the obvious sites of dams. The valley today is similar to the one described by Grinnell: a progression of wide floodplains alternating with narrower stretches bordered by desert hills. Today, however, this progression is punctuated by a series of large reservoirs (Lake

Table 1. Climatic Trends at Four Locations Along the Lower Colorado River*

Station	Elev. (m)	Temperature						Yearly Average Precipitation				Yearly Average Relative Humidity			
		Summer		Winter		Record Extreme**		Total cm		Days >0.25 cm		May–Sept		Oct–Apr	
		Avg. High	Avg. Low	Avg. High	Avg. Low	High	Low	May–Sep	Oct–Apr	May–Sep	Oct–Apr	0600	1800	0600	1800
Davis Dam	200	41	24	18	6	49	−6	3.10	7.54	3	10	24	15	37	26
Parker	130	41	24	21	4	51	−10	2.95	6.76	4	9	25	18	36	25
Ehrenberg	98	41	23	21	4	50	−8	3.23	5.74	4	8	34	16	39	22
Yuma Valley	37	40	21	21	3	48	−10	1.93	4.60	3	8	56	22	52	28

*Summer data are for 1 June–30 August and winter data are for 1 December–28 February. Temperatures are °C. Data are from Sellers and Hill (1974).

**Yearly average total days >32°C: Davis Dam, 165; Parker, 184; Ehrenberg, 180; Yuma Valley, 180. Yearly average total days <0°C: Davis Dam, 5; Parker, 19; Ehrenberg, 14; Yuma Valley, 18.

Mohave, Lake Havasu, and Imperial Reservoir) with fluctuating levels that inhibit the formation of a stable shoreline.

Besides the topographic features of the land and the meandering river channel, the most important physical feature of the lower Colorado region is its climate. By the time the river reaches Davis Dam, it has dropped to below 200 m in elevation and flows through one of the hottest desert regions in the world. This particular desert is often referred to as the Colorado Desert, but in reality is the edge of two larger deserts—the Sonoran to the east and the Mohave to the west and north. Most authorities list the Colorado Desert as a subdivision of the Sonoran Desert (Brown 1982). The Colorado Desert is hot and dry for much of the year, with summer temperatures exceeding 38°C for an average of 120 days each year (Table 1). Winter temperatures rarely fall below freezing for more than a few hours at a time. Precipitation is low, ranging from an annual mean of 6.5 cm in Yuma to 10 cm at Davis Dam, with a short midsummer rainy season ("monsoon") and brief, irregular winter storms. Perhaps more important, relative humidity is also very low (usually 5–20%), reflecting the effects of high temperatures and low rainfall.

Because of this arid climate, the lower Colorado River is a vital force, with its verdant floodplain valleys in sharp contrast with surrounding deserts. Plant or animal life existing in the floodplain must be tolerant of extreme heat, periodic flooding, and a constantly changing landscape.

RIPARIAN VEGETATION

Throughout the desert Southwest, a few plants are uniquely adapted to the narrow floodplains of seasonally fluctuating streams. These riparian plants exist where their roots can reach the water table and, therefore, occur only as far from the channel as the stream exerts its influence. This strip of vegetation essentially defines the floodplain of a river, and creates a marked contrast as a ribbon of green bisecting the desert uplands (Fig. 2).

The natural vegetation along the lower Colorado River was well described by Grinnell (1914). Belts of riparian vegetation stretched for many kilometers perpendicular to the river and filled the broad alluvial valleys. Numerous plant species existed in the riparian forests, but the most conspicuous and dominant were Fremont cottonwoods (Fig. 3) and Goodding willows. These occurred primarily on the first bottom and along braided channels in association with the understory shrubs seepwillow, mulefat, and occasionally other willows (e.g., coyote willow). Compared with honey mesquite and other upland trees, these plants are fast growing and relatively short-lived because they live in a frequently flooded environment. In fact, their existence was dependent

FIGURE 2. Riparian vegetation, contrasting starkly with surrounding desert uplands; Bill Williams River, October 1985. (Photo by K. V. Rosenberg.)

on the cycle of annual floods that created new silt beds for seed germination. However, these and other native riparian plants cannot tolerate prolonged inundation.

Where long-term flooding persisted, such as in backwaters, emergent marsh vegetation became established. This marsh vegetation consisted of cattails, bulrushes or tules, and, in the southern portions of the valley, common reed and giant reed (Fig. 4). On the drier sites, adjacent to the willow and cottonwood groves, arrowweed often formed dense monotypic belts. Where the floodplain of the first bottom had escaped inundation for several years, the rare screwbean mesquite grew in association with willows.

A very different type of riparian vegetation occurred on the second bottoms farther from the river. The dominant tree there was honey mesquite, which formed relatively open, monotypic woodlands. The roots of mesquite must find permanently moist soils to ensure survival, yet the tree itself apparently cannot tolerate inundations of even a few weeks. In addition to honey mesquite, several shrubs grew in dense clumps on parts of the second terrace, depending on the nature of the soil. Saltbush was the most conspicuous, occurring in sandy soils. Inkweed or pickleweed preferred denser, more alkaline soils. Both of these shrubs formed mats interspersing the mesquite woods and also lined the bases of mesas. Additional shrubs, such as quail bush and wolfberry, occurred locally as narrow belts between the first and second bottoms (Fig. 5).

The riparian vegetation of the floodplain provided a ribbon of greenery in the vast Colorado Desert and was the only forest-like vegetation for hundreds of kilometers. Each plant and animal species in this narrow belt was highly adapted to the hostile local environment in which it occurred. As will be described in the next chapter, these natural plant associations today bear little resemblance to what Grinnell described in 1914. The cycle of annual flooding has ceased, the effect of terracing is barely apparent, and the most productive land has either been inundated by reservoirs or developed for agriculture. Yet every element is still represented today, mostly as remnant islands of what were once widespread plant communities.

FIGURE 3. Cottonwood-willow forest. (Photo by B. W. Anderson.)

FIGURE 4. Extensive cattail marsh at Bill Williams Delta, July 1976. (Photo by K. V. Rosenberg.)

FIGURE 5. Honey mesquite woodland north of Ehrenberg, Arizona, April 1979, with wolfberry shrub (left) and saltbush (foreground). Note infestation of mistletoe on mesquite trees. (Photo by K. V. Rosenberg.)

FIGURE 6. Saltcedar habitat. (Photo by B. W. Anderson.)

The essential character of the existing riparian vegetation has also been altered by the appearance of the exotic tree, saltcedar (Fig. 6). Saltcedar dominates under conditions that characterize the modern valley: frequent fire, prolonged and unpredictable inundation, and increased soil salinity. The fragmentation and alteration of riparian habitats have resulted in dramatic changes in those animals that are dependent on these habitats. Although plant and animal communities have changed through time, the Colorado River continues to be essential to the vegetation and its wildlife.

2 A Recent History of the Lower Colorado River

GENERAL HISTORY OF HUMAN USE

The lower Colorado River has a long history of human use, from Native Americans, Spaniards, and Anglo-American fur trappers to modern-day corporate farmers and recreationists. Human dependence on and greed for water have brought about many conflicts with the natural environment. As refinement of river management has increased, so has modification of natural aquatic and terrestrial habitats.

Information on Native Americans' use of the fertile lower Colorado River Valley comes primarily from the diaries of Spanish explorers. These earliest written records, combined with studies by anthropologists and bioethnologists, provide insight into the cultures and habits of these riverine people, including Mohave, Cocopah, Chemehuevi, Quechen, and Hakhidhoma (Forde 1931; Castetter and Bell 1951; Forbes 1965; Crowe and Brinkeroff 1976; Kelly 1977). These people were dependent on the annual flooding of the river to provide irrigation and new fertile soils. Annual receding summer floods left a wet rich deposit of soil and organic material in which crops were planted. Honey mesquite pods were also an extremely important food because they contain carbohydrate- and protein-rich beans (Fig. 7). If annual floods were not productive and mesquite trees did not produce a heavy crop of beans, the indigenous people exerted greater pressure on native vegetation and wildlife by using fires to flush and capture rabbits, rodents, and other small mammals.

The Spaniards were primarily transitory explorers in this region who sought glory and gold and dispersed the word of God (Bolton 1936). Priests, such as Father Eusebio Kino, entered the Colorado River Valley during the late 1600s and early 1700s, bringing herds of cattle, sheep, horses, mules, and burros. Although the Spaniards persisted for many years, they did little to modify the lifestyle of the native people. Cattle and horses relish mesquite pods, which created competition with the Indians for this vital resource. Conflicts between the Spanish and Indians peaked in 1781 when the Indians attacked and burned the crude missions along the river near Yuma, killing most of the resident Spanish (Crowe and Brinkeroff 1976).

FIGURE 7. Honey mesquite pods, an important food for Native Americans and wildlife. (Photo by R. D. Ohmart.)

Fur trappers were the first Anglo-Americans to reach the lower Colorado River; they used the river illegally because at that time it was part of the territory claimed by the Mexican government. The Mexican-American War resulted in the acquisition of the lower Colorado River region by the United States Government in 1848. The 1853 Gadsden Purchase added the territory south of the Gila River which completed the present-day international boundary with Mexico.

The next 20 years brought various members of the U.S. "Army of the West" to visit and describe the Colorado River. Several of these explorers greatly contributed to our historical knowledge of both plant and animal life, including J. R. Bartlett (1854), W. H. Emory (1848; Calvin 1951), A. W. Whipple (1856; Foreman 1941), and J. C. Ives (1861). More and more people arrived on the river with the discovery of placer gold in 1862. The resultant increase in steamboat traffic placed great demands on cottonwood and willow trees for fuel. Steamboat use flourished until about 1890, after which demand for fuelwood decreased. By this time, most mature cottonwoods along the lower Colorado River had been eliminated, but large-scale natural regeneration of these groves continued after each annual flood.

John Wesley Powell was the first Anglo-American to describe to the American public both the natural beauty and potential for development of the Colorado River Basin. Powell's role in this region's history is quite ironic in that he was both a forerunner of the environmental movement and of the forces leading to water development. Powell's expedition through the Grand Canyon in 1869, along with his other explorations in the West, made him uniquely qualified to set policy for future development of the region during his tenure as director of the U.S. Geological Survey (Stegner 1953; Fradkin 1981; Reisner 1986). Unfortunately, Powell's inability to control the greediness of Western developers led to his political demise and to the eventual degradation of the Colorado River he so revered.

By the early 1900s, agricultural activities were booming around Yuma and in Imperial Valley, California. However, annual flooding, especially the disastrous floods of 1905 and 1907 that filled the then-dry Salton Sink, had devastated farming efforts. These setbacks generated public pressure on the government to control the river for human use. Water users wanted the Reclamation Service (now the U.S. Bureau of Reclamation) to assume responsibility for developing the river for power generation, water storage, and flood control. All of these needs could be met by a single solution: damming the Colorado River.

Laguna Dam, constructed in 1907, was the first water management structure to be built. When another large flood occurred in 1922, the demands of valley residents combined with pleas from thirsty Los Angeles, California, led to the authorization of Hoover Dam, completed in 1936. The stage was now set for total control over the wild and unpredictable flows of the lower Colorado River. A series of lesser dams followed, with Parker and Imperial Dams operational by 1938 and Davis Dam by 1954. These structures permanently changed the character of the lower Colorado River by ending the cycle of annual flooding that had shaped the valley over geological time.

Concomitant with water management activities, the first mitigation for habitat losses took the form of three U.S. Fish and Wildlife Service national wildlife refuges. Havasu and Imperial National Wildlife Refuges were established in 1941 to mitigate the effects of channelization and dam construction. Imperial National Wildlife Refuge was established primarily to manage and protect wildlife attracted to the backwaters formed after construction of Imperial Dam. Havasu National Wildlife Refuge is divided into three units (Topock Marsh, Topock Gorge, and Bill Williams River) to provide protection and food for waterfowl and other wildlife. Until 1965, all of Lake Havasu was in the refuge and it was called Havasu Lake National Wildlife Refuge. Topock Marsh, near Needles, was formed by man-made impoundments that backed water into a honey mesquite woodland, killing the mesquite trees. The standing "skeletons" that remained are now used as nesting platforms by herons and cormorants. Topock Gorge is one of the last remaining, relatively natural reaches of the lower Colorado River. In 1977, the lower Bill Williams River was added, as the Bill Williams Unit of the Havasu National Wildlife Refuge, to preserve native riparian and desert upland habitats. Cibola, the third national wildlife refuge, was established in 1964 to mitigate continued loss of fish and wildlife habitat involved in Bureau of Reclamation channelization projects. This refuge provided waterfowl with wintering habitats.

With floods controlled and irrigation water readily available, large stands of natural floodplain vegetation were rapidly converted to agricultural uses. Wide portions of the floodplain near Yuma, Blythe, Parker, and Needles were cleared during the 1940s and 1950s. The only large tracts of natural terrestrial vegetation remaining through the 1970s were on the five Indian reservations and three national wildlife refuges.

Native American communities followed the lead of the Anglo-Americans in bowing to economic incentives by developing their land for agriculture during the 1960s and 1970s. Much of the Mohave Valley was devoid of native vegetation by 1980, and large tracts between Parker and Ehrenberg continued to be cleared. Total agriculture production was about 120,000 ha by 1986. Most production is in alfalfa, cotton, and winter wheat, three crops that require large amounts of irrigation water.

An additional consequence of these recent practices is increased salinity of Colorado River water. Although the river and its tributaries flow over a series of salt beds that contribute relatively high natural salt loads, the condition is exacerbated by a combination of human-induced factors. First, large amounts of low-salinity water are now diverted near the headwaters of the Colorado River in the Rocky Mountains, increasing the concentration of salts reaching the lower basin. Next, millions of liters of water evaporate yearly from shallow reservoirs behind dams, concentrating salts in the stored water. Finally, as fields are flooded before planting, normal irrigation practices cause leaching (i.e., dissolution by percolation) of salts from the soil. This saline water then drains directly back into the river. High salinity has become a major problem for people as well as the wildlife in the valley today. According to international treaty, Mexico is entitled annually to 1.7 billion m^3 of water which must meet certain salinity standards (Ohmart et al. 1988). The United States must, therefore, engage in large-scale desalinization of Colorado River water before it reaches the border below Yuma. The most insidious result of this increased salinity is its contribution to the establishment and spread of exotic saltcedar in place of less salt-tolerant native plants.

In addition to agricultural development, the lure of mild year-round temperatures and an abundance of water for recreation has increased urbanization in many parts of the valley. Numerous trailer parks and various resorts now accommodate an annual migration of winter vacationers ("snowbirds") from northern states, as well as a growing number of year-round residents. Development of these communities has resulted in clearing additional riparian vegetation and filling emergent wetlands in areas where agriculture was not present.

Today the valley supports 200,000 people, mostly in Yuma, Blythe, Parker, Lake Havasu City, Needles, and Bullhead City. Numerous other small communities are dispersed throughout the agricultural valleys and the riverbank is lined with trailer resorts wherever accessible by road. For the present, the lower Colorado River has been tamed and molded. What was once a formidable barrier to human settlement now supports a thriving economy based on large-scale corporate agriculture and tourism.

Waterfowl and upland gamebird hunting are important recreational activities along the river today. Early agricultural practices initially provided ideal conditions for many game species. The abrupt edge of riparian vegetation against small isolated agricultural tracts, inefficient grain-harvesting practices, and predominance of grain crops provided invaluable wildlife habitat. Today, extensive farm tracts, "clean" farming practices, and shifts to crops such as cotton and lettuce have caused declines in once-valuable wildlife habitats. In addition, river management with dredged and riprapped channels has dried up old oxbow lakes and, in general, reduced the quality of the aquatic habitat for waterfowl.

Nevertheless, upland game and waterfowl hunting is still an important recreational activity along the river. Many southern Californians and some Arizonans spend a few days to a week engaged in these activities each year. For example, in 1984 hunters took over 350,000 gamebirds in the valley (Table 2). Mule deer, rabbits, and furbearers are also hunted heavily along the river.

VEGETATION CHANGES

Written accounts by explorers and missionaries from the 1600s to mid-1800s leave the reader with a vision of cottonwood and willow forests lining the banks of the lower Colorado River, except where bedrock formed the channel. Spring floodwaters, rich in silt and organic debris, spread water, new soil, and nutrients over the floodplain. Trees, shrubs, and vines were abundant, making travel along the river or attempts to cross it difficult. Wild grape, wolfberry, mistletoe, and other berry-producing plants provided a rich and varied food resource for wildlife. Raging floods uprooted thousands of hectares of forest vegetation in some

Table 2. 1984 Colorado River Gamebird Harvest*

Species	Total Hunters	Trips/Hunter	Total Harvest
White-winged Doves	5,000	9.3	30,000
Mourning Doves	8,000	5.4	200,000
Gambel's Quail	7,000	5.2	110,300
Ducks	1,000	5.4	10,000
Geese	1,000	5.4	400

*Data from Arizona Department of Game and Fish files.

years, but the flood-adapted riparian plants quickly reinvaded denuded areas. The river was dynamic, as was the vegetation that grew on its floodplain.

There is some controversy concerning the original number, extent, and duration of backwaters along the lower Colorado River north of the Mexican boundary. Ohmart et al. (1975) studied the dynamics of emergent wetland formation along the river, reviewing historical records and evaluating factors responsible for early marsh development. Early diaries (unpubl.) contain accounts of persons wandering for several days in wet areas choked with tules and other thick undergrowth, especially near Yuma. These accounts initially create the impression that these marshes were quite extensive and persistent.

However, historical study of the better-known and better-named backwaters suggests that these were few and of small size, with their total life span rarely more than 70 years and usually less (Ohmart et al. 1975). Another convincing perspective is offered by Grinnell (1914), one of the most prominent field biologists of his day. He led an expedition that floated the river from Needles to Yuma in 1910, before the construction of any major dams (Laguna Dam was built in 1909 but silted in within six months). Of backwaters, Grinnell (1914:72–73) stated:

> The river's habit of overflow would be expected to result in rather extensive tracts of palustrine flora. As a matter of fact, however, marshes were few and of small size. This was probably due to the rapid rate of evaporation of overflow water so that favoring conditions did not last long, and also to the rapid silting-in of such water basins as ox-bows or cut-offs. As a result there were either almost lifeless alkali depressions, or lagoons practically identical in biotic features with the main river. But in a few places there were well-defined palustrine tracts kept wet throughout the year, chiefly by seepage. They were marked by growths of tules, sedge, and salt-grass, sometimes the latter alone, and were usually surrounded by the arrowweed or willow association.

Prior to dam construction, the Colorado River was unpredictable in the amount of flooding and instream flow. This high annual fluctuation, combined with a constantly meandering channel and an arid climate, explained the short life expectancies of most backwaters. In addition, big floods carried heavy sediment loads that settled out as the floodwaters spread laterally into dense vegetation and then receded, expediting the filling and drying of many marshes. Finally, seepage or subterranean water entered the backwaters, primarily from washes that entered the valley from adjacent desert mountain ranges. In a climate where annual rainfall averages about 5 to 13 cm, seepage flows were too small to maintain permanent marshes.

During the brief heyday of steamboat traffic in the mid-1800s, virtually any tree large enough and close enough to the river was burned for fuel. However, the natural resiliency of riparian vegetation ensured that the cottonwood and willow trees would regenerate. The raging floods of 1905 and 1907, however, slowed this normally rapid regeneration. By 1910, Grinnell found the willow–cottonwood association thriving in the river bottomlands once again, and did not mention a conspicuous lasting impact from either the fuelwood cutters or prolonged floods.

Grinnell (1914:61) did, however, describe in detail the observable effects of the first dam, Laguna, on the lower Colorado River when he wrote:

> The water level has been raised conspicuously for at least ten miles, and we saw evidence of deepening of the first-bottom deposits and slowing of current for fully thirty miles, above the dam. The cottonwoods of the first-bottom had all been killed, evidently by the raising of the general surface around their trunks; and the mesquites and other vegetation of the second-bottom had all been drowned out, there thus being no trace of second-bottom conditions except for dead stalks. These were replaced by vast mudflats growing up to arrowweed. All of this change, of course, involved the birds and mammals of the area affected, in addition to the plant life.

Grinnell could not have guessed that what he had witnessed on a small scale would be an accurate prelude to the massive changes in the coming decades.

Two major events and their consequences have dictated the demise and, possibly, the eventual disappearance of the cottonwood and willow forests along the lower Colorado River. First, by 1936 Hoover Dam essentially stopped all threat of floods, except when heavy runoff from local rains brought floods from larger tributaries such as the Bill Williams River. Without floods, new rich alluvial seedbeds were no longer formed and the life-history cycle of the cottonwoods and willows was irreversibly changed. In addition, lakes that followed behind Hoover and other dams inundated thousands of hectares of riparian habitat. Of these rapid changes after Hoover Dam, Phillips et al. (1964:xv) commented:

> ... The river became a steady, clear-flowing stream that no longer annually overflowed its banks to create lagoons and silt flats. The building of this and other dams produced large lakes of clear, open water that drowned much excellent bird habitat. Most of the surviving river-bottom habitat has been cleared, leveled, and converted to farmlands. . . . Perhaps nowhere else in Arizona have the changes been more dramatic.

The second major event took place some time around 1920. An exotic species of tree, saltcedar, spread into the valley from the Gila River. Saltcedar, which has little value to native wildlife, found ecological conditions optimal for its spread and eventual dominance. In 1894, Mearns (1907) estimated that about 160,000–180,000 ha of alluvial bottomland between Fort Mohave and Fort Yuma were covered by riparian vegetation. As of 1986, total riparian vegetation comprised only about 40,000 ha, approximately 25% of the available bottomland estimated by Mearns (Anderson and Ohmart 1984; Younker and Andersen 1986). Roughly 40% of the area remaining in 1986 was covered by pure saltcedar; an additional 43% consisted of native plants mixed with saltcedar; 16.3% was covered by honey mesquite and/or native shrubs; and only 0.7% (307 ha) could be considered mature cottonwood or willow habitat (Ohmart et al. 1988).

The successful spread of saltcedar is an example of how an introduced species can optimally exploit an environment disturbed by humans, to the detriment of native vegetation. Initially, it became established in areas where native vegetation had been cleared and the land left fallow (Ohmart et al. 1977). Saltcedar has a high rate of seed production, with as many as 600,000 seeds per plant produced from April through October (Robinson 1965). This long period of seed production allows it to germinate well into fall, when most native trees are no longer producing viable seeds. Saltcedar has become dominant along the lower Colorado River by also being salt-, fire-, and flood-tolerant.

Where channelization and river-flow management have resulted in very little native plant regeneration, senescent stands of mesquite or willow have been replaced by saltcedar. In addition, soil and water salinity levels have risen dramatically in association with irrigation practices and evaporation from reservoirs. Native plants, except saltbush and quail bush, exhibit a low tolerance to saline soils. In contrast, saltcedar thrives under highly saline conditions.

Saltcedar is deciduous and, without floods, large amounts of leaf litter accumulate. Therefore, after 10 or more years fires almost become a certainty, especially during the hot and dry summer months. After a

FIGURE 8. Two years after a burn, with vigorous growth of saltcedar but no regeneration of native trees. (Photo by W. C. Hunter.)

fire, saltcedar and arrowweed quickly regenerate, whereas cottonwood and quail bush usually fail to return (Fig. 8). Thus, saltcedar will be the first to regenerate in stands of mixed vegetation, and through successive fires this species eventually displaces most native species. Currently, saltcedar is the numerically dominant tree along the entire length of the lower Colorado River.

Riparian areas, especially on Indian lands, are still being cleared for agricultural and residential developments. The last of the large continuous mesquite bosques was beginning to be cleared in 1984, with about 800 ha removed north of Ehrenberg. These activities have slowed with the recent decline in crop prices.

Channel straightening and armoring along most of the lower Colorado River was undertaken by the U.S. Bureau of Reclamation to increase the efficiency of water transport and to reduce riverbank erosion. Channel dredging (or deepening) lowered adjacent water tables, effectively draining most remaining backwaters. Cessation of floods also precluded the development of new backwaters. All of these activities have decreased the amount of vital instream organic materials used for food in aquatic habitats.

Ironically, the most recent cause of vegetation change is the same factor that was most essential to the continued health of the entire system; i.e., flooding. After 1935, the river did not overflow its prescribed channel until the summer of 1983, when controlled water releases from dams by the Bureau of Reclamation exceeded any previously recorded flows. The long duration of high flows during 1983, 1984, and again in 1986 resulted in the death of most of the remaining cottonwoods along the river. Native plant regeneration was limited by flood timing and high soil salinity, both of which favored saltcedar establishment. Even though some regeneration of cottonwoods and willows occurred, many more hectares were lost. Recent floodwaters have also covered many hectares of emergent vegetation with sediment and debris, while other marshes have been totally scoured of their vegetation. However, some marshes have benefited temporarily from inundation because emergent plants regenerated and spread quickly on new silt beds covered by shallow water. These silt beds eventually allow the reestablishment of terrestrial vegetation, most of which probably will be saltcedar.

To summarize the vegetational changes that have occurred, a floodplain that once was filled from end to end with expansive and impenetrable forests of cottonwood, willow, and mesquite has been converted, in little more than a century, to a largely treeless valley dominated by farms and towns. The relatively little remaining riparian vegetation exists in fragmented strips and islands, most being saltcedar. Sadly and ironically, some of the worst destruction, including the virtual elimination of cottonwood–willow habitats from the entire valley, came only in the last decade, a period during which our nation's commitment to conservation and research should have been strongest. The information was at hand, the consequences were obvious, but admonitions went unheeded.

ORNITHOLOGICAL HISTORY

THE ORNITHOLOGISTS

Unlike the dramatic vegetational changes described above, historical changes in the birdlife of the lower Colorado were only sporadically documented. The few ornithologists who visited the lower Colorado River Valley before dam construction limited their studies to specific geographical areas. J. G. Cooper (1869, 1870) worked in the vicinity of Fort Mohave. E. A. Mearns (1894) studied the area from Fort Yuma to the Gulf of California. H. Brown (1903), E. Coues (1866), H. S. Swarth (1914), and others added details on relative bird abundance and occur-

FIGURE 9. Wreck at edge of a whirlpool of Grinnell's expedition party on
11 March 1910 in the lower Chemehuevis Valley. (Photo courtesy of the
Museum of Vertebrate Zoology, University of California, Berkeley.)

rence before the river was dramatically altered. The best single source
of ornithological data from this era came from Grinnell (1914), whose
party systematically observed and collected plants and animals as they
floated the river from Needles to Yuma in the spring of 1910 (Fig. 9). His
keen powers of observation and uncanny perception of ecological patterns permitted a detailed synopsis of the status, habitat preference,
and life history of most Colorado River riparian birds. Grinnell's work
outlined many of the important relationships between birds and vegetation that we recognize today. His observations provide a sound basis for
comparison with our studies.

Few ornithologists visited the Colorado River after 1910, until Gale
Monson served as refuge manager at Havasu and Imperial National
Wildlife Refuges from mid-1943 to 1962 (except from 1944 to 1945,
during World War II). Monson documented important avifaunal changes
associated with extensive habitat alterations after the construction of
dams. He was the first to observe seasonal and annual variations in bird
species occurrence and to discover the true variety of migrants and
strays that could be found in the lower Colorado River Valley. His col-
lection of birds established numerous first-occurrence records for both
California and Arizona and added greatly to the understanding of migration patterns in this region. Although many of Monson's observations

are summarized in Phillips et al. (1964), we have been fortunate to draw upon his original field notes and personal recollections in compiling our species accounts (Chapter 7).

Since 1960, most additional knowledge of species' status and distribution comes from amateur field observers who sporadically visit the Colorado River Valley. Notable were the visits of R. Guy McCaskie and Arnold Small throughout the 1960s. These men documented the range expansions and declines of several bird species on the California side of the river, particularly around Laguna Dam and Earp, California. In 1971, bird observers in Yuma conducted the first organized Audubon Christmas Bird Count (CBC; see Appendix 1) in the valley at Martinez Lake. Most information from this period is available in the seasonal reports of *Audubon Field Notes* and *American Birds*.

The most recent era of ornithological discovery began in 1972 when our research team from Arizona State University began a long-term study of plant–animal habitat relationships along the entire lower Colorado River from Davis Dam to the Mexican boundary. For nearly 12 years, up to 15 professional biologists systematically surveyed all major riparian vegetation types, as well as agricultural land, marsh, and open-water habitats. Chapter 3 summarizes these efforts. Beginning in 1976, a detailed record of all migrant and rare bird species was kept and summarized in seasonal reports to the regional editors of *American Birds*. Also, two new CBCs were established at Parker and Bill Williams Delta (Appendix 1), and the region began to receive more frequent visits by birdwatchers from throughout Arizona, California, and southern Nevada. This most recent and detailed fieldwork builds upon the earlier accounts and serves as the basis for much of the information included in this book.

THE BIRDS

Although relative increases and declines of avian populations are discussed for individual species in Chapter 7, the following are some overall patterns in avifaunal change since the mid-1800s. These changes fall into three general categories: (1) bird species that have declined with loss of riparian habitats; (2) species that have increased with conversion of these lands to farmland, reservoirs, or marshes; and (3) species that have expanded their geographic range in recent years to include the Colorado River Valley.

Perhaps the most dramatic changes are in those species dependent on tall cottonwood–willow forests that originally filled the floodplain. Summer resident insectivores, such as Yellow-billed Cuckoos, Willow Flycatchers, Vermilion Flycatchers, Bell's Vireos, Yellow Warblers, Yellow-breasted Chats, and Summer Tanagers, were once considered common or abundant elements of the bottomland willow–cottonwood associations (Grinnell 1914; Swarth 1914). All of these species have greatly declined in number with the loss of these forests, and currently occur very locally within the river valley. Cavity-nesting birds, such as Elf Owls, Northern (Gilded) Flickers, Gila Woodpeckers, and Brown-crested Flycatchers, also have declined with the loss of tall snags often present in the older groves. Interestingly, the Brown-crested Flycatcher began to spread into the valley in the 1940s, but is presently declining with the other cavity-nesting species that are dependent on cotton-wood—willow habitats.

The invasion of saltcedar into large sections of the valley has also caused changes in the birdlife. Almost none of the cavity-nesting or other species dependent on cottonwood—willow habitat (i.e., habitat specialists) mentioned above occurs in these new saltcedar habitats in the lower Colorado River Valley. In contrast, birds primarily associated with mesquite (e.g., Verdin, Black-tailed Gnatcatcher, and Crissal Thrasher), as well as species that have broad habitat affinities (i.e., habitat generalists, such as Lucy's Warblers, Northern Orioles, and Abert's Towhees), use saltcedar to some extent. Only a few species have apparently adapted well to the spread of saltcedar along the lower Colo-rado River. Blue Grosbeaks reach high densities in this shrubby tree habitat, and Mourning and White-winged Doves nest abundantly in mature stands.

Although the increase in agricultural lands in the valley has had a negative impact on the breeding avifauna, many migratory and winter-ing species use these areas extensively. Geese, ducks, hawks, Sandhill Cranes, shorebirds, Burrowing Owls, Horned Larks, pipits, meadow-larks, blackbirds, and sparrows have all benefited from food available in agricultural areas. Some species, such as Cattle Egrets, Mountain Plo-vers, Whimbrels, Sprague's Pipits, and three species of longspurs, un-doubtedly visit the valley more frequently as the open habitats they prefer become more prevalent. Some riparian species also may benefit from agricultural—riparian edges that provide food as well as adjacent shelter and nest sites. Greater Roadrunners, Gambel's Quail, White-winged and Mourning Doves, Crissal Thrashers, and Abert's Towhees are numerous in these situations. However, certain predators and brood parasites such as European Starlings, Brown-headed and Bronzed Cow-birds, and Great-tailed Grackles also have increased, and they may in-terfere with nesting of riparian birds. We have witnessed a steady change in the bird species composition of the lower Colorado River Valley, and that change continues. As certain native breeding species have declined, many new wintering species and migrants have ap-peared and are increasing.

Many changes also are evident with development of open-water and marsh habitats. In these situations loons, grebes, ducks, herons, rails,

*Table 3. Waterfowl Species Seen in 1910 (Grinnell 1914)
and 1978 (Anderson and Ohmart 1988)**

Species	Grinnell February–May 1910	Anderson and Ohmart February–May 1978
Cinnamon Teal	1	207
Green-winged Teal	100–400	977
Lesser Scaup	100–400	116
Mallard	6	213
Northern Pintail	1	149
Northern Shoveler	4	37
Red-breasted Merganser	23	31
Ruddy Duck	4	6
American Wigeon	0	620
Barrow's Goldeneye	0	92
Bufflehead	0	226
Canvasback	0	3
Common Goldeneye	0	1,743
Common Merganser	0	591
Gadwall	0	95
Greater Scaup	0	4
Redhead	0	56
Ring-necked Duck	0	68
Surf Scoter	0	4
Total number observed	239–839	5,238
Total species observed	8	19

*Table from Ohmart et al. (1988).

shorebirds, gulls, and marsh-nesting passerines have benefited. Grinnell (1914:72–73) commented on the paucity of waterbirds in 1910: ". . . The little open water sometimes attracted a few transient ducks and mudhens, but so far as known no water bird outside of the Ardeidae remain to breed anywhere along the Colorado River" (Table 3). Among the many waterbirds occupying these habitats today is a race of the Clapper Rail termed the Yuma Clapper Rail, a federal endangered species. This rail spread north from the delta as productive marsh habitats developed behind impoundments and along irrigation canals where water levels were relatively stable (Ohmart and Smith 1973). Presently, the Yuma Clapper Rail population ranges between 700 and 1,000 individuals and occurs north to Topock Marsh. Similarly, Black

Rails were first reported in the marshes around Imperial Dam in 1969, and now number about 200 birds (Repking and Ohmart 1977).

The almost annual occurrence of rare ducks (e.g., scoters and Old-squaw), jaegers, Sabine's Gulls, and other typically oceanic species is associated with the formation of large lakes and deep channels now found along the lower Colorado River. Waterbirds dispersing from the Gulf of California, such as Blue-footed and Brown Boobies, Brown Pelicans, and Magnificent Frigatebirds, also are attracted to these large bodies of water. Again we are witnessing an ecological trade-off, with declines of native, riparian-breeding birds being offset by the establishment and expansion of species associated with open-water, marsh, and agricultural habitats.

In addition to locally induced changes in species composition, several birds have undergone, and still are undergoing, continental-wide range expansions. Most of these have spread into the Colorado River Valley from population centers to the east or south. A chronicle of these events begins with Grinnell discovering House Sparrows at the Needles railroad station in 1910. Hooded Orioles were already common locally around date palms planted in the previous century. Brown-crested Flycatchers were first noted at Bard, California, in 1921, and were found north to Parker in 1946. Inca Doves reached Yuma by 1942, and the first Northern Cardinal was found the following year at Parker Dam. The now-abundant European Starling was not present in the valley before 1946. By 1950, Bronzed Cowbirds followed the expansion of their preferred host, the Hooded Oriole. Anna's Hummingbird is the most successful species spreading from the west, reaching the lower Colorado River Valley by the 1940s. Perhaps the most famous avian range expander, the Cattle Egret, found the Colorado River by 1970, and probably is breeding now. The most recent invader, the Indigo Bunting, was first noted in 1974 and was widespread in small numbers by 1977.

The large diversity of bird species now found along the lower Colorado River is primarily a result of changes undertaken to "modernize" the river. Although some species not found in Grinnell's day are now common or increasing, immediate action needs to be taken to preserve the valley's original avifauna associated with pristine riparian habitats. Our attempts to study the entire lower Colorado River avifauna and recommendations for improved management of declining bird species are discussed in the next two chapters.

3   Current Research

As described in the previous chapter, the lower Colorado River Valley has experienced major changes in habitat and, ultimately, in bird species composition. Today, river flow is strictly controlled and managed for human benefit. Conflicts between human and wildlife needs are sharply evident. Those responsible for the future habitats of the Colorado River Valley include several state and federal agencies, Native American tribal councils, and many private landowners. The U.S. Bureau of Reclamation manages the availability of Colorado River water and simultaneously provides flood control for valley residents. Western water and power corporations, along with the Bureau of Reclamation, built and operate the seven dams that have so radically altered the lower Colorado River. In contrast, other government agencies such as the U.S. Fish and Wildlife Service, the Arizona Department of Game and Fish, and the California Department of Fish and Game are charged with management and conservation of wildlife that is affected by intensive river management.

If some of the floodplain's integrity as a natural system were to be conserved, a long-term study of wildlife–habitat relationships was necessary. Perhaps ironically, it was the Bureau of Reclamation, given a broad scope of management responsibility under the 1971 National Environmental Policy Act, that funded such a study in 1972 that spanned the next 12 years. With the leadership of Regional Director Edward Lundberg, and spurred by the new Regional Environmental Office under F. Phillip Sharpe, this agency provided funds for a comprehensive assessment of the riparian flora and fauna, entitled, "Vegetation Management for the Enhancement of Wildlife Along the Lower Colorado River" (Anderson and Ohmart 1984). Of primary interest to the Bureau was the value to wildlife of exotic saltcedar compared with native vegetation. There was a strong desire to manage these extensive tracts of saltcedar because of water loss via evapotranspiration, and because the dense saltcedar that now dominates much of the floodplain could exacerbate flood conditions by blocking high-water flows. The Bureau reasoned that if these saltcedar habitats could be modified in some ideal configuration to expedite flow and reduce water loss, their goals could be achieved while actually enhancing the wildlife of the valley.

Under the principal direction of Robert D. Ohmart at Arizona State University, and the field leadership of Bertin W. Anderson, research was underway by 1973. The Colorado River Project's first goal was to develop a solid information base from which to predict wildlife–habitat relationships in these managed riparian environments. Also, the discovery of ecological patterns and habitat management needs of wildlife would be of great value to resource managing agencies and the scientific community.

Twenty-three habitat types were defined based on the presence of six dominant plant species and the vertical configuration of vegetation within each stand. The first phase of study was a five-year inventory to quantify abundance and variety of birds and other wildlife species associated with each recognized habitat type. Birds became the central focus because they respond quickly to environmental changes and are excellent indicators of ecological health and change. Also, compared with all other vertebrate groups, birds are easiest to monitor in determining ecological relationships. Reestablishing native plant communities that would attract wildlife was the primary goal for the second phase of study, which involved intensive efforts to discover the best configurations, mixtures, and methods of cultivating native trees (Chapter 4).

To conduct the inventory, approximately 110 permanent census routes between Davis Dam and the Mexican border were established by cutting narrow (1-m-wide) trails (transects), about 0.8 to 1.6 km long, through stands of relatively homogeneous vegetation. Information on tree species composition, tree densities, foliage volumes, small mammal numbers, and other data were collected along these transects. Birds were censused on each transect three times monthly throughout each year by a team of trained biologists. This monitoring of bird populations in all riparian habitat types was nearly continuous from November 1974 through March 1984. However, after August 1979, the geographic scope of the inventory was restricted to the Cibola National Wildlife Refuge and Parker Valley. For most of this period detailed observations were made of avian foraging behavior, data were collected on the food resource base (insects, seeds, etc.), and almost 5,600 bird specimens were collected for dietary analysis.

Concurrent with the floodplain riparian study, bird and small-mammal inventories in normally dry riparian (desert washes) and nonriparian habitats were conducted at various intervals during the 10-year period. Desert washes at the periphery of the valley were censused along eight routes in 1977 and 1978. On a larger scale, 25 census routes were established in 5 agricultural areas in or near the Colorado River Valley between June 1978 and November 1980. Efforts in agricultural areas were concentrated on the Fort Mohave and Colorado River Indian reservations, Cibola National Wildlife Refuge, Wellton-Mohawk Irrigation District in the lower Gila River Valley, and Imperial–Coachella Valley in California (Anderson and Ohmart 1982a). The objectives of this work were to determine what value agricultural lands have for wildlife, especially birds, and to assess overall use of cultivated crops and features such as weedy margins, canals, flooded fields, feedlots, and rural residences.

Similarly, the value of marsh and open-water habitats that have formed largely since the placement of dams was assessed. About 30 marshes were inventoried monthly from May 1976 to July 1978. Marshes included artificial backwater levees below Parker, extensive marshes at Topock and Imperial Dam, and scattered natural marshes along backwater lakes and in flooded riparian habitats. In addition, up to 15 river and lake segments were censused between January 1977 and March 1981, with 5 segments monitored until November 1982. These included both relatively pristine and channelized sections of river, a backwater, lower Lake Mohave, and the north and south ends of Lake Havasu.

GENERAL HABITAT ASSOCIATIONS

Our most important finding was that each major habitat supports a relatively distinct assemblage of bird species. All these bird species occur in relatively predictable numbers from year to year and from season to season among habitats. However, species composition and abundance vary seasonally within each habitat according to specific foraging and nest-site selection, food resource levels, and seasonal status of the birds themselves.

COTTONWOOD–WILLOW

As noted earlier, the cottonwood–willow association was characteristic of the Colorado River Valley before settlement and even into Grinnell's day (early twentieth century). Although this habitat is now significantly reduced in area, it remains vital to a key segment of the region's birdlife. Our studies supported and confirmed what researchers had noted along other Southwestern rivers (Carothers et al. 1974); in terms of bird abundance and variety, mature cottonwood–willow forests are among the richest habitats in North America. It quickly became obvious that to enhance the valley for wildlife we must include restoration of this vegetation type.

Numerous migratory birds that either breed or winter in the Colorado River Valley prefer tall willows and cottonwoods over shorter or shrubby vegetation. These seasonal residents are largely responsible for the high diversity of birds in this habitat. Summer residents, such as Yellow-billed Cuckoos, Vermilion and Brown-crested Flycatchers, and Summer Tanagers are restricted largely to native cottonwood–willow stands; others, such as Bell's Vireos, Yellow-breasted Chats, and Northern Orioles attain their highest densities in this habitat. Three permanent resident cavity-nesters, Northern (Gilded) Flickers and Ladder-backed and Gila Woodpeckers, also reach their highest numbers throughout the year in this habitat type, as does the ground-dwelling Abert's Towhee. Today, this assemblage of birds can be found together only at the Bill Williams Delta, the last stronghold for what Grinnell noted as the most conspicuous element of the valley's avifauna. Here, on a June morning, the dawn chorus of these species, blending with the desert harmonies of Canyon Wrens and Common Poorwills and the marsh melodies of Common Yellowthroats and Song Sparrows, rivals that of even tropical forest regions.

The bird community in the Bill Williams Delta was studied intensively in 1977 and 1978. We found the breeding season peak to be in June and July, although some permanent residents began nesting as early as March and reared two or even three broods in a season. Even though temperatures were hottest in midsummer, this late nesting among the breeding resident species coincided with the annual emergence of cicadas, on which most of the birds feasted.

Outside of the breeding season, the Bill Williams cottonwood–willow groves continue to attract a changing variety of abundant bird species. As the summer breeders depart in August and September, common migrant tanagers, grosbeaks, flycatchers, vireos, and warblers take their place. By late fall, large throngs of Yellow-rumped and Orange-crowned Warblers and Ruby-crowned Kinglets move through the forest canopy consuming abundant aphids, leafhoppers, and other small insects. At the same time, in the tangled understory of dense saltcedar, dead branches, and wet leaf litter, House, Bewick's, and Marsh Wrens increase in abundance, along with a few Hermit Thrushes and Rufous-sided Towhees.

All these species remain common through the winter, unless an infrequent cold snap causes the trees to shed their leaves and insect populations to decline. In late January, cottonwoods are in bloom, and warblers and other insectivores flock to these trees to feed on nectar and the insects that are attracted to the flowers. These insectivores are joined by flocks of Lesser Goldfinches, usually mixed with Pine Siskins and a few American or Lawrence's Goldfinches, all feeding on the fresh blossoms and insects. A month later, the willows begin to bloom and the feeding flocks shift their activities accordingly. By mid-March, a bright green flush of new leaves clothes the trees in time to greet the first spring arrival of Ash-throated Flycatchers, Lucy's Warblers, and Northern Orioles. Besides this progression of seasonal residents and migrants, several uncommon wintering species are found regularly only in the tall cottonwoods and willows. These include Red-breasted Sapsucker, Brown Creeper, and Winter Wren.

Remaining tracts of willows or cottonwoods outside of the Bill Williams Delta attract portions of these species assemblages depending on tree maturity, grove size, and the amount of saltcedar or other shrubs present. Even sparse and isolated willow patches, however, are better habitats for birds than are pure saltcedar or sparse, stunted mesquite stands.

One important feature that separates mature cottonwood–willow habitats from other riparian vegetation is their structural complexity. Cottonwoods and willows typically grow to be the tallest trees in the valley, often up to 21–24 m under optimal conditions. Therefore, these mature stands provide both vertical and horizontal foliage layers that are absent in most of the other valley habitats. Such an increase in foliage diversity has been shown repeatedly in many ecological studies to support higher numbers of bird species. Along the lower Colorado River such structural complexity allows for additional cover from extreme summer temperatures that otherwise may interfere with the nesting of many late-breeding species.

As mentioned, it is the cottonwood–willow plant community that has declined most with modern river management. Consequently, birds strictly associated with this community are in serious danger of extirpation (Table 4). Due to poor planning by river management agencies, the mature grove in the Bill Williams Delta that we studied in 1977 and 1978 was subjected to two years of flooding from 1979 to 1981. Deliberate discharges into the Bill Williams River drowned 99% of the cottonwoods and about 66% of the willows in our study sites, although other parts of the Bill Williams suffered somewhat less damage (Hunter et al. 1987a). Five years after flooding, marsh vegetation became dominant and bird community composition changed accordingly. Many riparian forest-dependent species (Yellow-billed Cuckoos, Summer Tanagers, Abert's Towhees, and Northern Orioles) have declined, whereas many marsh species (Virginia Rails, Soras, Marsh Wrens, and Common Yellowthroats) have increased. At present, no large tract >20 ha, dominated by both cottonwoods and willows, exists in the entire lower Colorado River region. In particular, cottonwood has been virtually eliminated in natural stands, making our revegetation efforts, described in Chapter 4, all the more timely and critical.

Table 4. Estimated Population Changes in Seven Riparian Bird Species From 1976 to 1986 on the Lower Colorado River*, †

Species	Population Size			Percent Change		Overall Change
	1976	1984	1986	1976–1984	1984–1986	1976–1986
Yellow-billed Cuckoo	450	353	261	−22	−26	−42
Gila Woodpecker	883	690	561	−22	−19	−37
Northern (Gilded) Flicker	278	272	188	−2	−31	−32
Brown-crested Flycatcher	806	714	437	−11	−39	−46
Bell's Vireo	203	191	88	−6	−54	−57
Yellow-breasted Chat	997	970	700	−3	−28	−30
Summer Tanager	216	198	138	−8	−30	−36

*Table from Ohmart et al. (1988).
†All these species were common to abundant at the turn of the twentieth century.
Estimates are based on density data from 1976 to 1979 and total habitat size in 1976, 1984, and 1986 (see Chapter 6 *in* Ohmart et al. 1988). Bird density data from Anderson and Ohmart (1984).

HONEY MESQUITE

Honey mesquite habitats rank second to cottonwood–willow for bird abundance and variety. Honey mesquite generally dominates on the upper floodplain terraces, and is frequently the only riparian tree to form pure stands in which saltcedar is not an important component. Unlike the seasonal progression of bird species in the cottonwood–willow community, honey mesquite is dominated for much of the year by permanent resident insectivores such as Crissal Thrashers, Cactus Wrens, Verdins, and Black-tailed Gnatcatchers. In addition, Ash-throated Flycatchers reach their highest numbers in honey mesquite, although this species is usually absent from the valley in midwinter. One notable seasonal resident is the Lucy's Warbler, which arrives as the leaves and flowers begin to appear on mesquites in mid-March. Very high breeding densities can be found in optimum habitats during April and May, but each pair raises only one brood. Most Lucy's Warblers depart by mid-July. Most other birds in honey mesquite begin nesting very early (by March), but these permanent residents usually continue breeding through early summer and raise multiple broods.

Gambel's Quail maintain their highest winter and spring breeding populations in honey mesquite habitats. They feed on mesquite seeds and abundant desert annuals that occur there in most years. A few other typical desert species, such as Loggerhead Shrike and Black-throated Sparrow, are widely dispersed throughout sparse mesquite woodlands and avoid denser riparian vegetation.

Two botanical features of honey mesquite habitats attract seasonal residents, which add greatly to the overall composition of this bird community. One is mistletoe which parasitizes honey mesquite more frequently than other tree species in the area. Mistletoe clumps produce large amounts of berries that support a huge wintering population of Phainopeplas (Walsberg 1977; Anderson and Ohmart 1978). The Phainopepla is highly adapted for feeding almost exclusively on mistletoe berries during winter. This silky-flycatcher is unique among the valley's birds because it begins breeding in late winter and migrates out of the valley in May. Other frugivorous birds attracted to the mistletoe-infested mesquite woods in winter include small flocks of Cedar Waxwings, American Robins, and Western and Mountain Bluebirds. In addition, a small contingent of migrating Sage Thrashers arrives in February and March, when lone birds will take up temporary residence at individual mistletoe clumps. The Northern Mockingbird is the only permanent resident that feeds heavily on mistletoe, although Gambel's Quail, Gila Woodpeckers, and House Finches occasionally eat the berries.

The second important feature of honey mesquite habitats is the presence of several shrub species that form large patches in more open honey mesquite stands. Quail bush and saltbush are most common, providing perennial foliage for small insectivores during winter, including Verdins, gnatcatchers, and Orange-crowned Warblers. These shrubs also provide abundant food and cover for wintering granivores. Large roving flocks of White-crowned Sparrows predominate, often mixed with smaller numbers of Brewer's and Chipping Sparrows and Dark-eyed Juncos. Resident Gambel's Quail and Abert's Towhees feed and take refuge in these shrubby patches as well. Another shrub, inkweed, is common in only a few parts of the valley, but where it grows one also finds wintering Sage Sparrows (Meents et al. 1982). Large inkweed patches north of Ehrenberg and east of Poston, Arizona, are important wintering areas for this uncommon sparrow. The Parker CBC (Appendix 1) recorded a national high count for this species in 1978 and 1983.

HONEY MESQUITE–SALTCEDAR MIX

A mixture of honey mesquite and saltcedar occurs rather locally in the vicinity of Cibola National Wildlife Refuge and on the Fort Mohave Indian Reservation. Unfortunately, the floods of 1983 greatly reduced the honey mesquite component of this community near Cibola. This mixed-tree community supports avian species not found in either pure saltcedar or pure honey mesquite stands. Saltcedar forms a dense understory in these stands and adds significantly to summer insect production. Conversely, honey mesquite offers accessible foraging sites along with a well-developed, but patchy canopy layer.

An interesting bird found in this habitat is the Bell's Vireo, which historically was most closely associated with willow-dominated habitats. Although the Bell's Vireo is now very rare along the lower Colorado River and does not occur in all honey mesquite–saltcedar stands, this vegetation type represents its most important habitat outside the willow habitats of the Bill Williams Delta and near Needles. Similarly, Yellow-breasted Chats reach their highest densities in these mixed communities along the lower Colorado River mainstem.

Both the chat and vireo were once abundant and were two of five species that Grinnell (1914) considered characteristic of the willow–cottonwood association. These two species seem to require both a dense understory and at least a moderately tall canopy layer of vegetation. A honey mesquite–saltcedar mix is apparently adequate for these two summer-visiting insectivores and illustrates the potential importance of vegetation structure rather than tree species in determining the habitat preferences of certain bird species.

SCREWBEAN MESQUITE–SALTCEDAR MIX

Along the lower Colorado River, screwbean mesquite stands are frequently mixed with saltcedar. In contrast to honey mesquite, screwbean mesquite is rarely parasitized by mistletoe and grows so dense that few understory shrubs become established. Screwbean mesquites usually grow taller than honey mesquites, and the former often mix with a few isolated cottonwoods and willows close to the riverbanks.

The bird community in screwbean mesquite habitats is composed almost entirely of permanent resident species for much of the year. The general lack of perennial foliage, fruit, or seeds makes these areas among the least attractive of the riparian habitats to winter resident warblers, sparrows, and frugivores. However, in summer the tall canopy and scattered cottonwoods attract Northern (Gilded) Flickers, Gila Woodpeckers, Ash-throated and Brown-crested Flycatchers, Lucy's Warblers, and a few Bell's Vireos, Yellow-breasted Chats, and Yellow-billed Cuckoos.

Perhaps the most conspicuous avian feature of mature screwbean mesquite–saltcedar habitats is the tremendous density of nesting White-winged and Mourning Doves. These birds may place their loosely constructed nests as close together as 1 m throughout the dense canopy, and the din of their calls at first light may be deafening. In addition, during late summer and fall, seed pods of the screwbean mesquite ripen and fall providing an abundant food source for many wildlife species including deer and coyotes. In particular, large coveys of Gambel's Quail move into these woods from other riparian and desert areas to feed heavily on these seeds.

Screwbean mesquite habitats have increased in area with the stabilization and channelization of the river. Grinnell (1914) found screwbean mesquite to be rare and that it was concentrated primarily where the riverbed was very old or where backwaters had formed. At present, as decadent cottonwoods and willows are not naturally regenerated, screwbean mesquite in association with saltcedar is becoming more prevalent. Countering this trend, however, is the very recent clearing and burning of important screwbean mesquite stands, particularly on the California side of the river north of Blythe.

ARROWWEED

Arrowweed, another riparian vegetation type we recognized, is a 1.8- to 2-m-tall shrub frequently occurring in monotypic stands. The single, vertical stems grow so close together that stands are almost impenetrable. Large monotypic stands of arrowweed support few birds and attract only a few ground-foraging residents such as Mourning Doves, Gambel's Quail, and Abert's Towhees, as well as a few Verdins, Blue Grosbeaks, and wintering sparrows. Other species will feed in arrowweed if trees are available nearby.

Interestingly, Grinnell (1914) commented on the extensive tracts of arrowweed that formed the perimeters of many willow groves and that historically were the first plants to colonize the recently created silt beds and shoals. At that time, the only resident bird reaching peak abundance in arrowweed was the desert race of the Song Sparrow. Today, resident Song Sparrows are rarely found outside of marshes except in partially flooded willows or saltcedar, and during our intensive inventories this species was virtually never recorded in arrowweed.

SALTCEDAR

Monocultures of exotic saltcedar always support the lowest density and variety of bird species of any riparian habitat along the lower Colorado River except arrowweed. Most saltcedar stands are of short stature (<4.5 m tall) and are very dense, providing few choices for nesting or feeding sites. Bird species that occur in these habitats are generally permanent resident ground foragers or small insectivores. Cavity-nesting woodpeckers and flycatchers, as well as frugivores, are virtually absent in saltcedar habitats. Among the valley's summer residents, only Lucy's Warblers and Blue Grosbeaks do not seem to avoid pure saltcedar habitat. This habitat is largely devoid of birds in winter.

Several factors probably contribute to the scarcity of birds in saltcedar along the lower Colorado River. Although insects are often abundant during summer, the trees produce a sticky and salty exudate that may inhibit birds from foraging efficiently in the dense foliage. The absence of these birds from saltcedar along the lower Colorado River may be more complex than the mere inability to cope with the sticky exudate, because many insectivores nest and feed in saltcedar along the Colorado River within the Grand Canyon and in other river valleys farther east (see Hunter et al. 1988; Brown and Trosset 1989). During this study we observed that as summer temperatures become more severe from east to west across Southwestern deserts, certain migratory late-breeding birds become more specialized in their use of tall-statured, multilayered cottonwood–willow habitats (Hunter et al. 1987b, 1988). Perhaps the shrubby saltcedar cannot buffer the extreme summer heat, whereas the more moderate summer environment farther east allows a greater flexibility in bird use of lower-statured habitats. The extent to which each species uses saltcedar appears to be related directly to climate and the species' life-history characteristics, particularly residency status and breeding phenology (Hunter 1988). The interaction between geographical variation in environmental factors and life history in determining a species' use of available habitats is not a new concept and actually was described originally by Grinnell (1917a, b, 1927; James et al. 1984).

Notable exceptions to the above generalizations are occasional saltcedar stands, spared from fire long enough to attain heights >8–10 m, and tall stands of athel tamarisk. Athel tamarisk is a different saltcedar species that only reproduces vegetatively in the lower Colorado River Valley, and thus only occurs where it was purposefully planted. Although rare, mature athel tamarisk groves can nearly equal native vegetation in their value to some breeding birds. White-winged and Mourning Doves nest there abundantly, and these areas attract good numbers of such uncommon summer residents as Black-chinned Hummingbirds, Yellow-breasted Chats, and Summer Tanagers.

Ironically, after a saltcedar stand burns it is also temporarily more attractive to birds (Anderson and Ohmart 1984; Higgins unpubl. data). The opening of the canopy and presence of numerous snags, as well as emergence of abundant summer insects (such as the cicada), attract many birds in high densities that are not normally found in saltcedar.

These birds include aerial-foraging Western Kingbirds and Lesser Night-hawks, as well as the recently invading Indigo Bunting.

Saltcedar frequently occurs mixed with native riparian vegetation, especially willows and screwbean mesquite. Bird occurrence in these areas is determined generally by the dominant native tree species, and the effect of saltcedar in the understory is usually negative. Even a very few native trees or patches of native shrubs (e.g., saltbush or quail bush) scattered through what is otherwise a pure saltcedar stand will greatly enhance an area's value to birds. This was an important finding that served as the basis for many of the revegetation experiments, as discussed in the next chapter.

DESERT WASHES

Cutting through the desert hills and flats are arroyos, or washes, formed by surface-flowing water from an occasional torrential rain. These desert washes also support some riparian bird species that cannot survive in the harsh, dry uplands. Typical desert trees such as palo verde and ironwood frequently grow along these washes. These trees are taller and lusher than their conspecifics on drier upland soils. They are also parasitized by mistletoe, and wide spaces between the trees allow a lush growth of flowering shrubs, annual herbs, and grasses.

The combination of tall trees and low-growing vegetation attracts a diverse blend of desert and riparian bird species. The bird community is basically similar to that in honey mesquite habitats, including all the frugivores and wintering sparrow species. However, a few desert specialists such as Costa's Hummingbird and Black-throated Sparrow are more numerous in desert washes. A few riparian species, such as Crissal Thrasher and Abert's Towhee, also extend into these washes. On the Arizona side of the river, giant saguaro cacti grow along these desert washes to the edge of the valley. The full complement of cavity-nesting species are found where saguaros are present. This clearly illustrates how a single critical resource, such as sites for nest cavities provided by saguaros, can determine the distribution of a bird species or group of species.

MARSH

As we noted earlier, marshes and other aquatic habitats became important components of the valley's wildlife habitats after the construction of dams. Like terrestrial riparian habitats, present-day marshes vary in plant species composition and structure, as well as in their proximity to open water or trees. In general, marshes dominated by dense cattails or bulrushes support large numbers of breeding insectivores, rails, Least Bitterns, and other waders. Most of these species can be found in almost any marshy situation along the river. Marshes composed primarily of reeds or cane attract the fewest birds of any marsh type.

Nonbreeding birds (primarily wintering waterfowl, migratory shorebirds, and dispersing waders) prefer open marshes, especially those with exposed mud flats and sandbars. Where riparian trees are interspersed with marsh vegetation, other birds are added to the community. Trees also act as roosting or nesting places for herons and egrets. The only marshbird endemic to the lower Colorado River basin is the federally endangered Yuma Clapper Rail. Today, this rail is fairly common north to Topock Marsh. It prefers the tallest, densest cattail and/or bulrush marshes.

It was encouraging to find that artificial marshes behind backwater levees compare favorably with natural marshes as wildlife habitats. Marshes tend to evolve, through natural succession, into terrestrial types of plant communities. Without the maintenance of hydric conditions these areas lose their attractiveness to birds such as the Yuma Clapper Rail. It must be emphasized that without such management, increased channelization of the river will result in a decrease in marsh habitats because higher, swifter flows deepen the flow channel, lower the water table, and prevent the growth of emergent plants. Ironically, the marshbird community, which has enjoyed more extensive and stable habitats since historical times, is perhaps the most susceptible to immediate changes from current water-use policies.

OPEN WATER

Our surveys of open-water habitats confirmed the increased value of the lower Colorado River to waterbirds since river management began. Grinnell (1914) recorded only a few ducks and coots along the entire river in 1910. Today, at least 10 species of waterfowl, as well as American Coots and several species of grebes, can be considered common to abundant along the river in winter (Table 3). Waterbirds typically associated with oceanic or other deep-water habitats have probably benefited most. We have recorded loons, grebes (Western, Clark's, and Eared), goldeneyes, Buffleheads, mergansers, and gulls (Ring-billed and sometimes California), primarily on large lakes and in deep channels immediately below dams. On the other hand, many puddle ducks, Pied-billed Grebes, and American Coots are most numerous in unchannelized stretches of the river and in shallow backwaters that support emergent and submergent vegetation. Year-to-year abundances of several duck species, including Gadwall, American Wigeon, and Redhead, are largely determined by the local distribution of Sago pondweed beds. Pondweed cannot withstand swift currents, so it is adversely affected by channelization or by unusually high water levels. River segments least used by birds are channelized stretches away from dams.

In summer, very few birds are found in open-water areas. Marsh-nesting coots and grebes will venture out onto the river or lakes, and herons use riverbanks for feeding year round. However, summer is peak migration time for several species of shorebirds and terns. Postbreeding dispersers such as Brown Pelicans, boobies, or Magnificent Frigatebirds also occasionally visit the Colorado River in summer. All of these species must compete for space with the hordes of recreationists that flock to the river and lakes in summer to boat, fish, and water ski. Increased development of recreational areas along the river and increased pressure to channelize more of the river because of recent flooding will adversely affect many waterbird species.

AGRICULTURAL AREAS

Of all habitat changes experienced in the lower Colorado River Valley, extensive conversion of riparian areas to agricultural production is certainly the most dramatic. Our studies have shown that bird species using agricultural lands are usually very different from those that use riparian vegetation (Anderson and Ohmart 1982a). In fact, of all riparian residents only doves, Western Kingbirds, Yellow-rumped Warblers, and White-crowned and Brewer's Sparrows were found regularly more than 1.5 km from riparian tracts (Conine et al. 1978). Only the Western Kingbird breeds in both agricultural and riparian situations. However, along riparian–agricultural edges some riparian bird populations appear to benefit from the combination of increased food resources (from cultivated crops) and escape and nesting cover provided primarily by trees and shrubs. In particular, Greater Roadrunners, doves, Gambel's Quail, Crissal Thrashers, Abert's Towhees, and wintering sparrows and warblers were found in high densities along these edges.

Nearly all bird species using agricultural lands to any extent are migratory, and most of these only stay through the winter months. Among the few permanent resident, breeding species are Western Meadowlarks, Horned Larks, Killdeer, and Burrowing Owls. In addition, marsh-nesting Red-winged and Yellow-headed Blackbirds rely heavily on agricultural fields for food, and occasionally will establish breeding colonies in marshy canals or even in tall alfalfa fields. Where human residences provide tall trees and other cultivated plants, species such as Inca Dove, Western Kingbird, European Starling, Great-tailed Grackle, and House Sparrow are added to complete the agricultural breeding bird community.

During nonbreeding seasons the number of species in agricultural areas can be quite large. The number and variety of birds are definitely greatest where weedy margins and earthen canals are interspersed with cultivated fields. Margins may attract large flocks of wintering sparrows and are favorite feeding areas for Say's Phoebes, Loggerhead

Shrikes, and American Kestrels. Irrigation canals are used by grebes, cormorants, ducks, and herons. Bitterns, Green-backed Herons, rails, and Marsh Wrens may be found when marsh vegetation becomes established in canals.

Among the cultivated crops, alfalfa is most attractive to a variety of birds in winter. Western Meadowlarks and Savannah Sparrows are abundant in the taller alfalfa fields, and shorter alfalfa fields attract large flocks of American Pipits. Northern Harriers often hunt over these fields, and concentrations of geese or Sandhill Cranes may feed there as well. In contrast, dry plowed fields, cotton, and various truck crops (lettuce, onions, etc.) consistently support the fewest birds. However, large flocks of Horned Larks and Mountain Plovers occasionally may be attracted to dry plowed fields.

When fields are irrigated a new dimension is added to the agricultural landscape. A plowed and flooded field is the best place to find concentrations of migratory shorebirds, especially in late summer (see Ohmart et al. 1985). Here too, flocks of White-faced Ibis, Cattle Egrets, puddle ducks, gulls, pipits, and blackbirds congregate to feed on insects flushed out by irrigation water. In addition, doves, starlings, and blackbirds concentrate heavily at feedlots or sheep pastures. Finally, recently harvested grain fields may attract geese, cranes, doves, blackbirds, sparrows, and House Finches.

The abundant foods provided by agricultural habitats certainly benefit the wide variety of birds that uses them opportunistically. They only overwinter in these habitats so they may attain very high densities. If survival of these overwintering species has been higher in the past few years, then their population dynamics may be changing, with effects perhaps evident on their northern breeding grounds. However, species using agricultural areas may face serious problems involving massive spraying of pesticides as well as the constant push to line all canals with concrete and remove all weedy vegetation (Anderson and Ohmart 1982a). The future for many riparian bird species in agricultural valleys is also not optimistic. Unless a mosaic of native habitats is left among the developed areas, and the removal of remaining tall trees and weedy margins is discouraged, dramatic declines and extirpations of the original riparian avifauna will continue.

UNRAVELING ECOLOGICAL PUZZLES

Although the primary focus of our research was to delineate the relationships between various bird species and their habitats, our results were also important in addressing several major topics of ecological study. The large scope and extended time frame of our data base allowed a more complete evaluation of avian populations and community ecology than previous short-term studies on small plots. In particular, we wished to relate our empirical data on riparian birds in the Colorado River Valley to theoretical ideas regarding population regulation, habitat selection, and community organization. Our studies were conducted when several major controversies were brewing among ecologists, most notably whether interspecific competition for limited resources was a dominant factor influencing bird populations. These controversies, in part, influenced the interpretation of our results. Much of this work has already been published; this is a summary of our major findings.

POPULATIONS

As noted frequently throughout the individual species accounts (Chapter 7), most bird species undergo marked fluctuations in numbers from season to season and year to year. In general, population responses are different among groups of species depending on if they are migratory or on the types of foods they eat. Factors thought to regulate or limit populations also differ from group to group. Species that migrate to the lower Colorado River to breed show the most consistent population levels. Although timing of arrival and breeding densities differ among species, the peak density attained by a particular species in a given habitat is nearly identical from year to year. Length of the breeding season, number of broods reared, and relative use of various habitats are also very similar from year to year.

In contrast, migratory species that winter in this area are characterized by large annual variations in numbers. In any given year, highest numbers normally occur in November with declines noted through February. Decreases are usually not uniform among habitats and are associated with annual differences in winter severity (Table 5), average monthly temperatures, changes in plant phenology, and food abundance. These factors are summarized in a study of the wintering ecology of the Ruby-crowned Kinglet (Laurenzi et al. 1982). Similar responses for wintering frugivores and granivores are illustrated by studies of the Phainopepla (Anderson and Ohmart 1978) and the Sage Sparrow (Meents et al. 1982).

During our long-term study, we noted specifically that large numbers of kinglets, as well as Orange-crowned and Yellow-rumped Warblers, and lesser numbers of other wintering insectivores remained throughout the winter in mild years. During harsher winters, these populations declined dramatically through the season. Whether these declines represented heavy mortality or movements to other, less hostile regions could not be readily determined.

Terrill and Ohmart (1984) addressed this question with Yellow-rumped Warblers in southern Arizona and northern Mexico. They found that these birds continue to exhibit true nocturnal migratory

*Table 5. Winter Minimum Temperatures Along the
Lower Colorado River Near Poston, Arizona**

Winter	Nights With Frost (No.)	Lowest Temperature (°C)	Times < −4°C	Times < −7°C
1974–75	72	−8	14	1
1975–76	32	−8	10	2
1976–77	52	−6	18	0
1977–78	10	−2	0	0
1978–79	50	−9	9	4
1979–80	36	−5	6	0
1980–81	34	−6	9	0
1981–82	34	−6	7	0
1982–83	44	−6	6	0
1983–84	15	−4	0	0

*Poston is within the Parker Valley and is near the center of the intensive study area
where bird use of habitats was determined for many riparian species during winter.

behavior, orienting southward, through early January. Furthermore,
when birds disappeared from northern sites, increases were noted to
the south, with these apparent movements associated with relative
climatic conditions and insect availability. An important aspect of the
population ecology of many wintering insectivores, and probably other
species, is their flexibility in selecting a wintering site and their ability
to shift locations as local conditions dictate.

Permanent resident populations are influenced by both local summer
and winter environments. In general, most species attain their highest
densities in late summer and their lowest numbers in late winter. Declines through winter usually parallel those observed in wintering residents, especially among insectivorous species. The magnitude of the
decline and its distribution among habitats are dependent, in part, on
the severity of the winter. Most permanent residents begin breeding
early (by March) during times of increasing food abundance, and most
attempt to raise two or more broods in a season. Despite variations in
winter numbers, peak summer populations attained by most species
are very similar from year to year.

Among the factors thought to regulate bird populations, food supply
is often considered the most important. For some species, the relationship between population size and food abundance is clear. For example,
the abundance and distribution of Phainopeplas in the lower Colorado
River Valley is highly dependent on the presence of mistletoe fruit

(Anderson and Ohmart 1978). During the winter of 1978–79, when prolonged freezing temperatures (Table 5) killed virtually all exposed mistletoe, Phainopepla populations plummeted and remained low for the next two years. It was not until 1982 that berry production again approached pre-1978 levels; however, Phainopepla numbers did not begin to recover until 1983. Similarly, local reductions in insect availability appear to be responsible for declines in migratory bird populations during some winters. Some resident species, such as the Verdin, also seem susceptible to such resource "crunches." Others, such as the ecologically similar Black-tailed Gnatcatcher, can shift their foraging effort to more seasonally stable foods such as insects hiding in bark crevices or leaf litter.

In other cases, the limitation of food resources is less apparent. For example, food was found to be superabundant for insectivorous breeding birds in cottonwood–willow habitats (Rosenberg et al. 1982), with an annual emergence of cicadas making up the bulk of the available prey. For breeding birds, however, other factors may limit their ability to take advantage of these superabundant resources. For example, nest predation by snakes or other birds, brood parasitism by cowbirds, and climate have been shown to limit reproductive success in Black-tailed Gnatcatchers, Crissal Thrashers, and Abert's Towhees (Finch 1981, 1982, 1983a , 1984; Conine 1982). Abert's Towhees spend more time than expected in nest attendance, or merely perching, perhaps to guard against predators or parasites or to reduce their energy expenditure during extreme heat (Finch 1984). Also, the availability of nest sites may be more limiting than food supply. This has been demonstrated for Ash-throated and Brown-crested Flycatchers, which are cavity-nesting species (Brush 1983).

Another interesting suggestion is that birds may somehow regulate their own population sizes through socially mediated behaviors such as territorial defense. Stephen Fretwell (1969, 1972) proposed a set of predictions regarding species that should or should not exhibit such social regulation of populations. Essentially, two strategies could be followed by the birds. Socially regulated species would be expected to limit their breeding effort, usually to one brood, in spite of abundant resources for continued breeding. Individuals should maintain territories year round and should force their young to disperse to other, suboptimal habitats. In this way populations would seem to decrease well before the season of resource scarcity (winter); however, individuals holding territories would suffer very low mortality through the winter.

At the other extreme, some species may prolong the breeding season, raising several broods and remaining in extended family groups or non-territorial flocks through the late summer and fall. Such populations would exceed the carrying capacity of the habitat at the onset of re-source declines and, consequently, numbers would decline through the winter.

These predictions were tested by studying seven permanent resident species in the lower Colorado River Valley (Anderson et al. 1982). We found that these species exhibited a continuum between the two theoretical extremes predicted by Fretwell. Ladder-backed and Gila Woodpeckers showed all the characteristics of socially regulated species. Gambel's Quail, Verdins, and Cactus Wrens exhibited the least evidence for social regulation. Crissal Thrasher and Abert's Towhee were intermediate; they had prolonged breeding seasons but showed population declines before the onset of winter.

In the long term, each adaptation appears equally successful. These differences in population characteristics may help to explain the short-term, annual fluctuations seen in species such as Gambel's Quail and Verdin, compared with the relatively stable populations of the wood-peckers and the Black-tailed Gnatcatcher.

HABITAT SELECTION

As described earlier in this chapter, each habitat type in the lower Colorado River Valley supports a distinct assemblage of bird species. However, each species may occupy a variety of habitats in differing abundances, and these habitat associations may vary from season to season. The mechanisms by which birds select and fill available habitats have been a popular topic of ecological study (e.g., Cody 1985). We have undertaken a series of analyses to address several important unresolved issues.

Previous studies attempted to relate the composition of bird species in various habitats to easily measured features of those habitats. These included measures of the structure or configuration of the vegetation in vertical layers, usually using an index called foliage height diversity (FHD) (MacArthur et al. 1962; Cody 1978). Bird species were then arranged according to the foliage configuration of their primary habitats (James 1971; Shugart and James 1973; Anderson and Shugart 1974). Other investigators suggested that horizontal foliage distribution (patchiness) was an important habitat feature (Wiens 1974; Roth 1976). Finally, individual bird species may select habitats based on the presence of specific plant species (Balda 1969; Tomoff 1974; Franzreb and Ohmart 1978; Holmes et al. 1979). Nearly all of these studies attempted to predict the number or composition of bird species from particular habitat measures.

In our work we measured vertical and horizontal foliage diversity, density of vegetation in three vertical layers, and plant species composition of each of the study transects in an attempt to delineate which vegetation characteristics explained the habitat affinities of Southwestern riparian birds. The first analysis (Rice et al. 1983a) indicated that the vertical foliage profile was the most important variable for predicting the occurrence of a majority of riparian bird species. In general, areas with greater FHD supported a larger variety of birds, and individual species appeared to select habitats with either high or low FHD. The proportion of individual plant species, in particular cottonwood, willow, and honey mesquite, was also very important in predicting the presence of bird species, but only in combination with the FHD measure.

The interrelationship between FHD and plant species composition proved to be very complex, however. Areas with tall trees (e.g., cottonwoods or willows) have more foliage layers and, therefore, higher FHD than shorter-statured habitats. In addition, areas with mixed vegetation have higher FHD than monocultures. Therefore, it was possible for a site with many willow trees, which made up only part of the mixed vegetation, to have higher FHD and be more attractive to birds than a pure willow stand with relatively few individual trees. Our ability to predict bird species occurrence could be altered, depending on whether the absolute or proportional amount of a tree species present was considered.

In a second analysis (Rice et al. 1984), the absolute number of cottonwood, willow, and honey mesquite trees proved to be more important than any structural measure, including FHD, to the largest number of bird species. Surprisingly, if an index of habitat patchiness was considered, our ability to predict bird occurrences decreased. We believe, therefore, that most riparian birds in the lower Colorado River Valley select their habitats on the basis of the abundance of the dominant tree species present.

A few examples, however, illustrate the interrelationship between plant species and vegetation structure in the habitat selection process of particular birds. Summer Tanagers in summer, and Brown Creepers in winter, prefer areas with tall trees (≥ 10 m). Although most tall vegetation in the lower Colorado River Valley consists of cottonwoods or willows, saltcedar or other trees that attain this height may also attract these two bird species. Song Sparrows are associated with dense, low vegetation usually above water, regardless of the overall structure of the habitat or canopy species. Song Sparrows were therefore abundant in the flooded understory of closed canopied willow or cottonwood groves, as well as in open marshes and saltcedar (and, historically, in arrowweed). Similarly, the abundance of Abert's Towhees, especially in winter, was related to the overall density of the vegetation, regardless of its species composition or height (Meents et al. 1981). In contrast, Sage Sparrows were present only where inkweed was also present, regardless of the dominant vegetation type (Meents et al. 1982). Similarly, the presence of Phainopepla depended on the availability of mistletoe clumps. Therefore, Phainopeplas may occur in plant communities that may vary in vegetation structure.

Overall, using a combination of plant species composition and foliage configuration, we were able to predict the presence or absence of all bird species at a given site with 92–95% accuracy (Rice et al. 1984; also see Rice et al. 1986). Some bird species consistently were present on only a few sites, indicating high selectivity; others were distributed over nearly all study plots. In general, migratory species breeding in the Colorado River Valley showed more precise habitat selectivity than did permanent residents. During nonbreeding seasons, wintering and permanent residents showed equivalent levels of differentiation between habitats used and unused. Among permanent residents, most species were more selective in fall, winter, and early spring. Late spring and summer seasons were times of greater habitat flexibility for permanent residents.

The analyses discussed thus far have only considered the presence or absence of bird species at particular sites. We also examined relative abundances of birds in the range of habitats available, because most species occur in a variety of habitats. This measure, called habitat breadth, allowed us to evaluate more subtle changes in habitat affinities and relate these to changes in population size and to seasonal variability in vegetation characteristics and resource abundance (Rice et al. 1980).

Our strongest finding was that habitat breadths of most species narrowed during seasons of supposed resource scarcity. That is, permanent residents occupied fewer sites and were more concentrated in the most optimal sites during fall and winter. Similarly, wintering residents exhibited their broadest habitat use in fall, and habitat use narrowed through winter and the following spring. In addition, the narrowing of habitat breadth for many permanent resident species occurred before the harshest season (winter). Finally, we found that most permanent residents reached their peak abundances in the same types of habitats from season to season. That is, a particular species' habitat affinities were generally consistent throughout the year. Migratory breeders were even more consistent in their summer habitat preferences from year to year.

Temporal changes in habitat selection were closely related to the population levels previously discussed. For example, narrowing habitat breadths coincided with overall population declines for many permanent and wintering resident species. In addition, a concentration of individuals into the most optimal habitats before the onset of winter was seen most often in those populations thought to be socially regulated. Habitats used in early spring were those in which highest survival occurred the previous winter and may indicate the most preferred habitats for breeding. Additional habitats used after breeding may be suboptimal and were filled probably by young birds dispersing from natal territories. Many of these individuals disappeared the following winter; the quantity depended on the harshness of the season. Thus, after an exceptionally mild winter, individuals surviving in normally suboptimal habitats may remain to breed there, contributing to the annual variation in habitat breadth seen in many resident species. Similarly, in mild winters when seasonal resident insectivores appear in larger than normal numbers, these birds also occupy a larger range of habitats. Typically, after a maximum density is attained in the most preferred habitat additional individuals "spill over" into less and less optimal habitats as each one is filled. Thus, habitat breadth in these species is directly related to population size and, ultimately, to weather conditions and resulting resource levels. These findings support many of the predictions of Fretwell (1972) concerning bird populations in seasonal environments.

COMMUNITY ORGANIZATION

Ecological study has frequently been conducted at the community level where assemblages of plants and animals occur together in time and place. These assemblages share a set of dependencies and adaptations in a common environment that, theoretically at least, interact in some predictable way to influence each other's life histories. Joseph Grinnell was among the first ecologists to recognize the importance of "faunal associations," using the lower Colorado River Valley as an outdoor laboratory. Since that time, a large body of theory has been amassed concerning species assemblages. Birds have been a frequent subject of such studies because of their variety and high visibility compared with other animal groups.

Of particular interest to ecologists has been the study of community organization, or structure; that is, the delineation of those forces that determine the composition of an assemblage by establishing "rules" for the occurrence of various species. Such forces may be physical, such as those presented by climate and the geology of the landscape. They may also be biological, such as the influence that predators have on their prey or the subtle modes of competition among species for shared and limited resources. Competition has come to dominate the thinking of many recent avian community ecologists.

Most ecologists investigate community organization in largely indirect ways. On one scale, the overall composition of species assemblages may be compared to determine how measures, such as total bird density and number of species, vary with habitat quality and time of year. At finer levels, consideration of microhabitat use, food habits, and behavior may provide more insight into the occurrence patterns of bird species. In both kinds of studies, the degree to which species share, or overlap, in their use of habitat or resources is considered an important measure of interspecific interactions and the potential influence each species has on the presence of the other.

Our investigations of avian community organization along the Colorado River represent a synthesis of our knowledge of habitat affinities, population levels, and foraging ecology. We have attempted a few large-scale and some finer-level analyses. However, our work in this area is still incomplete and additional published results are forthcoming.

An assumption behind most indirect studies of community organization is that the patterns observed represent processes determined by biological interactions among species. The extent to which species may act independently of one another and to which random, or stochastic, processes may dominate over deterministic ones is at the crux of a major controversy in modern community ecology (Strong et al. 1984). We first addressed the role of stochastic processes by examining the turnover rates of bird species in our individual study transects over a four-year period (Rice et al. 1983b).

Overall, we found that the pattern of changes in species' status (between present and absent) at a given site was essentially random. Furthermore, this turnover rate was surprisingly high, with up to 40% of the total species list, across all transects, changing from year to year. Turnover rates were significantly higher for nonbreeding seasonal residents than for species that bred in riparian habitats. Breeding species showed the highest turnover rate between early and late summer (perhaps relating to postbreeding activities) and their lowest rate between winter and the following spring. These changes in species composition were not consistently related to the vegetation characteristics of any particular habitat type. In addition, estimates of population densities varied considerably at sites where particular species were consistently present from year to year.

Summer bird communities proved to be much more stable than communities during other seasons, exhibiting constant number of species, lower variation in density, and lower turnover rates between years. Even so, the locally breeding species exhibited turnover rates of 25–35% on specific transects. That both breeding and nonbreeding species exhibited random distributions of turnovers among transects further suggests that a large stochastic element to community composition may exist.

These results may tell us more about how ecologists study communities than about the processes of community organization. Nevertheless, the implications of these findings are potentially great. The uncertainty associated with measuring bird species composition at specific sites, even if merely an inherent bias in our sampling techniques, would be compounded in studies that did not cover broad geographic areas or long time spans. Any short-term study may isolate areas of high turnover or community stability by chance alone, and may lead to spurious conclusions regarding habitat affinities or community organization. We were able to overcome these difficulties by combining results from several transects within homogeneous habitat types.

In another broad-scale analysis, we related the community parameters of number of species, total density, and bird species diversity (BSD) to the various habitat characteristics discussed above (Anderson et al. 1983). As in the species-by-species analysis, we found that both the structure of the vegetation and foliage density were important predictors of the bird community measures, but the patterns were complex. In general, the total density of birds in an area was more closely related to habitat characteristics than was the number of bird species. Also, bird density and diversity measures were more closely related to vegetation features in the nonbreeding seasons than in summer. Habitats were most different from each other in terms of bird use in fall and winter. Breeding bird population size was best predicted by the density of the vegetation, especially in the canopy layers. A better predictor of bird diversity was FHD, especially in the nonbreeding season.

These results clearly indicate that relying on a single habitat measure (such as FHD) to evaluate its value to bird communities is unrealistic and potentially misleading. It is preferable to consider several independent aspects of plant composition and structure. These findings also suggest that bird community organization may be regulated in different ways in different seasons and for different subsets of the bird community, such as breeders and nonbreeders. In the breeding season, when bird community attributes were not well predicted by habitat features, the presence of a superabundant food supply may lessen the need for habitat discrimination by the birds.

In an earlier analysis (Anderson and Ohmart 1977), it was found that the amount of overlap in habitat use also varied among subsets of the bird community and among seasons. Smaller insectivores and ground-foraging species overlapped more in the habitats they occupied than did larger insectivores and, in particular, flycatchers. Furthermore, overlap within each of these groups was much higher in summer than in winter.

We also investigated how the seasonal availability of various food resources influenced overall community composition and, in particular, the relative proportions of permanent and seasonal residents in each major habitat type. In general, the food habits of permanent resident species were consistent among habitats and through much of the year. Bark- and ground-inhabiting arthropods, such as beetles and ants, always made up one-third or more of most species' diets. In contrast, seasonal residents fed mostly on temporarily available foods such as flying insects, seeds, or mistletoe berries which varied considerably in abundance among the habitats. Consequently, overlap in resource use between these two groups also varied seasonally and from habitat to habitat.

Dietary similarity between permanent and seasonal residents was highest in summer or late summer in all habitats. This was due largely to the residents switching temporarily to exploit seasonally abundant insects such as cicadas and grasshoppers. In winter, the overall diets of the two groups were most distinct, with few permanent residents able to exploit efficiently the abundant mistletoe berries or flying insects. Furthermore, these dietary differences were accentuated in the most seasonally variable habitats (in terms of food supply) such as cotton-wood–willow and honey mesquite. In screwbean mesquite woodland, however, very few seasonal foods are provided and the bird community is dominated for most of the year by permanent resident insectivores (except abundant nesting doves that feed in agricultural areas). The few wintering residents that occur in screwbean mesquite (e.g., wrens, flickers) are ecologically similar to the permanent residents; therefore, dietary overlap remains relatively high there, even in winter. However, the reduced seasonal productivity of food in screwbean mesquite is reflected in relatively low bird population densities, especially in winter.

Our findings suggest that many avian community attributes, including total population density and relative proportions of permanent residents, seasonal residents, and various foraging guilds, may be explained by the productivity and seasonality of specific food resources in each habitat. These may or may not be related to habitat structure or any other measurable characteristics of the vegetation. However, the observed changes in food use are probably related to the stronger habitat selection by birds in winter, rather than in summer.

Next, we consider several smaller-scale studies to help clarify some of our general conclusions. The first is a detailed study of the breeding bird community in mature cottonwood–willow forests of the Bill Williams Delta (Rosenberg 1980; Rosenberg et al. 1982). Thirteen dominant bird species were studied that varied greatly in size, bill shape, and general foraging habits. These included three woodpeckers, two fly-catchers, Yellow-billed Cuckoo, Summer Tanager, Northern Oriole, and Blue Grosbeak. However, when diets were examined we found much greater overlap than expected in morphological or behavioral differences among all but the smallest species. Eight of the 13 species fed heavily on cicadas, which emerged annually during the birds' peak breeding activities. The number of cicadas available was nearly 10 times what was needed to support the entire bird community's energy requirements. Therefore, instead of partitioning the resources to avoid interspecific competition, as would have been predicted by many theoretical works (e.g., MacArthur 1958; Cody 1974), these birds opportunistically shared in the exploitation of the most abundant foods.

In addition we noted little microhabitat segregation, with most bird species overlapping extensively in territory placement and foraging sites. The only competition observed was between the several cavity-nesting species for limited nest sites in cottonwood and willow snags (Brush 1983). In general, the spatial distributions of each species appeared to be independent of the other species present, again implying that partitioning to avoid resource-based competition was unimportant in this community.

Our second detailed study involved the population distribution and food habits of 7 specialized frugivores in honey mesquite woodland over 10 consecutive winters (for summary of minimum temperatures see Table 5; Anderson and Ohmart unpubl. data). The major diet of all species was mistletoe, which develops in clumps in fall and ripens through winter. The availability of mistletoe fruit depended largely on local weather conditions, with occasional freezing temperatures resulting in temporary reductions in fruit. During December 1978, a prolonged freeze killed most of the exposed mistletoe plants, and berry numbers did not recover for four years.

Two bird species maintained territories around the mistletoe clumps each winter. Phainopeplas migrated into the valley each fall and bred in early spring, and Northern Mockingbirds were permanent residents in the mesquite habitat, as well as in nearby towns and agricultural areas. The populations of these two species (Phainopeplas were always much more abundant) varied in parallel over much of the study period, with both demonstrating sensitivity to changes in food supply. Immediately after the freeze, in late winter 1978–79, both species declined sharply to the lowest levels seen in the 10 years of study. By 1983–84, Phaino-pepla numbers had recovered completely, but mockingbirds were still in relatively low abundance. During the winters of resource scarcity, neither species switched to alternative foods. Both became even more specialized in terms of habitats occupied and, therefore, their ecological overlap actually increased. Their territorial behavior and high level of interspecific aggression, however, helped to maintain spatial segregation at individual mistletoe clumps.

Up to five additional frugivorous species occurred sporadically in the mistletoe-infested mesquite woodlands. American Robins, Western and Mountain Bluebirds, and Cedar Waxwings invaded the valley in some winters from northern regions, occurring primarily in nomadic flocks. Sage Thrashers were somewhat more regular migrants through the area in late winter, with small numbers occasionally establishing temporary territories at mistletoe clumps. The timing and magnitude of occurrence of these five species were independent of each other and of local resource conditions or weather. Therefore, for the entire frugivore community much of the variation in species composition, as well as spatial and ecological overlap, was not explained by any measure of habitat quality, including food availability. Only during 1983–84 did the food requirements of the combined frugivore population exceed the amount of mistletoe available. This was due to the chance simultaneous occurrence of several invading species, high local Phainopepla population, and only a moderately recovered mistletoe crop.

These findings support the notion that resource limitation, and the resulting competition for food, may only occur intermittently, interspersed with periods of relative resource abundance (Wiens 1977). Even within this specialized foraging guild, a large amount of the "organization" was best explained by conditions outside our region that led to separate invasions by each species, as well as by the stochastic effects of local climate. Only for the two resident territorial species (Phainopepla and mockingbird) could interspecific aggression and local resource tracking be viewed as evidence for possible competitive interaction.

Our third example focuses on another foraging guild consisting of several small foliage-gleaning insectivores that occur in honey mesquite habitats (Hink, Anderson, and Ohmart unpubl. data). Two permanent residents, the Verdin and Black-tailed Gnatcatcher, reach peak abundance in late summer and decrease through fall and winter, as discussed earlier. Lucy's Warblers migrate to the Colorado Valley and breed along with the two permanent residents in spring, remaining only through July. In September or October, three wintering residents, the Ruby-crowned Kinglet, Orange-crowned Warbler, and Yellow-rumped Warbler, arrive and co-occur with the two permanent residents until early spring. Thus, peak populations of the permanent residents occurred during the only season (late summer) when seasonal residents were absent from the valley. In addition, total populations of all guild members combined were roughly equal at the beginning of winter and at the start of the breeding season. The proportion of permanent residents was much higher than that of seasonal residents during colder winters, whereas seasonal outnumbered permanent residents in mild years.

Species' diets and size of insects eaten changed considerably throughout the year. All species consumed large caterpillars when they became available in late spring, and then switched to tiny foliage insects such as aphids or leafhoppers in winter. Similarity in foraging behavior and diet was highest among the three species present during the breeding season (April to June), with no significant differences observed, and was lowest among the five species present in fall and winter. For breeding species, a 50-fold increase in potential insect prey biomass between winter and summer was accompanied by only, at most, a 5-fold increase in bird density.

Of particular interest was the divergence in foraging habits between the two permanent resident species, the Verdin and Black-tailed Gnatcatcher, after food availability began to decrease in late summer. Verdins fed primarily in the perennial foliage of various shrubs in winter and took significantly smaller insects than did gnatcatchers. At that same time, gnatcatchers primarily foraged on bark and ground substrates and sallied more for aerial insects. Despite these differences, disparities in prey size between winter and spring in both the Verdin and gnatcatcher were larger than the differences between the two species in winter. The kinglet and two wintering warblers also differed from each other in foraging method, substrate, and height in fall and winter, although sizes of prey eaten were similar within each season. This indicated that seasonal changes in food availability may be more important than interspecific partitioning in determining each species' pattern of resource use. The same conclusion was reached by Rotenberry (1980) after examining diet relationships of birds in Great Basin shrubsteppe habitats.

Again, these findings suggest that even within a relatively simple foraging guild, several ecological processes may be acting. The two permanent resident species may show some evidence of resource-based competition during seasons of reduced food abundance. The three wintering seasonal resident species remain ecologically distinct from the permanent residents; their population sizes, however, are directly dependent on local weather conditions and resulting food supply. During the breeding season, permanent resident species again opportunistically exploit the most abundant food, in this case caterpillars. They therefore overlap extensively with the seasonal resident Lucy's Warbler, whose breeding density is consistent from year to year and independent of the previous local winter conditions. An interesting question remains. Why are there more species of small, foliage-gleaning insectivores in the lower Colorado River Valley in winter than in summer?

Our final community-level study was of the 17 species of ducks that winter regularly in the various aquatic habitats in the valley (Anderson and Ohmart 1988). We were primarily interested in the ecological similarity among species within the various duck genera. This is the taxonomic level at which differences between species should occur, as predicted from ecological theory. We found extreme similarity in habitat use and food habits within genera. All dabbling ducks (genus *Anas*) tended to be associated with large amounts of submerged aquatic vegetation away from dams. They often occurred in mixed-species flocks and did not appear to segregate along the river. The pochards (genus *Aythya*) were found together in the same area as the dabbling ducks or near hydroelectric dams. Bufflehead and goldeneyes (genus *Bucephala*) were associated with discharge from hydroelectric dams and reservoirs, where aquatic invertebrate populations were highest. Finally, the two mergansers (genus *Mergus*) were also most common below dams.

Estimates of food resources (both animal and vegetable) suggest that no more than 0.5% of the total available food is necessary to support the entire duck population in any given year. Therefore, the various species probably do not need to partition these resources to co-occur. The overall similarity among species within genera is best explained by their common evolutionary history and resulting morphological constraints. The distribution and abundance of species along the river is probably related more to such factors as hunting pressure (predator avoidance) and the amount of frozen water north of our region, than to local resource conditions. Furthermore, the overall diversity of ducks in this community has been determined largely by human alteration of aquatic habitats, such as the creation of dams and reservoirs and the dredging and riprapping of the river. Thus, disturbance may also be an important factor in the organization of some bird communities.

GENERAL CONCLUSIONS

As we strive to simplify the patterns we observed in nature, we have found that the ecology of the lower Colorado River avifauna remains very complex, and still requires additional study. Indeed, our answers to many ecological puzzles have been superseded by additional questions. Even the best-designed, long-term study can provide only an empirical data base from which patterns can be extracted and compared. These set the stage for the next step in our scientific inquiry; the testing of specific hypotheses with controlled experiments and manipulations of species and habitats. Only through these more direct approaches can we hope to discriminate among the underlying processes that may act to structure bird communities.

Nevertheless, from the variety of analyses described in this chapter, several strikingly consistent conclusions have emerged. The first is the almost unequivocal implication that winter was the time of greatest ecological stress and, consequently, of the finest levels of community interactions. Winter weather conditions affected food supplies which, in turn, affected population size and distribution of both permanent and wintering seasonal residents. Breadths and overlaps of habitat and resource use among species during winter were usually at their lowest levels, suggesting that this might be the time that potentially limited resources would be partitioned. Permanent resident species exhibited several patterns of population regulation in relation to declining food supplies, but for most, differential overwinter survival was the key determinant of the optimal habitat for later breeding in spring.

In contrast, nearly all of our evidence points to the breeding season as being a time of nonlimiting food supplies, relaxed habitat discrimination and segregation among species, greatest ecological overlap, and increased opportunism by birds. Being freed from exposure to local winter conditions on the breeding grounds, assemblages of breeding seasonal residents were the most stable subset of lower Colorado River avifauna in terms of numbers of species and population density. These findings provide broad support for the ideas of Fretwell (1972) regarding winter limitation of bird populations and seasonal patterns of habitat selection. They are at odds, however, with the view that subtle differences among breeding species are evidence for resource-based interspecific competition acting in these communities.

Our second conclusion is that stochastic processes, involving weather and random turnovers of species at individual sites, may be a very important part of the patterns we observed. It is dangerous to assume that community-level patterns are necessarily the result of deterministic processes involving closely interacting species. We view this stochastic element not as evidence of chaos and lack of structure in avian communities, but rather as a reminder that important biological processes like competition may act only rarely or intermittently and may be exceedingly difficult for us to detect or study. Our work underscores the critical need for long-term studies of complex systems and especially the danger of making inferences from shorter studies conducted only during the breeding season.

Finally, it is apparent that each subset of the avifauna is subjected to very different selection pressures and life-history constraints and must be considered separately in any community analysis. Species assemblages co-occurring in space and time may be organized as completely independent units that superimpose on each other to form the entire community. Some groups of species may interact in ways that suggest strong biological processes; others in the same community may not. Only through careful attention to the natural-history attributes of each species can their role in ecological communities be assessed.

Our primary purpose for studying the avian communities along the lower Colorado River was to develop models for predicting the seasonal habitat use of each species. Although we may not completely understand the processes underlying these community patterns, we can still use our models to predict the value of managed and manipulated habitats (Anderson and Ohmart 1985b; Rice et al. 1986). These predictions guided our efforts to artificially revegetate lost riparian habitats in the lower Colorado River Valley. These efforts are discussed in detail in the next chapter.

4 Conservation Efforts

EXPERIMENTS IN REVEGETATION

LOGISTICAL BEGINNINGS

All of the data collected from 1973 to 1977 gave us a framework from which to develop the second phase of our research. In 1976, we began cultivating native habitats that hopefully would contain all the vegetational components necessary to support a large assemblage of birds (Anderson and Ohmart 1982b). Our research had shown that dense foliage, especially at higher canopy levels, and high overall vegetation patchiness were characteristics of habitats that supported most bird species. In addition, we found that standing dead snags proved to be important for many cavity-nesting species; both those species that excavate their own cavities and those that use existing cavities. The cottonwood–willow plant community in its most mature state provides these essential features, and is especially important to those birds that have become rare during the twentieth century. We also learned that honey mesquite and native shrubs were important in attracting a different subset of the lower Colorado River avifauna.

When we began, little information was available on cultivation of native riparian plant species. Our first attempt at native plant revegetation near Parker Dam in 1976 had little success, but this small experiment provided critical information that could be applied to other sites. We realized, from our experience and from consultation with agricultural agencies, that we needed to consider soil type, soil layering, salinity level, and depth of water table before planting. By using various watering regimes, we hoped to determine the range of conditions that different riparian plant species require. We selected experimental plots that exhibited a wide range of physical factors, then planted trees and shrubs under various tillage, salinity, and irrigation regimes.

In 1978, two sites near Cibola National Wildlife Refuge were selected for our second experimental revegetation project. Soil type and salinity were purposefully varied to determine the best growing condition for each plant species. One site encompassed 30 ha and was virtually barren dredge spoil (material dredged from the river channel) that contained a few widely scattered honey and screwbean mesquite, saltcedar, and blue palo verde trees as well as some Bermuda grass. This site had sandier soil than the second site, although there was considerable within-site variation in soil characteristics. The second site, located on the refuge, consisted of 20 ha of dense saltcedar and arrowweed with a few scattered, rather decadent willow trees. This site contained a hard clay soil. The refuge site offered the opportunity to test our ability to clear tenacious saltcedar permanently and to replace it with more productive native plant species.

THE REVEGETATION PLAN

We planted approximately 1,000 individuals each of honey mesquite, cottonwood, willow, and blue palo verde trees. Cottonwood and willow cuttings were taken from along the river and nurtured in a greenhouse until they were 0.6 m tall. Other trees were germinated from seeds collected from the area. Before planting, each hole was classified by determining soil type, layering, salinity level, and depth of the water table.

The basic design of the revegetation plots followed as closely as possible the configuration suggested by our avian inventory results. We combined these basic bird-habitat relationships in a computer model that would predict the bird assemblages present as each site developed. Accuracy in predicting avian occurrence on revegetation sites thus became the criterion by which we judged the success of our efforts.

GROWING NATIVE PLANTS

On the dredge-spoil site, cottonwoods, willows, palo verdes, and mesquites were planted in alternating rows (or patches). In this way lanes were formed, and vertical and horizontal heterogeneity was introduced as the trees grew. This important aspect of our design produced a patchwork effect in the horizontal plane, as well as vertical layers of shrubs (inkweed, quail bush, and wolfberry), short trees (mesquite and palo verde), and tall trees (cottonwood and willow).

While our planted trees and shrubs grew, other plant species also colonized the once-barren dredge spoil. Plants such as Bermuda grass, Russian thistle, and smotherweed responded quickly to the irrigation water. Once the planted trees were able to survive, these other volunteer plants became an important added dimension to the dredge-spoil site, because they formed an herbaceous layer that provided food and cover for many granivorous and some insectivorous birds.

On the refuge site, cottonwoods, palo verdes, honey mesquites, and willows were planted in a design similar to that used at the first site. Shrubs found naturally in heavy clay soils (quail bush, wolfberry, and inkweed) were planted in high densities because of the heavier and more saline soils. On the refuge site we effectively changed a virtual monoculture of saltcedar to a heterogeneous mixture of native shrubs and trees (Fig. 10).

Survival and growth for all tree species were dramatically increased when tillage was provided to the water table. Also, survival and growth were higher for all tree species in sandy soils than in dense soil types such as clay. However, in dense soil types, mesquite and palo verde trees were able to grow better than cottonwoods or willows. Salinity had a profound negative impact on growth and survival but, again, mesquite trees were somewhat more tolerant than cottonwoods or willows. At moderate salinity levels mortality of all tree species increased. In contrast, we found no such effect on native shrubs. Quail bush, inkweed, and wolfberry grew better in these soils than in sandy soils and were tolerant of extremely high salinity. In addition, these shrubs required only two weeks of irrigation if they were planted in January or February.

TESTS OF BIRD-HABITAT PREDICTIONS

The revegetated areas provided a unique opportunity to test our predictions about which bird species should occur and what densities they should attain, based on our habitat inventory data. We not only wished to test our ability to predict bird use of habitats, but we also wanted to monitor avian response to various levels of habitat quality as the habitats developed.

On the refuge site, we knew the bird species composition in the uncleared area because we had censused the site for 36 months before clearing it. We predicted an increase in birds attracted to quail bush, such as Gambel's Quail, granivorous passerines, Orange-crowned Warblers, Verdins, Blue-gray Gnatcatchers, Crissal Thrashers, and Abert's Towhees, all of which were rare or nonexistent on the site before saltcedar was cleared. In addition, we predicted that inkweed would attract Sage Sparrows. All of these predictions were confirmed within a year of revegetation (Anderson et al. 1989).

We conducted a three-year bird-banding study after revegetation of the refuge site to gain insight into site fidelity and microhabitat use of wintering species (Romano, Anderson, and Ohmart unpubl. data). Three species (Orange-crowned Warbler, Sage Sparrow, and White-crowned Sparrow) were netted most often and exhibited the most interesting patterns. Many banded individuals of all three species returned to the same or adjacent capture locations year after year, even though the vegetation at each location was changing differentially in quality during development (judging from our inventory; e.g., better shrub growth and more inkweed). Predicted bird use of specific 2-ha locations, based on our inventory data, was not confirmed even though predictions were confirmed for the entire 20-ha site. These data suggested that vegeta-

FIGURE 10. Aerial view of revegetation site on Cibola National Wildlife Refuge, one year after planting native trees and shrubs (1981). Note the mixture of vegetation already evident in contrast with surrounding saltcedar. (Photo by R. D. Ohmart.)

tion characteristics from the entire 20-ha site were important primarily in attracting these birds. Individuals of these species then became behaviorally tied to specific 2-ha locations within the site, regardless of differences in relative quality developing through time.

If each of the sampled microhabitats had been isolated 2-ha units, they probably would not have attracted the same species in the same numbers. Thus, total size of the habitat (or size of sampling areas used to determine avian use of a habitat) may be a very important consideration in understanding habitat selection by bird species (also see Engel-Wilson 1981 for a related analysis of summer riparian birds). This finding is also important in designing the size of revegetation plots and predicting their potential use by birds.

On the dredge-spoil site, which had sandier soil and lower salinity, all trees grew rapidly and survival was high. Shrubs grew very well in the less-sandy parts of the area. Again, we had censused this site before planting and were aware of bird species that were previously present. Based on our inventory data we predicted above-average densities, during winter, of seasonal and permanent resident insectivorous birds as well as granivorous passerines. These predictions proved correct.

We knew the trees would not be tall enough to attract cavity-nesting species until several years after planting. Since these species (e.g., Ladder-backed and Gila Woodpeckers, Ash-throated and Brown-crested Flycatchers, and Lucy's Warblers) constitute half or more of the locally breeding insectivores, we predicted that summer insectivore densities would remain low. During the fourth year after planting, one or two pairs of Lucy's Warblers bred successfully on the site. One pair of Ladder-backed Woodpeckers excavated a hole in a dead tree that had been present before planting. In 1981, we girdled several cottonwoods to produce snags to attract Gila Woodpeckers and Brown-crested and Ash-throated Flycatchers. By 1986, several Ladder-backed Woodpecker pairs and at least one pair each of Gila Woodpeckers, Brown-crested Flycatchers, and Ash-throated Flycatchers were nesting on the site.

In addition to determining how well we could predict occurrence and abundance of birds on our sites, we were specifically interested in attracting those species that had declined to very low population levels since the construction of dams along the river. We predicted that Yellow-billed Cuckoos and Summer Tanagers would arrive after some trees attained heights of 9 m. Trees attained this height early in the third growing season. As we had anticipated, two to three pairs of cuckoos bred successfully on the dredge-spoil site during the third and fourth years after revegetation. However, cuckoos did not return after the fifth year. Their absence was more reflective of an overall decline of the species in the lower Colorado River Valley after 1983 than an inadequacy of the revegetated habitat (see Yellow-billed Cuckoo species account). Summer Tanagers have not yet been confirmed to breed on either site; however, recently fledged young have been noted. Although planted vegetation did attract breeding cuckoos and visiting tanagers, this 30-ha site was apparently too small to maintain viable populations for either of these declining species. This conclusion is made in light of the reduced pool of individuals that could colonize revegetation sites and the increasing isolation of the sites themselves from native habitats since 1983 (Anderson et al. 1989).

In general, response of birds to the revegetated areas was almost immediate. Shrubs grew rapidly during the first growing season and, by the first fall following planting, insectivores as well as granivores were abundant on both sites. This pattern has continued, except during the breeding season (March–July) when densities remain lower than predicted. More recent observations show high breeding densities which include both primary- and secondary-cavity nesting species.

SUMMARY OF EXPERIMENTAL REVEGETATION RESULTS

Our experimental work indicated that it was possible to grow native riparian plants, but that the species planted needed to be suited to the conditions of the revegetation site. Although cottonwood, willow, and honey mesquite trees attract large numbers of many bird species, these plants do not thrive in dense, highly saline soils. Trees planted under such conditions will die or attain maximum heights of only a few meters. Mesquites grown in highly saline soil were more shrublike in appearance and provided a habitat no more attractive to wildlife than saltcedar. Undoubtedly, only salt-tolerant plant species such as quail bush, saltbush, and inkweed should be planted where soil salinity is a problem. Where salinity was not a problem, all native tree species exhibited high survival and rapid growth (2–3 m for cottonwoods and willows in each of the first three years) when planted with tillage to the water table.

Habitat heterogeneity was found to be very important in attracting bird numbers and variety. By planting rows or patches with different plant species, we produced a habitat that attracted bird species associated with each riparian plant community found along the lower Colorado River. Habitat size proved to be a critical factor, not only in providing the minimum amount of habitat needed to support population centers for certain bird species (especially those that are rare), but also in providing a large enough sample to test our bird-habitat predictions from empirical inventory data. Even though small plots <2 ha may produce good habitat, they do not provide the mosaic of habitats necessary to attract a large variety of species. Also, within-site variation indicates that many birds do not respond as predicted to specific small locations, but do respond predictably to a large enough site (≈20 ha).

Our revegetation efforts on both study sites were an unqualified success in attracting and supporting populations of most riparian bird species, as predicted. Within five years of planting, many cottonwood and willow trees attained heights of 15 m and the areas took on the appearance of natural forested stands (Fig. 11). These now stand as islands in a sea of degraded saltcedar, burned mesquite, and agricultural lands. Despite the quality of our revegetation sites, their increasing isolation from native riparian habitats apparently dilutes their value to the rarer bird species. This fact stresses the need to quicken the pace of revegetating large areas to reverse the decline of these species, especially in areas adjacent to remnant native habitats.

PROSPECTS FOR CONSERVATION

POTENTIAL FOR REVEGETATION

Our studies have shown that with careful planning, revegetating areas for wildlife can be accomplished in a relatively short time. Lack of natural regeneration and extensive conversion to farmland may lead to

A

B

C

D

FIGURE 11. Revegetation of cottonwoods on dredge-spoil site. A, four months after planting (1979); B, one and one-half years after planting (1980); C, five years after planting (1983); and D, eight years after planting (1986). (Photos by R. J. Dummer [A], J. Disano [B, C], and R. D. Ohmart [D]).

a situation in which revegetation is the only way to ensure the continued existence of those birds dependent on mature cottonwoods and willows.

During the last 15 years, several federal agencies have proposed revegetation projects to mitigate against unavoidable habitat and anticipated wildlife losses. However, because of low budgets, little knowledge and planning, and poor logistical support, these activities have resulted in virtually no success. Smaller (<10 ha) revegetation sites do not provide enough continuous habitat to support population centers necessary for recovery of those bird species nearing extirpation. Even our sites appear to be too small to maintain these bird species as they decline. Thus, widely separated small plots, despite the good intentions of the agencies supporting them, are not adequate to accomplish the goals for which they were planned. Although we encourage all revegetation efforts, we believe that only large-scale, well-planned projects will reap measurable benefits to wildlife.

The most likely way for revegetation to be implemented is as mitigation for future habitat losses. Effective revegetation should be guaranteed by any firm or agency involved in future development along the lower Colorado River. At present, construction firms or agencies are not required by law to mitigate effectively against habitat loss. However, an active consortium of conservation groups and agencies is helping to make sure mitigation efforts are more successful. The Bureau of Reclamation remains the financial leader in these mitigation efforts, but has funded many more failures than successes.

A multitude of potential revegetation sites are available along the entire lower Colorado River. These include other dredge-spoil sites and areas where saltcedar can be cleared and replaced with native vegetation, especially on national wildlife refuges. Most government agencies overseeing water and wildlife management are moving forward, ever so slowly, with proposals for revegetation while potential for reclaiming native habitats remains high. However, the California Department of Fish and Game has become aggressive very recently in its approach to revegetation along the entire river.

IMPORTANCE OF RECREATIONAL LAND

By far the largest tracts of land that can still be managed as wildlife habitat are under county, state, or federal government control. These areas have been set aside for many uses including recreation and wildlife habitat. Most of these outdoor activities would seem to be compatible with, and even enhanced by, the presence of natural greenery.

However, the management of these lands along the Colorado River needs review. Many of these "parks" are little more than paved parking lots for fishermen, boaters, and recreational vehicles. Most disturbing is the continued use of nonnative tree species, particularly eucalyptus and fruitless mulberry, in the development of recreational park land. We have found these trees to be virtually devoid of native birds. State and county park officials have the potential to restore cottonwoods and willows as well as mesquites and native shrubs. This would be beneficial to wildlife and would provide more aesthetic conditions, recreation (e.g., birdwatching), and much needed shade for parks that should be representing the lower Colorado River's natural heritage. Such a park is presently being constructed in Yuma.

The Yuma Crossing Park stretches for 3 km along the Arizona bank of the Colorado River at Yuma. It is composed of two elements—a recreational greenbelt area, and a historic interpretive area made up of 10 thematic sites that represent 400 years of Southwestern history. This park, spearheaded by the efforts of Gwen Robinson, a Yuma area native, is planned for enhancement of the unique historical, recreational, environmental, and cultural qualities of the site.

The environmental aspects of Yuma Crossing Park include enhancement and management of approximately 15 ha of existing riparian vegetation and 35 ha of proposed revegetated habitat. The overall sensitivity of park planning to fish and wildlife concerns has opened the door for involvement by the U.S. Fish and Wildlife Service, U.S. Bureau of Land Management, Arizona Department of Game and Fish, and California Department of Fish and Game. Strong support has been voiced from these agencies to assist in the development of habitat plans. There is a consensus that this effort could serve as a model for future habitat rehabilitation projects along the lower Colorado River.

The most important federal government-controlled lands are the national wildlife refuges which, as their mandate dictates, represent the best-managed habitats for wildlife. A balance has been reached on these refuges between managing for waterfowl (by providing protected aquatic habitats as well as grains from agricultural areas for food) and maintaining populations of many terrestrial bird species (by preserving much of the remaining native riparian vegetation). The three refuges, Havasu, Imperial, and especially Cibola, have been instrumental in the recent increases in geese, cranes, and other gamebirds. In addition, the largest populations of wading and marshbirds, including the federal endangered Yuma Clapper Rail, are located within Havasu and Imperial National Wildlife refuges. Virtually all remaining mature cottonwood–willow stands in the lower Colorado River Valley are located on national wildlife refuges. Consequently, damage to any of these areas may result in serious harm to the overall populations of bird species using them.

Despite their mandate to manage for wildlife, the U.S. Fish and Wildlife Service has been involved in several recent incidents in which regard for wildlife has been outweighed by political expediency. An example comes from Cibola National Wildlife Refuge where, in 1981, several fires were accidentally started during river management activities by the Bureau of Reclamation, consuming large tracts of honey mesquite. These habitat losses were never mitigated for by the responsible agency. A second example involved landswapping of portions of Havasu National Wildlife Refuge to a private developer, allowing him commercial access to Lake Havasu. This "Jop's Landing incident" set a dangerous precedent by allowing wealthy developers to bargain and gain rights to develop once-important federally owned wildlife areas.

The most recent, flagrant example of refuge mismanagement was the deliberate and prolonged flooding of the last remaining cottonwood–willow forest on the Bill Williams Delta Unit of the Havasu National Wildlife Refuge between 1979 and 1981. Although this action was the work of the U.S. Army Corps of Engineers, the U.S. Fish and Wildlife Service stood idly by while the Corps, in ignorance, drowned this vital habitat. Ironically, the cottonwood groves could easily have been saved or even improved with planned releases. Although wildlife refuges continue to be important for conservation and preservation of natural resources along the river, these concerns are too often given low priority when financial gain and water management are involved.

THE ROLE OF PRIVATE LANDOWNERS

In many parts of the lower Colorado River Valley, the only tall vegetation remaining is that surrounding human habitations. Landowners and private developers can therefore greatly increase the value of these areas to birds through their efforts to preserve or restore natural vegetation. However, many landowners are unaware of these values, and their operations are too often detrimental. There are notable exceptions, and on close inspection these have confirmed the increasing importance of residential plantings to a variety of bird species (Rosenberg et al. 1987).

The first example comes from Willow Valley Estates, a mobile home and trailer community in the Mohave Valley south of Bullhead City. This development is somewhat unique in that corner lots have been left vegetated and the emphasis has been to plant native tree species, particularly cottonwoods and willows. The result has been the creation of a natural oasis in this largely agricultural valley. During the breeding season, the number of bird species in Willow Valley Estates compared favorably with those in natural cottonwood–willow habitats we studied. Species that benefited most were Gambel's Quail, Black-chinned Hummingbird, Western Kingbird, Northern Mockingbird, Northern Oriole, and House Finch. Lucy's Warbler, Summer Tanager, and Blue Grosbeak were also present in small numbers. Most encouraging was successful breeding by all of the valley's cavity-nesting species, including the uncommon Northern (Gilded) Flicker and Brown-crested Fly-catcher. In addition, nesting Yellow Warblers and American Robins were found. The Yellow Warbler is near extirpation as a breeding species in the valley, and the American Robin is a very recent breeder in desert lowlands of the region.

In winter, bird abundance and variety were again very high at Willow Valley Estates, with sparrows, juncos, and small insectivores benefiting most. Several uncommon wintering species were also found, such as Brown Creeper, Red-naped Sapsucker, and Black-throated Gray Warbler. Willow Valley Estates illustrates the compatibility of human development with wildlife habitat. Homeowners enjoy the presence of their avian neighbors and the abundant greenery adds to the aesthetic appeal of this community.

A second example is the Clark Ranch, north of Blythe, where many cottonwoods and willows were left intact and additional trees were planted along irrigation canals and alfalfa fields. This area serves as a refuge for many nesting species that have otherwise declined or are rare due to loss of riparian habitat. These include at least one pair each of Yellow-billed Cuckoos, Gila Woodpeckers, Northern (Gilded) Flickers, Vermilion Flycatchers, Brown-crested Flycatchers, and Summer Tanagers. One willow grove supports a small rookery of Great Blue Herons and Great Egrets. In addition, many more unusual species have been found on the ranch, such as nesting Cassin's Kingbirds and Northern Parulas, as well as a large diversity of very rare transients including Thick-billed Kingbird, Prothonotary Warbler, Common Grackle, and Painted Bunting. With its riparian vegetation confined to the borders of alfalfa fields and the riverbank, the Clark Ranch provides habitat for wildlife without interrupting ranch operations. Indeed, the owners of this ranch have enhanced their lives through their interactions with nature.

Many bird species will live and breed near humans if suitable habitat is provided. With careful planning, stable populations of certain sensitive species can probably be maintained or reestablished in areas where little native vegetation remains. Other species may not adapt to these altered habitats, and such areas can never fully replace natural riparian communities. In addition, the possible negative effects of increasing populations of species associated with agriculture, such as Brown-headed and Bronzed Cowbirds, European Starlings, and House Sparrows, need further study.

Willow Valley Estates and Clark Ranch serve as models for present and future land developers. Revegetation techniques can be easily employed where ranch and farm machinery are readily available and where irrigation water may be provided at little extra cost. The importance of riparian vegetation for bank stabilization (along the river and dirt-lined canals) and for providing shade should also be noted by future landowners and developers. Rapid growth rates and resistance to disease and insects should also make native cottonwoods highly preferable to slower-growing exotic trees.

PROSPECTS FOR THE FUTURE

There is still time to recover a small portion of what has been lost. This can be achieved by revegetation, wise management of public lands, and education of private landowners. Large-scale recovery of the entire lower Colorado River ecosystem, however, can be realized only with federal help, and possibly only through court action. Legislation has been timely (National Environmental Protection Act, Endangered Species Act, etc.), but federal agencies have not responded as these laws dictate.

Legislation that perhaps could be effective, if enforced, would allow classification of plant communities as federally endangered. This type of legislation could be an important first step in preserving and restoring entire ecosystems or even portions of, for example, cottonwood–willow and honey mesquite habitats along the lower Colorado River.

After viewing the rapid and almost complete demise of the aquatic and riparian habitats along the lower Colorado River in just 50 years, it would be difficult for anyone to be optimistic about the next 50 years. Rapid habitat loss will continue until private citizens and environmental groups exert enough pressure on state and federal agencies and elected officials to address the problem. Unless this pressure becomes focused and organized soon, it will be too late for the few remaining natural resources that struggle to survive.

5 Patterns of Bird Distribution and Status

With historical and ecological patterns outlined in the previous chapters, we now turn our attention to the differential status and timing of occurrence of the birds themselves. Despite extensive habitat degradation, the lower Colorado River Valley remains a region of significant ornithological interest. Its geographic location as a migration corridor, a lowland wintering ground for many northern species, and an oasis-like magnet for a huge variety of birds visiting the arid Southwest have made this region an outstanding outdoor laboratory for students of bird distribution, migration, and geographic variation.

Bird distribution patterns in the lower Colorado River Valley are complex and dynamic. Breeding activity extends from January through September. Wintering birds may arrive as early as August and linger through May. Migration through the valley may occur in any month of the year, and rare birds may occur at any time. Thus, at any given time, birds at a certain location may consist of breeding, wintering, migrating, and stray individuals. Similarly, at any given location the avifauna can be expected to change markedly from month to month or season to season.

Adding to this complexity are those species exhibiting noticeable geographic variation. Different populations of the same species may vary in their seasonal status. An example is the Northern Flicker. One form (Gilded) is a breeding permanent resident, another (Red-shafted) is a common transient and winter resident, and yet a third form (Yellow-shafted) occurs only as a rare straggler. Understanding the seasonal status of species and their recognizable subspecies is a continuous and exciting aspect of Southwestern ornithology.

PERMANENT RESIDENTS

Roughly 60 bird species are considered permanent residents in the lower Colorado River Valley, with at least one nonmigratory population. In general, these species are typical of the Sonoran Desert region, being largely associated with desert scrub or riparian habitats throughout their ranges. For species such as Black-tailed Gnatcatcher, Crissal Thrasher, and Abert's Towhee, the valley is a major center of abundance;

others, such as Northern Cardinal, are at the periphery of their ranges and are limited in numbers. Another set of permanent residents is associated with aquatic habitats. In general, these species are wide-ranging and familiar inhabitants of wetlands elsewhere in North America and include various herons, egrets, American Coot, and Killdeer. A few additional residents, such as House Sparrow and European Starling, are exotic species. Still others, such as Inca Dove and Great-tailed Grackle, have recently expanded their range and are increasing numerically as a result of human activities.

Permanent residents live in an environment that is changing throughout the year. Compared with migratory species, permanent residents tend to be generalists in terms of habitat and diet, although some have limited foraging habits. Permanent resident insectivores tend to feed mostly on the ground or on tree limbs and trunks where insect populations are presumably more stable year round. However, a few foliage-gleaning species, such as the Verdin, persist throughout the year by foraging on perennial vegetation in winter after the trees have dropped their leaves.

Populations of permanent residents exhibit marked seasonal fluctuations, with numbers usually highest in late summer and lowest in late winter. Winter mortality and dispersal of young birds both contribute to population declines. Many permanent residents begin breeding early in spring (or even in late winter) and often raise two or more broods during the prolonged nesting season.

SEASONAL RESIDENTS

Many bird species reside in the lower Colorado River Valley for only part of the year. Length of stay varies from only three months in breeders such as Yellow-billed Cuckoo, to nearly nine months in several wintering species. In some species, purely transient individuals confuse the status of those populations establishing seasonal residency in the valley.

Approximately 28 species regularly migrate to the lower Colorado River Valley to breed in spring or summer. A few others have reached the region or have bred only on rare occasions. Breeding seasonal residents tend to be habitat specialists. Half are dependent on mature riparian habitats, especially stands of cottonwood or willow, and an additional five species are associated with marshes and other wetlands. The Phainopepla is unique among the seasonal breeding species in that it arrives in the valley in fall, breeds early in spring, and exits before the onset of summer. Most other seasonal residents arrive on their breeding grounds in April or May and depart in August. These species normally raise only one or, rarely, two broods during their short stay.

About 80 species can be considered wintering seasonal residents that arrive mostly from the north and west. Nearly half of these are aquatic species attracted to the reservoirs, river channels, and marshes of the Colorado River. The remaining species flock into agricultural and riparian habitats, often in large numbers. In addition, many species that only irregularly visit the valley have been recorded primarily in winter. Finally, individuals of some otherwise purely transient species, such as Osprey, Solitary Vireo, and Black-throated Gray Warbler, will overwinter in small numbers.

The relatively mild climate and abundant food resources of the lower Colorado River Valley combine to provide a subtropical wintering ground for a wide variety of migrants. In some years, species such as Spotted Sandpiper, Say's Phoebe, Marsh Wren, American Pipit, Orange-crowned Warbler, as well as Vesper, Savannah, and White-crowned Sparrows have been tallied in higher numbers on Colorado River CBCs than in any other North American location (Appendix 1). The valley is also an important wintering area for Sandhill Cranes, White-throated Swifts, Ruby-crowned Kinglets, Yellow-rumped Warblers, and Sage Sparrows. Red-tailed Hawks, Northern Harriers, American Kestrels, and Loggerhead Shrikes now winter abundantly in agricultural areas, and Cooper's and Sharp-shinned Hawks are common where riparian vegetation persists.

Wintering waterfowl numbers are small compared with other regions, but the lower Colorado River usually hosts an excellent variety of species for an inland region. The three national wildlife refuges in the valley support a modest population of Canada Geese and smaller numbers of Snow Geese, as well as varying numbers of Mallards, Northern Pintails, Green-winged Teals, and a few other puddle duck species. Away from these refuges (and the grain they provide), duck numbers vary greatly in time and place depending on severity of the winter north of Arizona, water levels in the river, bank characteristics (channelized or unchannelized), presence of marsh vegetation, or disturbances. Populations of Gadwall and American Wigeon follow the distribution of Sago pondweed beds and may be scarce in years of high water flows. Among the diving ducks, Common Goldeneye, Bufflehead, and Common Merganser are always the most numerous, with smaller flocks of other species primarily concentrated close to dams. Several species, such as Canvasback and Ruddy Duck, are inexplicably scarce compared with their numbers elsewhere in Arizona or the Salton Sea in southern California.

In general, populations of wintering seasonal residents are characterized by large fluctuations from year to year. Unlike permanent resident populations, wintering seasonal residents may be greatly influenced by climatic or resource conditions outside the region, making it difficult to discern true population trends. For example, during unusually harsh winters, certain species (primarily granivores, frugivores, raptors, and diving ducks) may appear in larger than normal numbers, having moved from colder regions in the mountains or from the north. Conversely, some insectivorous species winter in large numbers only during mild years, while others may only attempt to winter at such times (e.g., Solitary Vireos). Still other species appear only during "flight years" that are tied somehow to food abundances over broad geographical areas.

In addition, there is evidence that some species may make midseason adjustments in their winter quarters in response to sudden changes in local weather or food supply. These moves may be in the form of continued migratory behavior, as in some small insectivores, or as massive invasions into more suitable habitat, as in berry-eating waxwings and thrushes. Other examples are midwinter influxes of northern gulls, diving ducks, and Bald Eagles. Thus, the notion of a static winter range for many species may be inappropriate, as winter in the lower Colorado River Valley is very much a time of flux.

Seasonal residents make up an ecologically diverse lot. However, one general tendency is for these birds to use resources that are only temporarily available and not effectively exploited by permanent residents. In winter, for example, many seasonal residents rely on mistletoe berries, abundant weed seeds, or emerging aquatic insects. Breeding seasonal residents are almost entirely insectivorous, foraging primarily by gleaning from foliage or by flycatching. It is uncommon for either summer or winter seasonal residents to feed on seasonally stable bark- or ground-inhabiting insects, as do most permanent residents. However, when food is seasonally abundant, such as during the annual cicada emergences in midsummer, both permanent and seasonal residents may share in the feast, with little evidence of competitive interactions among species. The ecological requirements of each seasonal group may be quite distinct, although this subject begs further study.

TRANSIENTS

It is the passage of migrants that contributes most to the constantly changing variety of valley birdlife. About one-third of all species recorded along the lower Colorado River are transients that linger in the region for only a short time. Migrants arrive from nearly every direction

FIGURE 12 (pages 75–78, following). Seasonal distributions of common landbird migrants in the lower Colorado River Valley. Each graph compiles all sightings from 1976 to 1983, from throughout the valley, in weekly intervals. N is the total number of individuals recorded during this period.

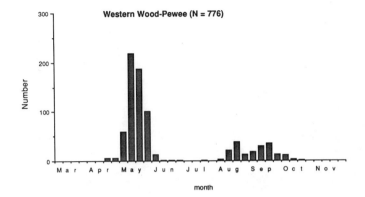

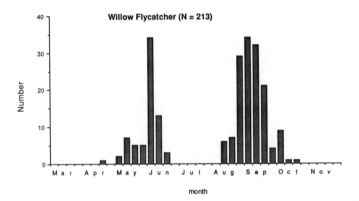

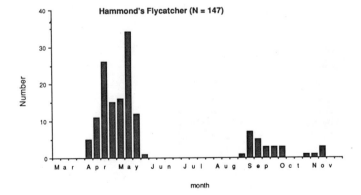

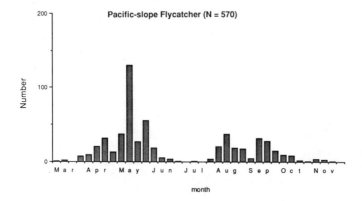

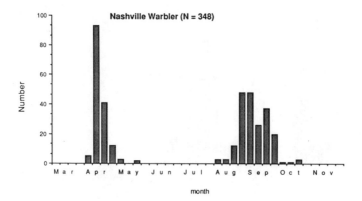

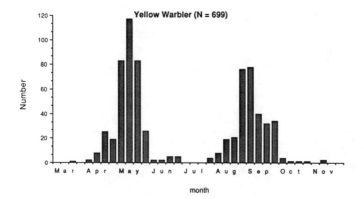

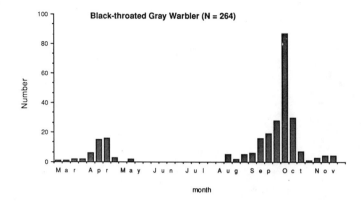

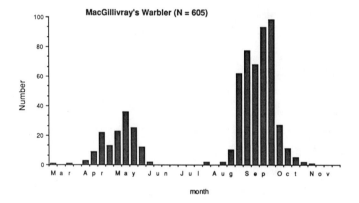

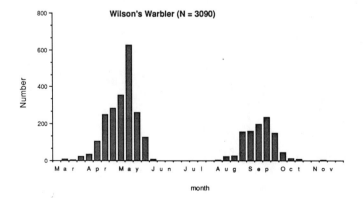

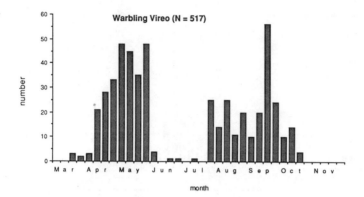

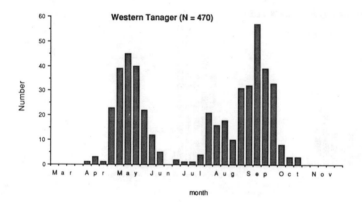

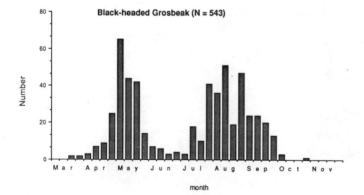

and we estimate that 70% of the roughly 440 North American migratory species have occurred in the valley at one time or another.

First signs of spring come early, with the arrival of migrating Cinnamon Teal, Costa's Hummingbirds, and Violet-green Swallows in January. A steady trickle of early migrants follows the warming temperatures through February and March, and by mid-April the major movement of shorebirds and passerines is under way. May is the peak month for both variety and magnitude of spring migration through the valley. Several species, such as Pacific-slope Flycatcher, Western Wood-Pewee, Warbling Vireo, Wilson's Warbler, and Western Tanager, are still regularly migrating through early June. Willow Flycatchers may appear in mid- or even late June. The seasonal distributions of the most common landbird migrants are illustrated in Figure 12.

Overlapping with late northbound migrants are the first southbound flocks of water- and shorebirds, consisting of Great Basin-nesting species such as White-faced Ibis, American White Pelican, Wilson's Phalarope, Black-necked Stilt, and Caspian, Forster's, and Black Terns. The migration patterns of these species are illustrated in Figure 13. These are soon joined by the first swallows, ducks, passerines, and other shorebirds, and fall migration is in full swing by late July. A steady progression of different species continues through the fall months with some still migrating in December.

While the number of species migrating through the lower Colorado River Valley is impressive, the number of individuals involved varies greatly among species and often appears surprisingly small for most. In general, the magnitude of fall migration is larger than in spring. This appears to be related more to the abundance of immature birds making their first trip south than to a change in migration routes. However, certain flycatchers and warblers may be more numerous in spring (Fig. 12).

Several factors may contribute to the frequent inconspicuousness of bird migration. First, the lower Colorado River Valley includes immense tracts of relatively wide and continuous wetland and terrestrial vegetation that are infrequently visited by ornithologists. Thus, the concentrating effect commonly observed in isolated desert oases or small riparian stands does not occur. Consequently, the bulk of migrants may pass unnoticed. Second, the weather patterns in the region are characterized by a lack of major storms or fronts for most of the year. It is not unusual, for example, for April and May to pass without a cloudy day or night. Under these conditions, most nocturnal migrants can pass over the Southwestern deserts easily without stopping to forage or rest. The infrequent storm that forces thousands of birds into the valley hints at the magnitude of migration occurring overhead. One such storm occurred on 9–10 May 1977. On those days, roughly 650 migrants of 26 species were found near Parker and the Bill Williams

Delta, including the extremely rare Black Swift, Prothonotary Warbler, Worm-eating Warbler, and Ovenbird.

The value of the lower Colorado River Valley to groups of migrants is highly variable. The river is not a major flyway for waterfowl, except perhaps for Northern Pintail and Green-winged Teal in early fall. Only a few waterfowl species are transients that do not normally winter in the valley. Greater White-fronted Geese were formerly common as spring and fall migrants. They now pass through only in fall (late September to mid-October), in much reduced numbers. Cinnamon Teal may be common anytime between February and September, but delineation of their north- and southbound migration seasons is clouded by occasionally summering or even breeding individuals. During the same period, the rarer Blue-winged Teal passes through in very small numbers. The only other primarily transient species, the Red-breasted Merganser, may be present in any month, but concentrates its passage in November, March, and April.

Migration of raptors is poorly defined and often inconspicuous. Individuals of many species undoubtedly move through the valley, but patterns are obscured by varying numbers that remain through winter. Only Swainson's Hawks and Ospreys can be considered true transients; a few individuals of both species pass through each spring and fall. In addition, a major movement of Turkey Vultures is occasionally noted in March and October, when single flocks of up to 750 have been recorded.

Shorebirds consistently follow the river valley during migration, but numbers are unimpressive compared with flocks that concentrate at the nearby Salton Sea. Migratory flocks usually contain nesting species from the Great Basin and Great Plains, occasionally including large concentrations of Black-necked Stilts, American Avocets, Marbled Godwits, Long-billed Curlews, or Wilson's Phalaropes (see Fig. 13). Among Arctic-nesting long-distance migrants, only Long-billed Dowitchers, Greater Yellowlegs, Western and Least Sandpipers, and occasionally Red-necked Phalaropes can be considered common at any time. Most other species appear regularly in small numbers, however, and the variety of birds seen in a season is often high. Shorebirds are attracted most frequently to irrigated agricultural fields, especially those recently plowed and flooded. This may be due, in part, to frequent human distur-

FIGURE 13 (pages 81–84, following). Seasonal distributions of migrant waterbirds nesting primarily in the Great Basin. Each graph represents all sightings from 1976 to 1983, from throughout the valley, in 15-day intervals. N is the total number of individuals recorded during this period. Note the early "fall" passage of many species in July or even June, often with no gap in records through early summer.

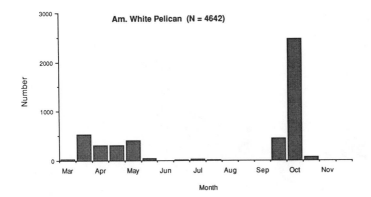

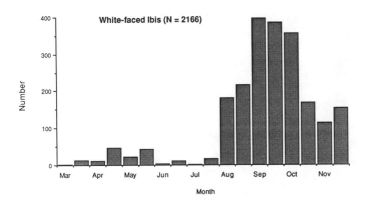

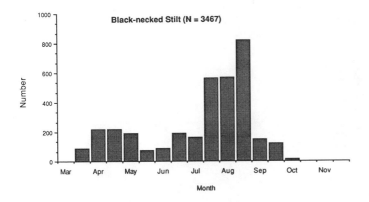

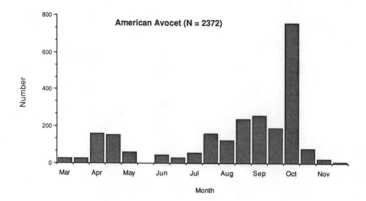

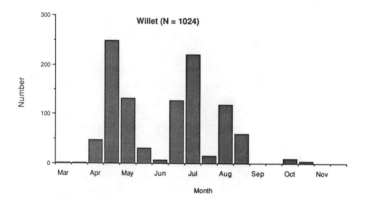

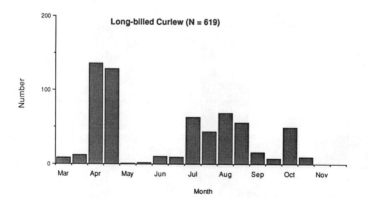

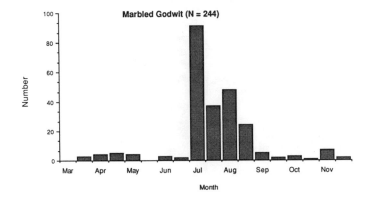

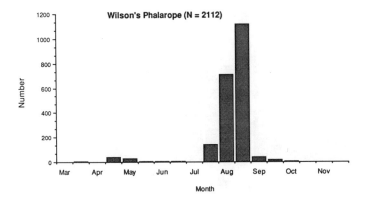

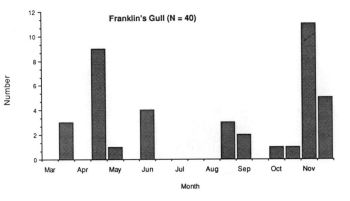

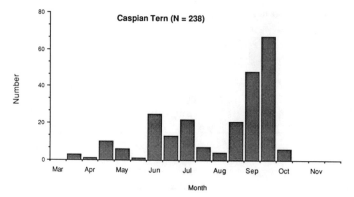

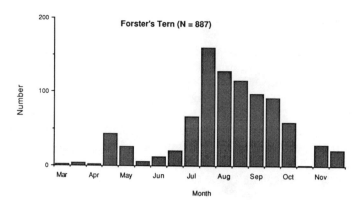

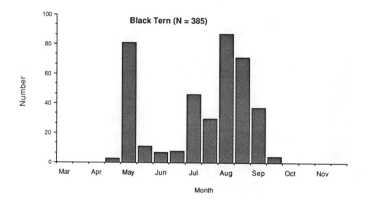

bance at riverine marshes and sandbars, but apparently these fields provide adequate food.

Swallows are perhaps the most conspicuous avian group during migration. We estimate that several million Tree and Barn Swallows use the valley during migration, with much smaller numbers of other species in the mixed flocks. Peak movement of Tree Swallows is always about a month earlier than that of Barn Swallows, and the latter's peak is compressed into a much shorter period. When large flocks of swallows are present, they frequently roost at night in extensive marshes such as those at Imperial Dam and Topock Marsh. There they exit their roost sites in never-ending numbers to forage by day over riverine habitats and agricultural fields. The return of these birds to their roosts at dusk is a remarkable sight. Huge flocks may congregate in a tight swarm that spirals erratically over the marshes with a nearly deafening whir of wings. As darkness ensues, this whirring stream of swallows will suddenly plunge and vanish into the cattails, leaving a deathly stillness.

As noted earlier, migration of other passerines is often light and variable. The most numerous species are those traveling primarily from the Sierra Nevada or Rocky Mountain regions to wintering areas in western Mexico. In both spring and fall these include Pacific-slope Flycatchers, Western Wood-Pewees (wintering in South America), Warbling Vireos, Nashville, Yellow, MacGillivray's, and Wilson's Warblers, Western Tanagers, Black-headed Grosbeaks, and Lazuli Buntings (see Fig. 12). Hammond's Flycatchers are common only in spring and Willow Flycatchers only in fall.

These migrant songbirds usually seek the tallest trees available, such as cottonwood or willow groves or cultivated trees around residences. Here they often feed incessantly throughout the day and frequently travel in mixed-species groups from one patch of greenery to another. It is not uncommon, however, for individual migrants to take up temporary residence in a single grove or tree, remaining for several days before departing.

Most records of out-of-range species are from the migration seasons. It is apparent that many species stray regularly from their normal migration routes and occur predictably at points where other migrants concentrate, such as desert oases and coastal peninsulas. The origin and fate of these "vagrants" remain unresolved; however, various theories exist to address these ornithological issues. A few individuals of some species may have alternative routes that are regularly followed, so they may not be vagrants at all. Other birds may be truly disoriented, traveling for example from east to west instead of north to south.

Isolated records from the lower Colorado River Valley may take on more meaning than mere "accidental" occurrences when put in a larger perspective of regional patterns. For example, the consistent appearance of Northern Parulas in late May, Eastern Kingbirds in early September, and Black-throated Green Warblers and other Eastern warblers in October and November is part of recently discovered patterns throughout the Southwest. Even single records of Hooded and Kentucky Warblers in June and Louisiana Waterthrushes in late July fit nicely into wider patterns of occurrence. Indeed, it is through the recognition of these patterns, and the growing interest in finding unusual species, that the avifaunal list from the lower Colorado River Valley has grown by roughly 15% in the past 15 years.

ORIGINS OF MIGRANTS AND VISITORS

Analysis of the avifauna of the lower Colorado River Valley reveals a web of directional movements from a variety of regions. For some species the region of origin is readily apparent. For many others it is only by studying the geographical or subspecific variation exhibited by collected specimens that their point of origin can be determined. Thus our knowledge of migration sources and pathways is still incomplete.

We are fortunate in that the lower Colorado River Valley has attracted the attention of several excellent avian scientists with an eye for geographic variation and temporal change. In particular, the work of Gale Monson and Allan R. Phillips contributes most to our knowledge of source regions and timing of migration, as summarized in their regional accounts for Arizona (Phillips et al. 1964; Monson and Phillips 1981). The collections of Joseph Grinnell and others from California are similarly summarized in Grinnell and Miller (1944). Although these sources do not always agree, they allow the same general conclusions. We rely heavily on these works in this chapter and in the species accounts (Chapter 7).

Most nonbreeding species arrive in the lower Colorado River Valley from locations in western North America as far north as Alaska. The Rocky Mountain and Sierra Nevada regions are the source of many regularly occurring landbirds. Most waterfowl and shorebird species migrate from either the Arctic tundra or wetlands of the northern Great Plains or Great Basin. Shorter movements occur in the form of mass dispersals or irruptions by some species from montane areas in Arizona and adjacent states.

Since it is within the broad region of western North America that many species exhibit their most noticeable geographic variation, we can discern more subtle migration patterns and make the following generalizations. First, in species that have distinct breeding populations in the mountains of Arizona, migrants from these areas are almost unknown in adjacent lowland regions such as the lower Colorado River Valley. Examples include Cordilleran Flycatcher, Hermit Thrush, American Pipit, Black-throated Gray Warbler, and Yellow-rumped Warbler. Second, there is a tendency for the earliest passerine migrants, at least in spring, to be bound for breeding grounds along the southern Pacific Coast, where the start of the corresponding fall migration is also early. Breeding birds from the Rocky Mountains usually pass through later in the season and the latest migrants are often from Alaskan populations. Subtly variable Wilson's Warbler specimens from the lower Colorado River Valley illustrate this quite clearly. The third generalization is that for species that are both common migrants and winter residents, the wintering populations usually originate far to the north in Alaska. More southern montane breeders comprise the bulk of the transients seen in early fall and late spring. American Pipits and White-crowned Sparrows are most notable in this regard.

Several typically Eastern species and subspecies have been found along the lower Colorado River. Most are long-distance migrants with breeding ranges extending westward across Canada, and many appear with some regularity in small numbers throughout the West. There is also probably a westward movement by birds from the southeastern and central United States which brings species such as Northern Parula and Scissor-tailed Flycatcher to the lower Colorado River Valley almost every year.

Southern subtropical areas contribute another set of bird species. A few Mexican landbirds have occurred primarily in fall or winter, including Ruddy Ground-Dove, Broad-billed Hummingbird, Greater Pewee, Thick-billed Kingbird, Rufous-backed Robin, and Varied Bunting. More conspicuous is the postbreeding dispersal by waterbirds such as Brown Pelicans, Magnificent Frigatebirds, and Wood Storks via the Gulf of California. Grinnell (1914) commented on the numbers of southern wading birds that visited the lower Colorado River drainage before the construction of dams. In fact, Wood Storks were much more common and widespread during this period. These birds apparently concentrated in summer, as the annual floodwaters were receding, to feast on stranded fish in drying backwaters and pools. So large and predictable a feast this must have been that the ancient Colorado River (and probably other Southwestern rivers) may have figured prominently in the evolution of this northward dispersal in many Mexican-breeding waterbirds. Of course, without annual flooding and subsequently stranded fish, such behavior today would not be very advantageous. However, we still see a shadow of this traditional dispersal, with individuals of nearly every southern wader appearing in recent years. Even periodic mini-invasions of Roseate Spoonbills occurred as recently as 1977. Dams and their large reservoirs have probably increased the attractiveness of the lower Colorado River to more water-based species such as pelicans, boobies, and frigatebirds.

We should note here that the Salton Sea has acted as a magnet to southern waterbirds in recent decades and continues to host a variety of species that now rarely, if ever, reach the lower Colorado River. In fact, the source of visiting storks, frigatebirds, and other rarities such as Black Skimmers and Laughing Gulls is now most likely the Salton Sea rather than the Gulf of California. This is supported by the preponderance of recent records of these and other species in the Palo Verde– Cibola region, which is the shortest overland route between these two aquatic habitats.

The appearance along the lower Colorado River of largely oceanic bird species has apparently increased in recent times because of the creation of large reservoirs and deep-water channels below dams. An overland migration route between the middle Pacific Coast and the Gulf of California is probably used by several species such as Brant and Heermann's Gulls, as evidenced by their multiple occurrences along the river and at the Salton Sea. Sabine's Gulls and jaegers visit the valley's lakes on their journey from the Arctic to more southerly oceans, and others such as Pacific Loons and Barrow's Goldeneyes now winter regularly on these lakes. Laysan Albatrosses, found in the Yuma area in 1981 and 1988, undoubtedly originated far from land in the Pacific Ocean.

6 Finding Birds in the Lower Colorado River Valley

This chapter offers guidelines to birdwatchers or ornithologists who visit the lower Colorado River. First, 10 of the most productive areas are described on a trip from north to south (Fig. 14). Numbers in parentheses refer to map locations. Distance references are in English units of measure, with metric in parentheses for vehicle odometer use. Table 6 is a calendar that illustrates the annual cycles of resident and migratory species composition, and predicts when certain rare species are most likely to occur.

BEST BIRDING SPOTS

DAVIS DAM–MOHAVE VALLEY

Located on Arizona Highway 95 on the Arizona–Nevada border, Davis Dam and the lower portion of Lake Mohave are among the best places to search for northern and oceanic waterbirds between late November and March. Such rarities as Pacific Loon, Horned Grebe, Barrow's Goldeneye, Oldsquaw, and Hooded Merganser are seen there almost annually. Up to seven species of gulls have been observed in a single winter. The spillway of Davis Dam is best viewed from Sportsman's Park (1), on the Nevada side below the dam. The lake may be checked from the top of the dam or from Katherine Landing (2), on the Arizona shore of Lake Mead National Recreation Area. A large flock of White-throated Swifts often forages over the river below Sportsman's Park.

The riparian habitat along the Nevada shore below Davis Dam (Fort Mohave area) is home for several breeding species, including Elf Owl, Gila Woodpecker, Brown-crested Flycatcher, and Bell's Vireo, that are rarely found elsewhere in Nevada. Recently recorded rarities from this area include Tropical Kingbird (June), Palm Warbler (October), Grace's Warbler (winter), and Painted Redstart (winter). In Arizona, Mohave County Park just below the dam (3), is perhaps the northernmost regular nesting site for Inca Doves and Bronzed Cowbirds. At Bullhead City and adjacent resort communities, tall trees and other plantings are attractive to migrant landbirds. Backwater levees and exposed sandbars, where accessible, can be checked for ducks and gulls in winter and for shorebirds during migration.

South of Bullhead City, most of the Mohave Valley has been cleared recently for agricultural use. Flooded fields in this area often attract large flocks of geese, ducks, and Sandhill Cranes in winter. A large winter roost of American Crows has formed there in recent years. Many hawks and other species typical of open country can be seen as well. Remaining patches of native riparian vegetation may be checked for common residents, and Willow Valley Estates (4), along Arizona Highway 95, provides habitat for some uncommon breeding species, including Vermilion and Brown-crested Flycatchers, American Robins, Summer Tanagers, and occasionally, Yellow Warblers.

NEEDLES–TOPOCK MARSH

Needles, like other desert towns, is an oasis of tall trees and ornamental plants that attracts a wide variety of migrant and resident landbirds. One productive area is along River Road in the vicinity of Needles Municipal Golf Course (5). Patches of willow and mesquite trees, trailer parks, and the golf course's tall cottonwoods and marshy ponds have produced such rare species as Northern Cardinal, breeding Bell's Vireo and Yellow Warbler, a wintering Yellow-throated Warbler, and a Scissor-tailed Flycatcher (paired with a Western Kingbird). North of Needles, about 12 mi (19 km) on River Road, is Soto Ranch (6) which supports the last stand of mature honey mesquite on the California side. This 54-ha area is currently the most reliable place to locate Elf Owls in California (5 pairs in 1987). The area also supports the now very rare Bell's Vireo and Yellow-billed Cuckoo.

Across the river, in Arizona, is the Topock Marsh Unit of Havasu National Wildlife Refuge. Topock Marsh may be entered by turning right (south) after crossing the bridge from Needles. This refuge protects extensive marsh and riparian habitats that host many of the valley's resident breeding species in summer. Birds of particular interest include colonies of Double-crested Cormorants, herons, and egrets (flooded trees), Least Bitterns and Yuma Clapper Rails (marshes), Yellow-billed Cuckoos and Bell's Vireos (willows), and Indigo Buntings (burned areas). From November through March numerous backwaters,

FIGURE 14. Maps of birding spots. Numbers correspond to highlighted localities in text. A. Davis Dam–Mohave Valley (1–3). B. Fort Mohave Indian Reservation (4–6). C. Needles–Topock Marsh (7–8). D. Lake Havasu City–Upper Lake Havasu (9–11). E. Bill Williams Delta–Parker Dam (12–15). F. Parker Valley–Colorado River Indian Reservation (16–21; 23–24). G. Blythe–Palo Verde Valley (22; 25–29). H. Cibola National Wildlife Refuge (30–32). I. Imperial National Wildlife Refuge (33–34). J. Imperial Dam–Laguna Dam (35–40). K. Yuma and vicinity (41–42).

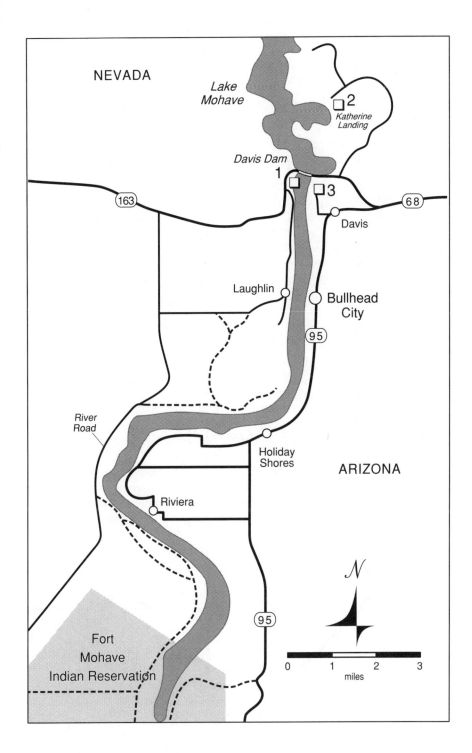

FIGURE 14A. Davis Dam–Mohave Valley.

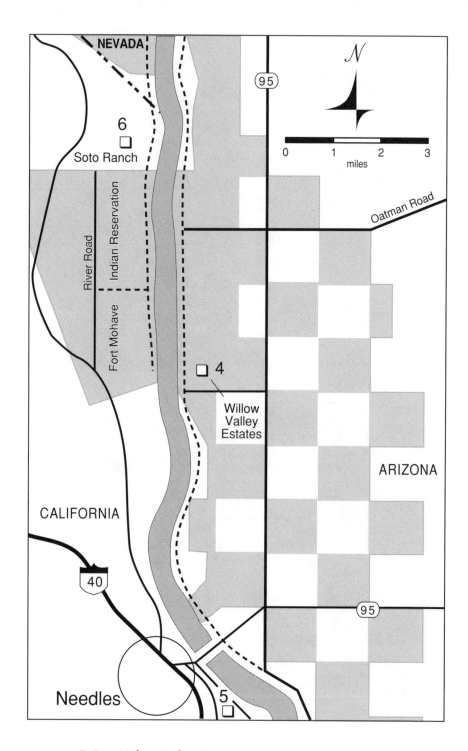

FIGURE 14B. Fort Mohave Indian Reservation.

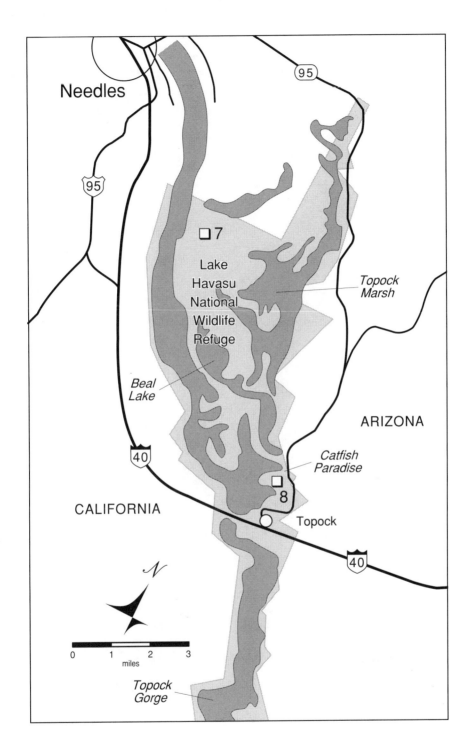

FIGURE 14C. Needles–Topock Marsh.

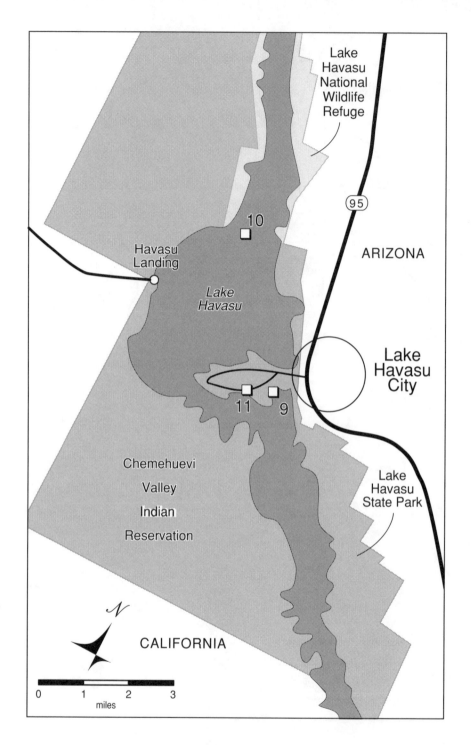

FIGURE 14D. Lake Havasu City–Upper Lake Havasu.

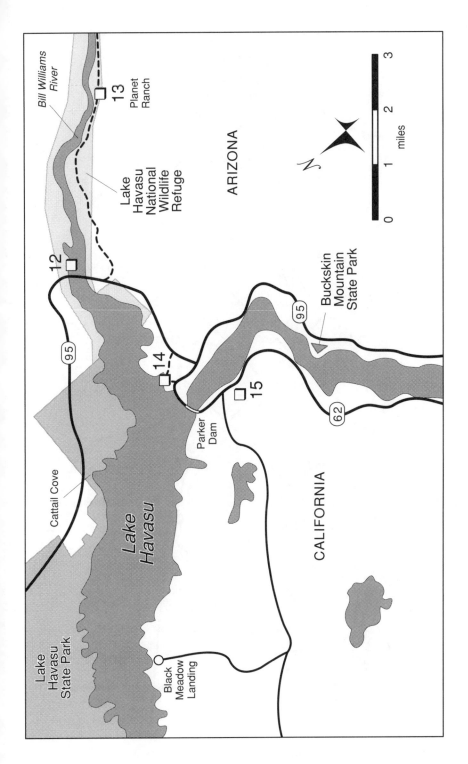

FIGURE 14E. Bill Williams Delta–Parker Dam.

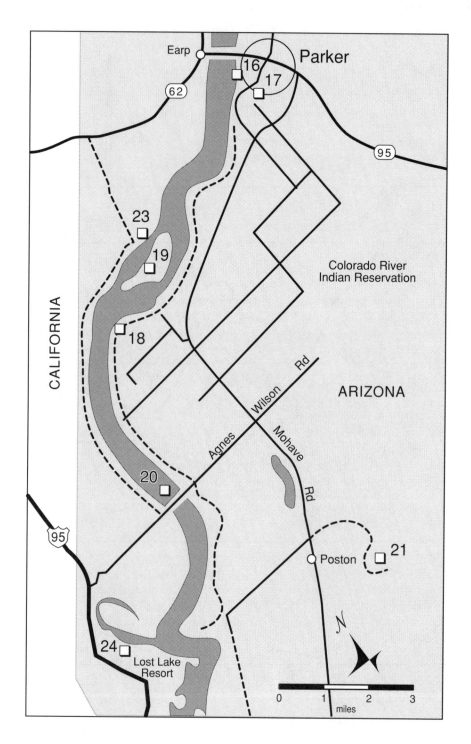

FIGURE 14F. Parker Valley–Colorado River Indian Reservation.

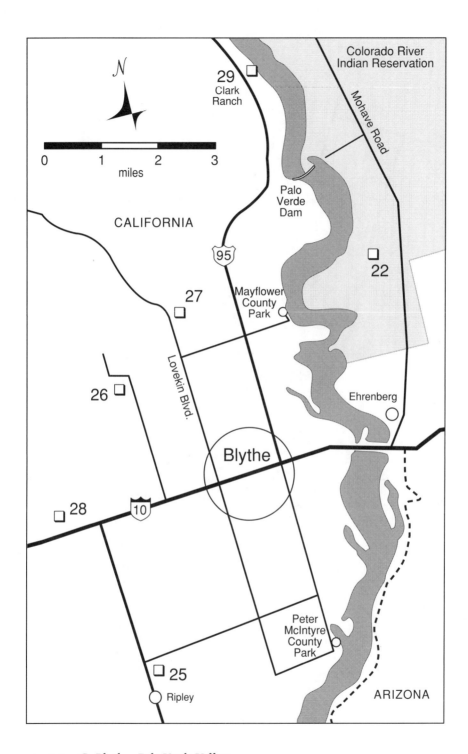

FIGURE 14G. Blythe–Palo Verde Valley.

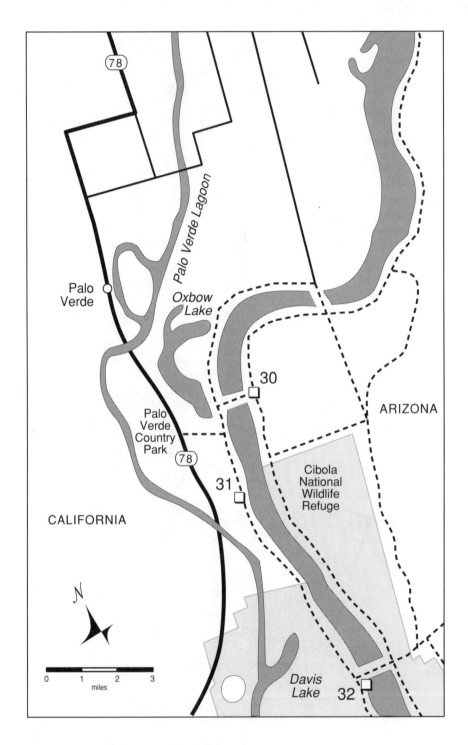

FIGURE 14H. Cibola National Wildlife Refuge.

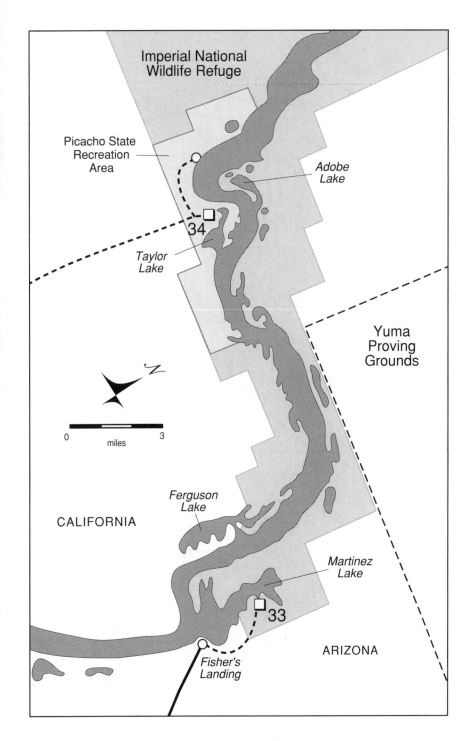

Imperial National
Wildlife Refuge

Picacho State
Recreation
Area

Adobe
Lake

34

Taylor
Lake

Yuma
Proving
Grounds

CALIFORNIA

Ferguson
Lake

Martinez
Lake

33

ARIZONA

Fisher's
Landing

0   miles   3

FIGURE 14I. Imperial National Wildlife Refuge.

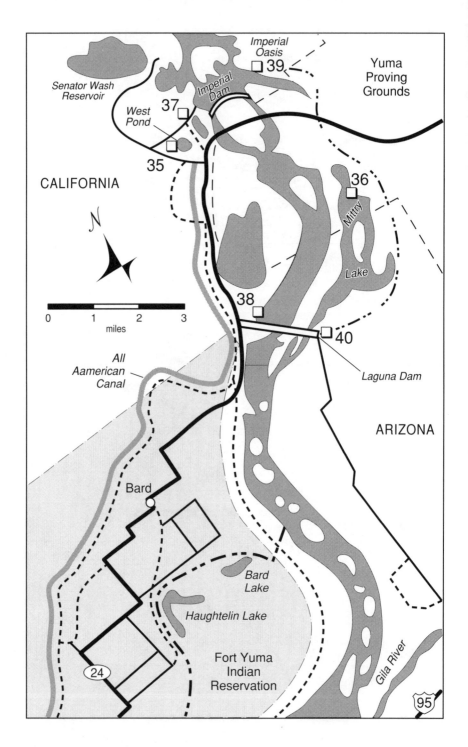

FIGURE 14J. Imperial Dam–Laguna Dam.

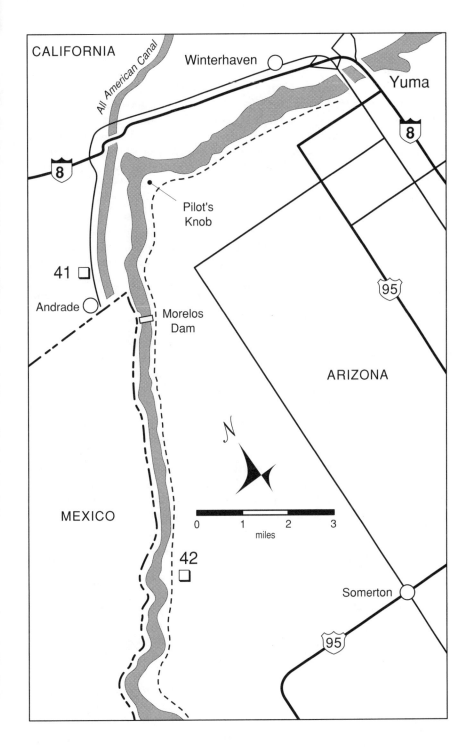

FIGURE 14K. Yuma and vicinity.

ponds, and cultivated fields attract varying numbers of ducks, geese, and occasionally swans. Nearly every waterfowl species found in the valley has been recorded in the marsh at least once. An observation tower overlooks a popular feeding area at the Topock Farm (7), where information on other access points can be obtained. The large flocks of geese seen from the tower often contain a few Ross' Geese or perhaps a rare blue-morph Snow Goose. Refuge checklists, maps, and hunting regulations can be obtained from: Havasu National Wildlife Refuge, P.O. Box A, Needles, California 92363.

A unique large stand of exotic athel tamarisk (8) flanks Arizona State 95 just north of the Topock Exit off Interstate 40. Lucy's Warblers (abundant breeders) and Summer Tanagers, which do not normally occur outside of mature cottonwood—willow habitats, are consistently present in summer. Rare but regular visitors in winter have included Long-eared Owl, Red-breasted Sapsucker, Brown Creeper, and Winter Wren.

LAKE HAVASU CITY–UPPER LAKE HAVASU

Lake Havasu City, on the desert shore of Lake Havasu, attracts a multitude of human visitors to view the world-famous London Bridge in its new home and to boat, water ski, and fish. The birdwatcher who endures these crowds and their bewildered stares may be rewarded with some of the most exciting waterbird watching in the Southwest. Across London Bridge, Lake Havasu State Park's marinas, beaches, and coves offer views of many species from shore, especially loons, grebes, and gulls. The best way to see these birds is to rent a boat at Pittsburgh Point Marina (9) and visit the north end of Lake Havasu (10) which is part of Havasu National Wildlife Refuge. Western and Clark's Grebes are abundant year round, and in late summer downy young are seen dodging boats along with their parents. This is one of the finest places in North America to observe the subtle differences in plumage and behavior between these two closely related grebes, as they breed here side by side.

"Pelagic" trips to this area in August and September have encountered boobies, frigatebirds, jaegers, up to five tern species, and many shorebirds. Early October is the peak time for Greater White-fronted Geese and Sabine's Gulls. Wintering birds on the lake often include Pacific Loons, Horned Grebes, rare diving ducks (e.g., Oldsquaw and Greater Scaup), and gulls.

In addition to the lake itself, the landscaped area around London Bridge should be checked for migrants or rarities. The Lake Havasu City sewage treatment ponds (11) across from the marina are attractive to ducks, shorebirds, and pipits. Two Rusty Blackbirds wintered there in 1982–83. At the Lake Havasu City Golf Course, we have observed Ring-billed Gulls hovering and eating the fruit of cultivated date palms.

FIGURE 15. Bill Williams Delta, September 1976, showing extensive marsh, riparian forest, and surrounding desert mountains. (Photo by K. V. Rosenberg.)

BILL WILLIAMS DELTA–PARKER DAM

Between Lake Havasu City and Parker, State Highway 95 crosses the mouth of the Bill Williams River and winds above the Bill Williams arm of Lake Havasu to Parker Dam. This portion of the lake, as well as the extensive marshes and riparian habitat upstream along the Bill Williams River, are part of the Havasu National Wildlife Refuge (Fig. 15). Together with agricultural fields on the Planet Ranch, trailer resorts below Parker Dam, and the surrounding desert cliffs and washes, this area offers perhaps the largest variety of birdlife of any place in the lower Colorado River Valley. More than three-fourths of the valley's species have been seen there.

The riparian tracts of the Bill Williams Delta include the largest remaining cottonwood–willow forests in the valley. Local breeding birds such as Yellow-billed Cuckoos, Northern (Gilded) Flickers, Brown-crested Flycatchers, and Summer Tanagers are common, and wintering wrens (up to seven species), kinglets, and warblers are particularly abundant. In summer, look for rare-but-regular species such as Zone-tailed Hawk, Common Black-Hawk, Elf Owl, and Indigo Bunting. In winter, Wood Duck, Brown Creeper, Winter Wren, Golden-crowned Kinglet, Black-and-white Warbler, and Swamp Sparrow may be found. The delta is a magnet for rare migrants and visitors throughout the year, but especially during peak migration periods in May, September, and October.

To visit the riparian groves, turn east onto a well-graded dirt road, 0.25 mi (0.4 km) south of the Bill Williams bridge (12). This road winds through desert hills before dropping into the Bill Williams floodplain and eventually reaching Planet Ranch (13); access on this road may be limited during flood conditions. Planet Ranch is now owned by the city of Scottsdale, Arizona, and permission from the ranch or the city should be gained before birding there.

Cattail marshes of the delta itself are home to nesting Virginia Rails, Yuma Clapper Rails, Least Bitterns, Marsh Wrens, and Common Yellow-throats. In winter, Canada Geese (and occasionally Snow and Ross'), dabbling ducks, and Long-billed Dowitchers often congregate near the bridge.

The Bill Williams arm of Lake Havasu supports large numbers of Western and Clark's Grebes. Their courtship dances may be seen in late spring and summer, and young birds are evident in late summer and fall. A wide variety of other waterbirds have been seen on this part of the lake, including Red-throated Loon, Red-necked Grebe, Brown Booby, Barrow's Goldeneye, Red Phalarope, Parasitic Jaeger, and Heermann's Gull. Boats can be rented at Havasu Springs Marina (14).

Downstream from Parker Dam, the deep-water spillway and channel are attractive to loons, diving ducks, and gulls in winter. This is an excellent place to look for Greater Scaup, Barrow's Goldeneye, and Hooded Merganser. The federal government residential park (15), just below the dam, is a good place to see Inca Doves and Bronzed Cowbirds on the California side of the river. Vermilion Flycatchers, presently rare in California, also have bred there recently. This and numerous other trailer resorts along the "Parker Strip" attract many migrant and wintering landbirds. In addition, recently recorded rarities in this area have included Greater Pewee, Eastern Phoebe, Worm-eating Warbler, and Clay-colored Sparrow. Although nearly any area with trees may be attractive to birds, local reaction to birdwatchers varies from very

friendly to extremely hostile. Permission should be sought before entering any private park.

PARKER VALLEY–COLORADO RIVER INDIAN RESERVATION

At Parker, take Agency Road to the riverfront residences, Parker Oasis
(16), where tall cultivated trees attract a large number of migrants and
rarities. Northern Cardinals are rare residents both here and across the
river at Earp. Also on Agency Road is the Parker Sewage Treatment
Plant (17), which is a protected resting site for ducks and shorebirds.
Just south of the sewage ponds is a small date and pecan orchard where
woodpeckers, warblers, orioles, and tanagers often concentrate in fall.
Lewis' Woodpeckers have taken up residence in this orchard during at
least four different winters.

Most of Parker Valley has been cleared for agricultural use, except for
a strip of dense screwbean mesquite, saltcedar, and scrubby willows
adjacent to the river. The levee road south of Parker follows the riverbank and passes a series of small backwater levees (18) where marshes
have formed. This is a good place to see herons, shorebirds, rails, and
other marshbirds. The deep channel in the vicinity of Deer Island (19)
often hosts large numbers of Common Goldeneyes, Buffleheads, and
Common Mergansers in winter.

Agricultural portions of the valley can be traversed on a network of
paved roads between Parker and Poston. Hawks, Say's Phoebes, Horned
Larks, American Pipits, Loggerhead Shrikes, and many sparrows are
abundant there in winter. Alfalfa and Bermuda grass fields may reward
keen observers with Sprague's Pipits or Chestnut-collared Longspurs.
Check bare plowed fields for Mountain Plovers and longspurs in winter
and for Whimbrels and other shorebirds in spring and late summer.
Irrigated fields are especially attractive to shorebirds in July and August. A flock of Sandhill Cranes winters in fields west of Poston and can
be easily viewed from the Agnes-Wilson Road bridge (20) as they fly to
their roost at sunset.

East of Poston (21) and north of Ehrenberg on Mohave Road (22) there
still are extensive tracts of honey mesquite woodland. Easy access to
these bosques provides an opportunity to see most common desert and
riparian species, including abundant Phainopeplas in winter and Lucy's
Warblers in spring. Wintering Sage Sparrows are common where
patches of inkweed shrubs grow along the roadside or at the mouth of
desert washes.

On the California side of the river, Big River RV Park (23) and Lost
Lake Resort (24) should be checked for migrants and rare landbirds. The
only riparian habitat on the California shore is restricted largely to the
area below Agnes-Wilson Road and contains primarily unproductive

saltcedar. However, there are small willow groves along the levee road north of the bridge. Look for Northern Cardinals in dense vegetation adjacent to the river and Indigo Buntings at burned or other open areas with scattered willows in summer.

BLYTHE–PALO VERDE VALLEY

To the north and south of Blythe, along Interstate 10, is the most extensive agricultural land on the California side of the river. As at Parker, many open-country species are abundant and rarities such as Mountain Plover, Sprague's Pipit, and longspurs should be looked for in winter. Plowed and irrigated fields, especially near Ripley and Palo Verde (25), are excellent places to look for shorebirds, White-faced Ibis, Cattle Egrets, and visiting Wood Storks in summer and fall. Occurrences of Wood Storks, as well as Laughing Gulls, Least Terns, and Black Skimmers around Palo Verde indicate that this may be a frequently used dispersal route from the Salton Sea.

Migrant landbirds can be found where there are tall trees such as at Blythe Golf Course (26), the pecan orchard north on Lovekin Boulevard (27), and Mayflower County Park. Citrus orchards west and south of Blythe (28) support huge nesting populations of Mourning and White-winged Doves, as well as the valley's only nesting Lark Sparrows. About 14 mi (23 km) north on U.S. Highway 95 is the Clark Ranch (29), a small oasis of native trees and marsh that is a working ranch maintained as a sanctuary for wildlife. At least one pair each of Yellow-billed Cuckoos, Summer Tanagers, Gila Woodpeckers, Brown-crested Flycatchers, and Vermilion Flycatchers can often be found there in summer. The long list of unusual species discovered there in recent years includes Thick-billed Kingbird, Prothonotary Warbler, Common Grackle, and Red-shouldered Hawk, as well as nesting Cassin's Kingbird, Northern (Baltimore) Oriole, and Northern Parula. Birdwatchers are warmly welcomed.

CIBOLA NATIONAL WILDLIFE REFUGE

South of Palo Verde, 6,651 ha of riparian and wetland habitat on both sides of the river are protected on Cibola National Wildlife Refuge. Cibola was established primarily for waterfowl management and has become an important wintering area for Canada Geese, Northern Pintail, and Sandhill Cranes. The main river channel through the refuge has been straightened and extensively riprapped, representing one of the most heavily managed reaches of the lower Colorado River. However, the original channel is maintained as a relatively pristine backwater and, together with Cibola Lake, provides shelter for many of the area's waterbirds.

Most of the riparian vegetation on the refuge is relatively unproductive saltcedar. However, until very recently, small stands of willows and cottonwoods sustained rare breeding birds, such as Yellow-billed Cuckoos and Summer Tanagers, and attracted many migrants. Dusky-capped Flycatcher, Mountain Chickadee, Ovenbird, Blackpoll Warbler, and Rufous-backed Robin are among the rare landbirds recorded on the refuge in recent years.

To reach Cibola Refuge, turn onto a marked river access road 1 mi (1.6 km) south of Palo Verde. After reaching the river levee either cross the north bridge (30), 0.5 mi (0.8 km) to the north, and turn right onto the east-side levee road until the refuge is reached, or proceed south on the west-side levee road. North of the refuge along the west side of the river, you will pass one of our experimental revegetation sites (31) where native trees and shrubs were planted in 1977 (see Chapter 4). Presently, this site is one of the largest stands of cottonwoods along the entire lower Colorado River. Breeding Yellow-billed Cuckoos were attracted there in the early 1980s. Rare migrants such as Thick-billed Kingbird, Black-and-white Warbler, and American Redstart also have been found recently on the site. This revegetation site harbors most of the representative riparian bird species found along the lower Colorado River and is always worth a visit.

About 11 mi (18 km) south of the access road, both levee roads reach the south bridge (32). This area, between November and March, is an excellent spot to witness the exhilarating morning flights of Canada Geese from Cibola Lake to the alfalfa fields in which they feed. Look carefully for the noticeably smaller Cackling Geese that may occur in the flocks, as well as for the rarer geese species that have all wintered at Cibola on occasion.

Along the main river channel loons, frigatebirds, pelicans, terns, and migrating shorebirds may be found at appropriate times of the year. Where isolated backwaters are accessible, Least Bitterns, Yuma Clapper Rails, and other marshbirds can often be observed easily. A large winter roost of American Crows has formed in recent years; some of these birds can be seen from Highway 78, west of the refuge. Note that the Arizona–California state boundary follows the old river channel in this area, so in many parts of Cibola National Wildlife Refuge both sides of the main river channel are in Arizona.

Further information and permission to enter limited-access bird-watching areas can be obtained from: Refuge Manager, Cibola National Wildlife Refuge, P.O. Box AP, Blythe, California 92225.

IMPERIAL NATIONAL WILDLIFE REFUGE

Although access is limited (except by boat), this refuge, encompassing 10,706 ha, offers an excellent variety of wetland and riparian habitats. In contrast to the stretch at Cibola, the Colorado River at Imperial has not been channelized and a trip by boat will hint at the pristine beauty observed by early explorers. Numerous backwater lakes support large numbers of wintering cormorants, coots, ducks, and geese, as well as nesting Least Bitterns, rails, and other marshbirds. The most recent record of Fulvous Whistling-Duck was of 17 birds at Martinez Lake during the winter of 1985–86.

Where small cottonwood or willow groves persist, Great Blue Herons and Great Egrets nest in mixed colonies. A few Yellow-billed Cuckoos, Northern (Gilded) Flickers, Brown-crested Flycatchers, and Summer Tanagers also may be found in summer. The remainder of the riverside vegetation is mostly dense saltcedar or reeds. Yellow-breasted Chats and Blue Grosbeaks are common in summer and Abert's Towhees are abundant year round.

The refuge headquarters (33) and Martinez Lake may be reached by car by turning west off U.S. Highway 95, 22 mi (35 km) north of Yuma. A small network of roads enables the visitor to see most habitats typical of the refuge. The only other access is at Picacho California State Park and Taylor Lake (34). Drive north 20 mi (32 km) from Winterhaven, California, on Picacho Road. Further information may be obtained by writing: Refuge Manager, Imperial National Wildlife Refuge, P.O. Box 72217, Martinez Lake, Arizona 85364.

IMPERIAL DAM–LAGUNA DAM

Between these two dams is a network of canals, spillways, ponds, and lakes that provides an excellent opportunity to view water- and marsh-birds. This is an excellent area to look for Mexican species such as Brown Pelican, Olivaceous Cormorant, boobies, and rare herons. Some rare species from the north such as Pacific Loon, Barrow's Goldeneye, White-winged Scoter, and Thayer's Gull have occurred there as well. Extensive marsh habitat near Imperial Dam, at West Pond (35) and at Mittry Lake (36) support Least Bitterns, Yuma Clapper Rails, and other marsh species. Black Rails are locally common in shallow marshes along West Pond and Mittry Lake.

From the west end of Imperial Dam, a raised levee separates West Pond from a series of settling ponds (37) and offers a good view of both. Large numbers of Ring-necked Ducks, Lesser Scaup, smaller numbers of Canvasbacks, and an occasional Greater Scaup winter on the settling ponds. Concentrations of egrets and Spotted Sandpipers often feed along the concrete dikes. Herons and shorebirds sometimes congregate in the shallow spillway directly below the dam. From Imperial Dam, take a side trip to Senator Wash Reservoir where loons, grebes (occasionally including Horned), and goldeneyes are found in winter.

Just north of Laguna Dam is the abandoned trailer community of Shantytown (38), set in a grove of tall mesquites, scattered willows, and cottonwoods. Although many of the taller trees have been cut, many riparian birds are common and several uncommon breeding birds such as Yellow-billed Cuckoo and Summer Tanager still may be found on occasion. Inca Doves, Bronzed Cowbirds, and Northern Cardinals also have been seen, as have several rare migrants.

On the Arizona side of Imperial Dam is Imperial Oasis (39), where food, ice, and gasoline can be purchased. A gravel road heads south along the shore of Mittry Lake to Laguna Camp (40) at the east end of Laguna Dam. This road passes desert washes, willow groves, and re-vegetation sites, while offering several views of the large lake and marsh. Look for wintering Bald Eagles and reintroduced Harris' Hawks. A Northern Jacana was found there in the summer of 1986. South of Laguna Dam the paved road passes through orchards and agricultural land, crosses the Gila River, and ends at U.S. Highway 95.

YUMA AND VICINITY

Having the longest history of human settlement in the valley, the Yuma area is largely agricultural land with very little remaining riparian veg-etation. With most of its water diverted at Imperial Dam, the Colorado River becomes little more than a sluggish stream, except when fed by floodwaters from the Gila River or releases from upstream dams. Yuma is a well-vegetated oasis, supporting quite a few native bird species and attracting many migrants.

North of Winterhaven, on the California side of the river, County Road 24 winds through date, pecan, and citrus orchards and agricultural fields towards Bard and Laguna Dam. This area historically has pro-duced many significant bird records but has remained largely un-searched in recent years. A Ruddy Ground-Dove was found at Bard in 1988, however. Check for rare summer residents or unusual migrants in patches of cottonwoods arising from seeps along the All American Canal. Along this same stretch the levee road parallels the river chan-nel, which is lined by dense reed and scrubby patches of riparian trees. Concentrations of cormorants, herons, egrets, rails, and other marsh-birds may be found, depending on water levels. West and south of Win-terhaven, the Mexican border is reached at Andrade (41). Rows of tall athel tamarisks just north of town are very attractive to migrants and the All American Canal is choked with marsh vegetation. This is one of the few sites where Ruddy Ducks have bred in the valley.

In Arizona, a network of paved roads traverses the agricultural land between Yuma and the Mexican border at San Luis. As in other agri-cultural valleys, open-country species are abundant and tall trees may attract migrants or strays. The levee road can be accessed at several points and remnant riparian patches along the river can be checked (42). This area is being restored through mitigation by the Bureau of

Reclamation and Arizona Department of Game and Fish, and is among the first sites where recently invading Black-shouldered Kites may have become resident.

Northeast of Yuma, U.S. Highway 95 enters the lower Gila River Valley, passing through more agricultural land with rows of tall tamarisks. About 1.5 mi (2.4 km) north of the Gila River crossing, Dome Valley Road turns east and zigzags towards Wellton and Tacna. Although technically outside the lower Colorado River Valley, this area's proximity to Yuma and its high potential for birds warrants its inclusion. Irrigation of the surrounding agricultural fields provides ideal habitat for waterbirds. Large concentrations of migrant ibis, shorebirds, and ducks use the area, as do visiting Brown Pelicans and Wood Storks. The region's largest rookery of Cattle Egrets also has formed recently along this part of the lower Gila River.

The only tall trees in this area are athel tamarisks. These trees, until many were cleared in 1985, supported breeding Lucy's Warblers and a few Summer Tanagers. This stand still provides an oasis for rare migrants and wintering residents, including such rarities as Winter Wren, Hutton's Vireo, and Williamson's Sapsucker.

Table 6 provides a brief synopsis of the most conspicuous or significant events in the annual cycle of Colorado River bird species. These include arrival, departure, and breeding activity of permanent and seasonal residents; the first, last, and peak appearance of common migrants; and various rarities to be looked for during particular seasons.

Table 6. Calendar of Avian Activity in the Lower Colorado River Valley

Month/Day		Residents	Migrants	Rarities
January	1	Rough-winged Swallows increase Crissal Thrashers start singing	First Costa's Hummingbirds and Cinnamon Teal	Oldsquaw
	10	Influx of Bald Eagles Influx of bluebirds, waxwings, robins		Northern Shrike
	20	Violet-green Swallows arrive at Parker Dam		Rare gulls
	30			
February	1	Lowest populations of permanent residents		Eastern Meadowlark
	10			Greater White-fronted
	20	Gambel's Quail begin "cow" calls Ash-throated Flycatchers increase Cliff Swallows arrive	Red-breasted Mergansers	Goose Rusty Blackbird
	28	Turkey Vultures arrive		Bohemian Waxwing at Lake Mohave
March	1	Black-chinned Hummingbirds arrive Phainopeplas begin breeding	Peak Tree Swallows Mountain Plovers	
	10	Last Canada Geese and Sandhill Cranes depart Western Grebes begin courtship Lucy's Warblers arrive Lesser Nighthawks arrive	Sage Thrashers	

Month/Day		Residents	Migrants	Rarities
March	20	First Western Kingbirds arrive Northern Orioles arrive		
	30	Bell's Vireos arrive Clapper Rails increase	First Black-headed Grosbeaks, Lazuli Buntings, Wilson's Warblers	Franklin's Gull
April	1	White-winged Doves arrive, Red-shafted Flickers depart	peak Nashville Warblers	Elf Owl
	10	Most Ruby-crowned Kinglets, Yellow-rumped Warbers, and sparrows depart	California Gulls, shorebirds	Common Black-Hawk Brant
	20	Summer Tanagers arrive Yellow-breasted Chats arrive	Swainson's Hawks Caspian Tern, Willets	
	30	Brown-crested Flycatchers, Blue Grosbeaks arrive	peak Barn Swallows	
May	1	Most species in peak breeding activity	Peak warblers and flycatchers	
	10	First Gambel's Quail broods	Vaux's Swifts	Eastern warblers
	20		Peak Western Wood-Pewees, Western Tanagers Swainson's Thrushes	
	30	First broods of many riparian species		Northern Phalarope Northern Parula, Scissor-tailed Flycatcher

10		Willow Flycatchers; Last spring migrants	Rose-breasted Grosbeak; Tropical Kingbird
20		First "fall" Willets; First Black-necked Stilts	Least Tern
30			Red-eyed Vireo
July 1	Second brood of many riparian species	Peak Marbled Godwits; First Tree Swallows	Black Skimmer
10	Most Lucy's Warblers depart	First Warbling Vireos, Western Tanagers; First Black-headed Grosbeaks	
20		Peak Forster's Terns	Magnificent Frigatebird; Brown Pelican
30	First Chipping and Lark Sparrows arrive		
August 1	Most Northern Orioles depart	Peak shorebirds	Whimbrel; Short-billed Dowitcher
10	First Northern Harriers, accipiters arrive; Belted Kingfishers, Orange-crowned Warblers arrive	First Yellow and Nashville Warblers	Wood Stork
20	Most White-winged Doves depart; Peak densities of most permanent residents; First Savannah and Vesper Sparrows arrive	First Northern Pintails, Green-winged Teals; Rufous Hummingbirds; Peak Black Terns	Northern Waterthrush

Month/Day		Residents	Migrants	Rarities
August	30	Most Brown-crested Flycatchers and Yellow-billed Cuckoos depart First House Wrens, Lincoln's Sparrows arrive		
			Peak White-faced Ibis, Wilson's Phalarope	Jaegers (any or all species) Eastern Kingbird
September	1	First Sage Sparrows arrive	Common Terns, Willow Flycatchers Red-necked Phalaropes	
	10	First American Pipits arrive	Peak Western Tanagers Peak Warbling Vireos	Dickcissel
	20	White-crowned Sparrows arrive Yellow-rumped Warblers arrive	Peak MacGillivray's Warblers	Sanderling
	30	First Sandhill Cranes arrive	Caspian Terns Greater White-fronted Geese, large Turkey Vulture flocks	Cassin's Kingbird
October	1	Most Ruby-crowned Kinglets and Phainopeplas arrive	Peak Barn Swallows Peak American Avocets and American White Pelicans	Sabine's Gull
		Red-shafted Flickers and sapsuckers arrive	Peak Black-throated Gray Warblers Last common fall migrants	Red-breasted Nuthatch
	10	Last Summer Tanagers depart Dark-eyed Juncos arrive		Eastern warblers
	20	First Snow and Canada Geese arrive		
	30	Influx of wintering raptors		Evening Grosbeak
November	1	Common Mergansers arrive		Painted Bunting

Date	Events	Species
November 10	Most Snow Geese arrive; Common Goldeneyes arrive	Heermann's Gull
20		Surf Scoter, Barrow's Goldeneye; Pacific Loon, Horned Grebe
	Lawrence's Goldfinch (during flight years)	Red-breasted Mergansers
30		Tundra Swan, Ross' Goose
December 1		Golden-crowned Kinglet; Sprague's Pipit
10		Greater Pewee; Thick-billed Kingbird
20	Anna's Hummingbirds begin to breed	Rufous-backed Robin; Longspurs
30		Golden-crowned Sparrow

7   Species Accounts

The species accounts cover 400 species that have occurred with certainty in the lower Colorado River Valley through 1989. Most are documented by specimens or photographs, and these records are noted in the text. The remainder are sightings by competent observers, with accompanying written details reviewed by either us or the appropriate state review committee. Because so many recent sight records (since 1984) are our own, or those of close colleagues, we feel that a very incomplete picture of the regional avifauna would result if these were not incorporated into our summary. It should be noted that virtually all sightings and other records from 1942 to 1962 are those of Gale Monson, who generously shared his knowledge and time. Other persons to whom we are indebted for field notes or individual records are listed in the Preface.

Those additional species reported with insufficient evidence for documentation are treated as hypothetical in Appendix 2. A few very recent records (1988–90) may not have been received or reviewed in time to be treated in the text. For the convenience of our readers, species are ordered following the American Ornithologists' Union Checklist of North American Birds (1983 and supplements 1985, 1987, 1989). Although there remain many additional arguments for taxonomic revision, we leave these to other authors (see especially Monson and Phillips 1981). Species accounts are organized into as many as five sections, depending on their applicability to each species, as follows.

STATUS

This section begins with a brief description, and documentation when necessary, of relative abundance, seasons of occurrence, and distribution within the valley. The relative abundances we use correspond with those in the bar graphs in Appendix 3 and may be defined as follows: *Common*—always present in moderate to large numbers in proper habitat; *Fairly common*—always present, but in small numbers, in proper habitat; *Uncommon*—occurs locally or patchily and in small numbers in the given habitat and season; *Rare*—occurs annually during the period indicated, but usually in very small numbers, or extremely local in distribution; *Casual*—occurrence is sporadic and normally unexpected, species outside their usual ranges.

Thus, common and fairly common species should be seen daily by most observers, and uncommon species may be seen almost daily with some diligent searching. Rare and casual species are much less predictable, and all sightings of these should be carefully documented. *Max:* refers to the maximum number of individuals of a species observed in one day (not necessarily all at one place) either by a single observer or group effort, such as on a Christmas Bird Count.

Changes that have occurred in the species' status through time or comparisons with regions outside the valley are then briefly discussed. The status of identifiable races or comments on taxonomy also are treated in this section. Comparative information for Arizona was taken primarily from Phillips et al. (1964) and Monson and Phillips (1981). Statements about southern California came principally from Garrett and Dunn (1981). These sources were consulted for virtually every species and because of general space limitations we do not specifically cite them in the individual records. However, we wish to acknowledge them fully here. Similarly, most statements concerning the subspecific identification of specimens, or about taxonomy, were taken from Grinnell and Miller (1944), Phillips et al. (1964), Monson and Phillips (1981), Rea (1983), and Phillips (1986). Whenever our discussions are contrary to these published accounts, or are from other sources, it is duly noted.

HABITAT

This second major section describes important habitats used by the species in the valley. These are usually listed in decreasing order of importance. Changes in habitat use through time (seasonal and annual) and comparisons with habitat use in areas outside the valley are commented on when appropriate. Relative abundance in various habitats also will be given when appropriate. These density estimates will be expressed as "birds/40 ha," which is equivalent to birds/100 acres.

BREEDING

Data on breeding times, nest locations, clutch size, and number of broods are treated only as they apply to the lower Colorado River Valley. Where we have extensive data on the breeding biology of a species, these are summarized.

FOOD HABITS

In this section, we summarize the food habits of each species, based on our specific information from the lower Colorado River Valley. Behavioral accounts are based on quantitative and qualitative observations made during our fieldwork. Diet information comes from stomach samples from common species collected from 1977 to 1980. Percentages given usually refer to percent volume of stomach contents. We discuss habitat and seasonal differences in diet whenever possible.

COMMENT

This category is used sparingly for miscellaneous information or for interpolation of information from previous sections.

The following abbreviations are used frequently throughout the species accounts: BWD = Bill Williams Delta; CBC = Christmas Bird Count; LCRV = lower Colorado River Valley; NWR = National Wildlife Refuge; photo = photograph; spec = specimen.

ACCOUNTS

Red-throated Loon (*Gavia stellata*)

STATUS A casual transient and winter visitor; may prove to be of regular occurrence. One seen 21 November 1947 near Bullhead City, one 31 March–2 April 1978 on lower Lake Havasu (photo), one present from 23 December 1980 to at least 21 February 1981 on upper Lake Havasu, and two 8 February 1986 at lower Lake Havasu. Also one summer record, 8–15 June 1948 on Lake Havasu.

This loon is extremely rare inland in the West. It has not been recorded at the Salton Sea.

Pacific Loon (*Gavia pacifica*)

STATUS Recently a rare but regular winter visitor and transient, primarily in the northern parts of the valley. Most records are from Davis Dam and upper Lake Havasu from late October (spec, 27 October 1949) through March, but seen south to Mittry Lake 19 April 1941 (first Arizona record; Arnold 1942) and Senator Wash Dam 27 February 1977. One remained through summer at Davis Dam to at least 17 July 1979, and one seen 13 August 1954 at Lake Havasu (may also have summered locally).

There were only five records before 1976, but since then it has been observed each winter, with up to eight on Lake Havasu in late fall 1980. It has been noted with increasing frequency at other inland bodies of water in the Southwest as well.

Common Loon (*Gavia immer*)

STATUS An uncommon winter visitor and migrant from October to early May, casually remaining through summer. Sometimes seen in small groups on Lake Havasu and Lake Mohave in November and April. Possibly increasing as a wintering species. Max: 24; 10 April 1954, at Lake Havasu.

Individuals that linger in spring often acquire their full breeding plumage. On rare occasions its eerie cries have been heard at Lake Havasu, chilling the hot desert air.

HABITAT All loons prefer deep water areas such as channels below dams or large reservoirs. They forage for fish primarily in protected coves or along rocky shorelines. These birds have undoubtedly increased.

Yellow-billed Loon (*Gavia adamsii*)

STATUS A casual winter visitor; one found 24 December 1989 on Lake Havasu was seen repeatedly, at least through February 1990. (Not shown in bar graphs.)

Recently this loon has been found on a number of inland reservoirs throughout the West. There is only one additional record for Arizona.

Least Grebe (*Tachybaptus dominicus*)

STATUS A casual visitor from Mexico; nested (six adults and three young) at West Pond near Imperial Dam in October 1946. Also seen at West Pond, 14 May–22 June 1955.

These records suggest that a few may have been resident in the area throughout that period. This common tropical grebe ranges north normally to southern Sonora, Mexico, but regularly occurs north to ponds along the Arizona border. There is a recent record from the Salton Sea, 19 November–24 December 1988.

HABITAT This is a rather shy and retiring grebe that usually seeks marsh or other dense vegetation. Look for it in heavily vegetated canals and ponds in southern portions of the valley.

Pied-billed Grebe (*Podilymbus podiceps*)

STATUS A locally common resident and breeder throughout the valley; more numerous and widely dispersed in winter, with an influx of northern migrants. Max: 158; 29 December 1945, Imperial NWR.

HABITAT This ubiquitous marshbird may be found at almost any aquatic area during much of the year. It is most common in backwaters, marshes, and canals. Although less common on large reservoirs, a few usually occupy protected coves, especially in winter.

BREEDING This grebe is usually solitary and quite secretive, but breeding pairs often announce their presence with surprisingly loud cuckoo-like howls that emanate from dense cattail marshes in spring. The

FIGURE 16. Red-necked Grebe on lower Lake Havasu, 23 March 1981; a first record for Arizona. (Photo by M. Kasprzyk.)

downy, pin-striped young have been noted as early as late March. Two broods may be raised in a season.

Horned Grebe (*Podiceps auritus*)

STATUS A rare, but regular winter visitor and transient from early November (exceptionally, 27 October 1957, spec) to mid-April. Most often seen at Davis Dam, Parker Dam, and Lake Havasu but has occurred as far south as Imperial Dam and Senator Wash Reservoir. Max: 8; 22 November 1985, Lake Havasu.

This species may be increasing as a winter resident because of the establishment of large reservoirs and deep river channels near dams. Occasionally individuals occur along other parts of the river during migration. They usually do not associate with the more numerous and gregarious Eared Grebe. All solitary grebes on open water should be closely scrutinized.

Red-necked Grebe (*Podiceps grisegena*)

STATUS A casual visitor. One in breeding plumage photographed 23 March 1981 on lower Lake Havasu (Fig. 16), and a possible sighting above Davis Dam, 28 January 1984.

These represent the only records for Arizona, although this species has been found on the Nevada portions of Lake Mead. The species remains unrecorded at the Salton Sea, and there are only two inland records for southern California.

Eared Grebe (*Podiceps nigricollis*)

STATUS A fairly common transient and winter resident from September through May. Winter numbers vary considerably from year to year, with occasional large concentrations (up to 2,000) present on Lake Havasu. More numerous in spring than fall, with peak migration in late April and early May. Individuals in breeding plumage occasionally are seen during June and July, but breeding has not been observed. Max: 3,375; 25 April 1982, Lake Havasu.

In fall, migrants appear during the last days of August and numbers accumulate throughout the season. However, fall flocks on Lake Havasu rarely exceed 200–300 birds, and these may disappear after early December in some years. In spring, large numbers may arrive suddenly in the latter half of April on Lake Havasu. Some birds linger late into May; e.g., 40 on 29 May 1979, and 120 on 27 May 1982.

Away from Lake Havasu only single birds or small groups (2–10) are usually encountered, but up to 100 have been noted in the Yuma area in winter. Use of lower Colorado River reservoirs as a migration or winter stopover is undoubtedly a relatively recent phenomenon. Grinnell (1914) did not record this species during the spring of 1910. However, as with many other waterbirds, its numbers on the lower Colorado River pale when compared to the huge rafts found on the nearby Salton Sea.

HABITAT Large flocks of several thousand may be present on Lake Havasu in some years, with scattered individuals in nearly every section of the river at those times. A few are seen occasionally in marshes, backwaters, sewage ponds, and canals.

Western Grebe (*Aechmophorus occidentalis*)

STATUS A recent local breeder on Lake Havasu, with numbers augmented by transient and wintering birds from late August to March. Large winter concentrations occur north to Lake Mohave; uncommon south of Parker Dam. One breeding record away from Lake Havasu; a pair with young at Topock Marsh, 3 September 1984. Max: 1,138; 15 November 1981, Lake Havasu (note that 6,803 *Aechmophorus* grebes were counted 21 December 1979 on lower Lake Havasu, but the two species were not differentiated).

The status is complicated by recent recognition of two distinct "morphs" now considered separate species (see Ratti 1979; Nuechterlein 1981; Storer and Nuechterlein 1985). Population data at Lake

Havasu from 1981 to 1983 allow these tentative conclusions. Western Grebes, which predominate in most North American populations, represent the bulk of Lake Havasu's wintering birds (averaging 1,000–1,500), as well as nearly all records of migrants elsewhere in the valley. They also constitute about 35% of the breeding population.

Range expansion of these grebe species has paralleled changes in aquatic habitats after placement of Parker Dam in 1938. After Lake Havasu filled in the mid-1940s, wintering grebe numbers began to increase; however, they remained scarce before 1960. Cattails first appeared on the lakeshore in the early 1950s, and in 10 years extensive marshes began forming at the shallow north end of the lake and at BWD. The combination of emergent vegetation as a substrate for their floating nests and large open lake as an arena for their elaborate courtship displays offered ideal nesting habitat. Breeding was first reported in 1966, and 100 pairs were noted at Lake Havasu by 1973. Estimates of 250 pairs were made in 1977, and 450 pairs were counted in 1982. Unfortunately, early records and counts of grebes on Lake Havasu did not differentiate the two currently recognized species, so it is impossible to describe the historical appearance of each species separately. Both species were common at least by 1980.

It is a marvel to see these birds thrive among the hundreds of jet-powered boats and water skiers who also find suitable habitat on Lake Havasu in summer. The grebes have an uncanny way of diving at the last moment to escape oncoming boats, and they allow close approach if not directly threatened. Although few grebes are killed, heavy summer boat traffic may cause delayed breeding in some years.

Away from Lake Havasu, individuals are noted most frequently in fall and winter. Occasionally, nonbreeding birds are seen in summer. Nesting may have occurred earlier at Topock Marsh and should be looked for elsewhere. Winter numbers in the Yuma area are normally low (5–15 birds); however, up to 350 were on Martinez Lake in November 1977.

HABITAT An open-water species, the Western Grebe is found almost exclusively on large reservoirs. Breeding grebes seek protected coves with marshy shorelines.

BREEDING They breed, along with Clark's Grebes, primarily in cattail marshes at the north end of Lake Havasu. Courtship dances begin in March and downy young have been noted from July through November. Nests are floating masses of dead reeds, usually anchored to bulrushes. Clutches consist normally of three to four eggs, but up to six eggs have been attended by adults. Eggs often lie in water that has seeped into the nest.

As in other mixed colonies, interbreeding by the two species on Lake Havasu is rare. In 1982, only 5% of grebes observed were intermediate

FIGURE 17. Clark's Grebe with young on Lake Havasu, August 1978. Note the white face extending above the eye, and the all-pale (yellow) bill. (Photo by K. V. Rosenberg.)

between the two species or were in mixed pairs. Because both species have colonized this area simultaneously in the past 20 years, this situation offers potential for further investigation of the ecological and behavioral differences exhibited by Western and Clark's Grebes.

Clark's Grebe (*Aechmophorus clarkii*)

STATUS A locally common resident and breeder at Lake Havasu. Two also found on the 17 December 1988 Yuma CBC. No certain records elsewhere at present. Max: 890; 23 November 1981, Lake Havasu.

Until recently, this species was considered a color morph of the Western Grebe. Its historical and present status in the LCRV away from Lake Havasu is uncertain. Clark's Grebes are the predominant form in Mexico; they also occur in "mixed" colonies in the southwestern United States north to Oregon. They seem to be largely resident on Lake Havasu, with numbers remaining somewhat stable at 500–700 birds, making up about 65% of the breeding population from 1981 to 1983.

HABITAT This is an open-water species, but it may use the deeper parts of the lake, whereas Western Grebes use coves and shallower water. Much work is still needed to explain these spatial differences (see Nuechterlein 1981).

BREEDING All breeding takes place at the north end of Lake Havasu. Nest-site characteristics, clutch size, and hatching rates do not differ between the two grebe species. More detailed studies, however, may determine if any differences have been overlooked.

An interesting behavior seen only in Clark's Grebe is that many parents carry their newly hatched young to the Bill Williams arm at the south end of Lake Havasu, some 50 km from where they were hatched (Fig. 17). All Western Grebes and the remaining Clark's Grebes stay near the north end of Lake Havasu with their young. The reason for the southward movement of Clark's Grebe families, or whether it occurs every year, remains a mystery.

Laysan Albatross (*Diomedea immutabilis*)

STATUS A casual visitor. One found alive on a Yuma street, 14 May 1981 (Fig. 18) and another found dead along the Gila Gravity Main Canal east of Yuma, 18 July 1988 (photo of spec). The first bird was released and later died at San Diego, California.

These records are not without precedent. Another inland record from the California desert and recent sightings from the northern Gulf of California and Salton Sea (all in May) suggest that this species may occasionally move north into the Gulf and seek an overland route back to the Pacific Ocean.

Least Storm-Petrel (*Oceanodroma microsoma*)

STATUS Accidental visitor. One seen 17 September 1976 at Davis Dam after tropical storm Kathleen.

This storm also transported this species to the Salton Sea (> 200) and several other inland Southwest localities. Two other individuals were on Lake Mohave, just north of our study area, on 12 September 1976. For an analysis of the effect of tropical storm Kathleen on bird occurrences in this region, see Kaufman (1977).

Blue-footed Booby (*Sula nebouxii*)

STATUS A casual postbreeding wanderer from Mexico in late summer and fall. Recorded in 1953, 1954, 1959, 1971, and 1977 north to Lake Havasu. At least two individuals have remained through winter.

Years of occurrence correspond with larger invasions of boobies to the Salton Sea and coastal southern California. The lack of records along the Colorado River in other such years (e.g., 1969) is probably due to lack of observers.

HABITAT Boobies seek open water, such as reservoirs and deep river channels around dams, and they may frequent marinas.

Brown Booby (*Sula leucogaster*)

STATUS A casual postbreeding wanderer from Mexico in late summer and early fall. Recorded in 1943 (photo), 1946 (spec), 1953, 1958, 1973 (photo), and 1977 (photo) north to Lake Havasu. One individual remained at Martinez Lake from 5 September 1958 to 7 October 1960, and a bird lingered on Lake Havasu from 19 August until early December 1977.

As with the preceding species, records correspond to occurrences at the Salton Sea. These two boobies may occur in the same year or in separate years, and their dispersals to our region are not necessarily related. The 1943 and 1946 records were the first for the United States.

American White Pelican (*Pelecanus erythrorhynchos*)

STATUS An uncommon transient (but sometimes in large flocks) primarily from March through May and from late September through October. Definite migrants have been noted as early as late February in spring and late July in fall. Individuals or small transient groups are rarely seen during summer months. However, a flock of 250 was near

FIGURE 18. Laysan Albatross found on a Yuma street, 14 May 1981; a first record for Arizona. The photograph was taken at San Diego's Sea World, where the bird was sent for rehabilitation. (Photo courtesy of G. McCaskie.)

Imperial Dam in July 1958. Rare and irregular in winter with most records of 1–3 birds from the Yuma area, but up to 100 have wintered at Cibola Lake. Max: 3,200; 4 October 1956, near Laguna Dam.

Although the seasonal status of this species has remained unchanged, the number of birds passing through the LCRV has declined considerably since the 1950s. Largest flocks noted in recent years have been of 300–500 birds, both in spring and fall (see Fig. 13, p. 81). Consistent passage of large flocks during the first week of October in most years suggests that the bulk of the migrants may come from one or a few breeding colonies. At least one individual recovered at Imperial NWR had been banded at Yellowstone National Park. This species winters abundantly at the Salton Sea.

HABITAT This species most often is seen in flocks migrating over the river or lakes. They occasionally rest in marshes or on sandbars in the river.

Brown Pelican (*Pelecanus occidentalis*)

STATUS A rare but annual postbreeding wanderer from Mexico in late summer and early fall. Individuals occasionally linger through winter and into the following spring.

This pelican is most frequently seen around Imperial Dam, but individuals have occurred north to Davis Dam and even to Lake Mead. Virtually all records are of lone immatures, undoubtedly dispersing from breeding colonies in the Gulf of California, or perhaps via the Salton Sea. Up to four have been seen together in Dome Valley east of Yuma, August–September 1979.

HABITAT In the LCRV they prefer large open-water areas such as near dams. They occasionally frequent marinas where they may become quite tame. Numerous weak or dying birds have been recovered, attesting to the difficulty they have in finding food away from marine habitats.

Double-crested Cormorant (*Phalacrocorax auritus*)

STATUS A fairly common local breeder in secluded backwaters at Topock Marsh and Imperial NWR. Also a common transient and winter resident, often in large flocks. Winter numbers vary greatly from year to year. Max: 2,314; 19 December 1987, Yuma CBC.

Although a few may be seen year round, peak migrations through the valley are from late March through April and in October. Most wintering birds have arrived by November, but there is occasionally a midwinter influx from the north.

HABITAT This species requires flooded snags or live riparian trees for nesting and is, therefore, sensitive to recent habitat alterations involving loss of trees. Away from breeding colonies, cormorants occur along the entire river channel and are occasionally seen on small ponds and agricultural canals.

Olivaceous Cormorant (*Phalacrocorax olivaceus*)

STATUS A casual visitor from Mexico, perhaps becoming regular in recent years. First recorded at West Pond near Imperial Dam 13 April 1971, and probably the same individual 22–23 April 1972 and again 7 April 1973.

This species occurred annually from 1978 to 1982 as follows: 27 January 1978 at Cibola Lake, 20 December 1979 to late January 1980 at Mittry Lake, 7 September to 7 October 1981 at West Pond (photo), and 18 June 1982 at Lake Havasu. All records are of single birds in flocks of Double-crested Cormorants, making them easily overlooked. Unlike the pattern for most waterbirds dispersing from Mexico, most (if not all) of our records are of adults. Recently this species has reached the Salton Sea and Lake Mead.

Magnificent Frigatebird (*Fregata magnificens*)

STATUS A rare summer visitor from Mexico, occurring almost annually from 1971 to 1981 (first observed in 1954). Most records are between 30 June and 8 September north to Lake Havasu.

Although most sightings are of lone immatures that probably represent a "normal" dispersal pattern from the Gulf of California, perhaps via the Salton Sea, there has been one notable exception. Associated with tropical storm Kathleen, four frigatebirds were seen together at Tacna, east of Yuma on 10 September 1976. The following day, six (five adults) flew over the desert west of Blythe. A single bird that flew south over Davis Dam, 17 September 1976, probably also was associated with the storm. There has been speculation that some Arizona records (Monson and Phillips 1981) and, in particular, some storm-blown birds (Kaufman 1977) may be Great Frigatebirds (*F. minor*); therefore, all wayward frigatebirds should be carefully identified to species if possible.

HABITAT Frigatebirds are usually seen flying high over the river or lakes. They can undoubtedly traverse much of the valley in a relatively short time.

American Bittern (*Botaurus lentiginosus*)

STATUS An uncommon and local winter resident from late August to early May. Recent summer records of calling birds, such as at Martinez

Lake in 1978 and BWD in 1980, suggest breeding but this is uncon-firmed. Max: 11; 20 December 1976, Parker CBC.

HABITAT This species is solitary, secretive, and usually only detected when flushed from dense marsh or wet riparian areas.

Least Bittern (*Ixobrychus exilis*)

STATUS A locally common breeder from April through September, with a few present after late February and in October. Uncommon in winter around Imperial Dam and south to Yuma (up to 5 on CBC); rare farther north. Max: 41; 13 July 1978, Imperial NWR.

Visual encounters with this secretive species are rare. The observer who is familiar with its varied and rail-like calls, however, will appre-ciate its true abundance. In July, recently fledged young will occasion-ally perch in the open, affording good views. By counting calling birds we have estimated the breeding density of Least Bitterns to be about 40 birds/40 ha in some marshes. This species is listed as a Species of Spe-cial Concern by the California Department of Fish and Game.

HABITAT Largest populations are in extensive cattail or bulrush marshes, such as at Topock and near Imperial Dam. A few are found in a variety of other marshy sites elsewhere throughout the valley, includ-ing ponds and along agricultural canals.

Great Blue Heron (*Ardea herodias*)

STATUS A fairly common year-round resident and local breeder throughout the valley. Usually seen singly or in small groups, except when at rookeries or, rarely, in migratory flocks. The breeding popula-tion is augmented by migrants from the north and west in winter. Max: 150; 18 December 1982, Yuma CBC.

Grinnell (1914) found them abundant in 1910, with many active col-onies present along the entire river. Today only a remnant of the origi-nal population persists, and nesting colonies are few. Besides the pres-ence of abundant suitable nest sites (cottonwoods), the availability of fish regularly stranded by fluctuating river levels was thought to pro-mote large numbers of this and other heron species before construction of dams.

HABITAT For much of the year they are seen feeding solitarily on river-banks, mud flats, marshes, and occasionally in agricultural fields or along canals. Tall riparian trees are usually required for nest sites, al-though a few colonies have formed on protected rock outcrops, such as "Heron Island" in the Bill Williams arm of Lake Havasu.

BREEDING Rookeries form in late February and nesting continues through May and June. There are usually 10–15 nests/colony, except at Topock Marsh where up to 75 pairs have nested. Although cottonwoods are usually used for nesting, tall cultivated trees such as eucalyptus are sometimes used. Such a rookery in a residential area at Parker is evidence of this species' tolerance of human activity in the valley, as long as suitable nest sites are available. Birds at Heron Island are even known to nest in saguaro cacti. Elsewhere, as remnant cottonwoods vanish, there is an increasing tendency for isolated pairs to attempt nesting away from rookeries.

Great Egret (*Casmerodius albus*)

STATUS A fairly common year-round resident and local breeder. Concentrations occur near breeding sites and occasionally flocks are seen elsewhere, especially in spring. Max: 550; 31 July 1952, Imperial NWR.

In contrast with the preceding species, Great Egrets were encountered only once by Grinnell in 1910, and no nesting was noted. The first evidence of nesting was in 1947 at Topock Marsh (Monson 1948a). Historically, anecdotal evidence indicates that this species was common as a postbreeding visitor from the south.

HABITAT The breeding population is limited by availability of tall riparian trees for nest sites. Away from breeding colonies, egrets gather at a variety of sites to feed, especially in marshes, backwaters, and below dams. This species visits agricultural fields more often than other resident waders, except Cattle Egrets.

BREEDING Like the Great Blue Heron, this species breeds colonially at rookeries in tall willow or cottonwood groves. Where the two species nest together, Great Egrets begin their breeding activity later and have been seen to displace herons from specific nest sites. At present, fewer than 10 rookeries are known. All are small (2–16 nests), except at Topock Marsh where as many as 144 pairs have nested in drowned mesquites.

Snowy Egret (*Egretta thula*)

STATUS A fairly common resident and very local breeder in small numbers. Transients occasionally concentrate in spring and early fall. Most numerous from Imperial Dam south, where flocks of up to 150 have been recorded recently. Uncommon to rare in winter in northern parts of the valley, where formerly it was more numerous year round. Max: 490; 30 July 1959, above Laguna Dam (California Swamp).

Interestingly, this egret was not seen by Grinnell in spring 1910, although it may have been present as a postbreeding visitor in summer. A change in status took place in the next 30–40 years, with nesting first noted at Topock Marsh in 1947 (Monson 1948a).

HABITAT This egret seeks protected backwaters, marshes, or canals for feeding and only rarely visits cultivated fields. At Imperial and Laguna Dams the many spillways and settling ponds are favored foraging sites, and it is here that largest numbers still gather.

BREEDING Definite nesting by this species has been noted only sporadically. It nests most often at Topock Marsh and Imperial NWR. At these two locations, they join other species to form mixed colonies in riparian groves or flooded snags. An exceptionally large colony at Topock Marsh built up to 141 nests in 1950, but had disbanded by 1953.

Little Blue Heron (*Egretta caerulea*)

STATUS A casual visitor from Mexico. Three records: an immature in calico plumage found dead in May 1976 at the southern tip of Nevada, and adults seen 9 May 1979 in Dome Valley and 19 July 1979 at Andrade. An odd heron at BWD, 14–15 August 1976, was reported as a hybrid between this species and the Tricolored Heron.

The pattern of spring and summer dispersal fits well with occurrences elsewhere in the interior Southwest. They are rare but regular visitors to the Salton Sea during these periods and remained to nest there once (1979).

Tricolored Heron (*Egretta tricolor*)

STATUS A casual visitor from Mexico in fall. Recorded 13 October 1954 at Martinez Lake and 8–30 September 1955 at Imperial Dam.

These dates correspond with the primary period of occurrence in Arizona. However, dispersal to the California coast and Salton Sea has been noted chiefly in winter and spring.

Reddish Egret (*Egretta rufescens*)

STATUS A casual visitor from Mexico. Three records from Imperial NWR: 30 September 1954–3 March 1955, 19 November 1955, and 2 September 1960. Also, immatures at Lake Havasu, 4–9 September 1954 (spec), and at Imperial Dam, 11 February to mid-March 1979 (photo).

The 1979 bird was frequently seen to feed characteristically in the shallows below Imperial Dam, in close association with a locally wintering immature Brown Pelican. This behavior may indicate a feeding relationship between these two species where both are common, as along the coast of western Mexico. Being a bird of tidal flats and shallow bays, this egret's northward dispersal is primarily to the coast of southern California. It is among the rarest of the southern herons and egrets inland in the Southwest.

Cattle Egret (*Bubulcus ibis*)

STATUS At present, uncommon and irregular year round but increasing slowly; occasionally seen in large flocks. Most numerous in southern parts of the valley (especially near Yuma), but small flocks regularly winter near Palo Verde and Parker. Max: 500; 7 July 1980, northeast of Yuma.

This famous invading species was first recorded along the river on 23 August 1970, when four were seen at Parker. That same year, breeding was first noted at the Salton Sea. However, numbers in our area never attained levels recorded elsewhere, and the species' presence here remains unpredictable. The success of these egrets in the LCRV has possibly been limited by a scarcity of suitable rookery sites. The presence of large numbers in breeding plumage near Yuma (Dome Valley) in 1980 and Cibola in 1984 suggests that breeding is at least imminent in these areas.

HABITAT This species is found primarily on agricultural land, especially in cattle pastures. They are seldom seen on riverbanks or in marshes with other herons and egrets.

Green-backed Heron (*Butorides striatus*)

STATUS A fairly common summer resident from March through September in the southern half of the valley; less numerous farther north. In winter, uncommon around Imperial Dam and Yuma and rare but regular farther north. Max: 14; 18 December 1978, Parker CBC.

This species has increased as a wintering bird throughout Arizona during the past 50 years. Grinnell (1914) did not find Green-backed Herons before 24 April in 1910 and considered it a common transient with no evidence of breeding. Young of the year were noted, however, in summer 1902 by Stephens (1903).

HABITAT This is a solitary heron, most often encountered feeding silently, sitting motionless on a riverbank, or when flushed from a backwater or agricultural canal.

BREEDING This species is not colonial, although pairs are frequently seen together when breeding. Their stick nests are in trees over or very near water; thus, dense willows are preferred.

Black-crowned Night-Heron (*Nycticorax nycticorax*)

STATUS An uncommon to fairly common year-round resident throughout the valley. Usually occurs singly or in small groups but occasionally is seen in large concentrations at roost sites. Its nocturnal habits make this species' status difficult to monitor. Max: 200; 29 December 1953, Topock Marsh.

HABITAT This heron nests and roosts colonially in dense riparian trees, including mature tamarisks, and occasionally tall cattail marshes. The birds venture out at night to feed in open marshes and along lakeshores, riverbanks, and irrigation canals. Their silent and sluggish appearance by day belie their noisy and active nocturnal foraging antics.

Yellow-crowned Night-Heron (*Nyctanassa violacea*)

STATUS A casual visitor from Mexico; a subadult seen 17 April 1973 at Imperial Dam.
This species' affinity for coastal mangroves probably makes it less prone to wander inland than other wading birds. However, because of difficulty in identifying immature night-herons, the true dispersal patterns may be harder to detect.

White Ibis (*Eudocimus albus*)

STATUS A casual visitor from Mexico. One definite record of an adult 4–5 April 1962 at Martinez Lake (photo). Other records are an old report of one seen March 1914 at Palo Verde and a probable sighting of two adults, 7 October 1982, north of Ehrenberg.
Most other Arizona reports of this species are in spring. However, three records at the Salton Sea have been in midsummer.

White-faced Ibis (*Plegadis chihi*)

STATUS An uncommon to fairly common transient in small flocks from March through May and late July through October. Rare transient throughout June and early July when, as with the other Great Basin-nesting species, north- and southbound migrations seem to overlap (see Fig. 13, p. 81). Also, a rare but regular winter visitor in southern parts of the valley.
An exceptionally large concentration of 5,000 roosted in Dome Valley east of Yuma on 9 September 1979. Although more scarce than in former years, this species may now be wintering more frequently, even north to Topock Marsh. For example, ibis have wintered at Cibola NWR each year since 1976, with up to 50 present in December 1981. In addition, up to 16 were found near Imperial Dam in December 1978,

and at Parker individuals were found in 1979, 1981, and 1983, with a group of 3 found there in 1982.

HABITAT This species prefers flooded fields where it frequently congregates with other wading birds and shorebirds during migration. Others may be found in marshes, along lakeshores, or flying in loose lines low over the river.

Roseate Spoonbill (*Ajaia ajaja*)

STATUS A very rare and erratic visitor from Mexico in summer and fall. Recorded in 1942, 1959, 1969, 1973, and 1977. At least one bird remained through winter until 21 March 1974 at Palo Verde. Max: 21; 18 June 1973, northeast of Yuma.

Most records are from near Yuma, but they have ranged north to Parker Dam (six birds, 24 June 1973) and Topock (7 September 1977). Years of occurrence correspond with similar invasions to the Salton Sea and central Arizona, all involving only immature birds. The 1977 flight also reached Phoenix, Las Vegas, and the coast of southern California.

HABITAT In the LCRV spoonbills seek protected backwaters and marshes.

Wood Stork (*Mycteria americana*)

STATUS A rare but regular postbreeding visitor from Mexico in summer and early fall, rarely north to BWD. Formerly more widespread, numerous, and occurring all year. Max: 230; 16 July 1959, near Laguna Dam.

Storks were described as common or abundant in many of the historical accounts of bird life along the lower Colorado River (e.g., Coues 1866; Cooper 1869; Swarth 1914). Brown saw flocks at Yuma in January and March 1903 (Phillips et al. 1964) and Grinnell (1914) noted a flock of 12 near Laguna Dam, 21 April 1910. Since that time, this species has not been found earlier than June, and since 1974 all records are between 9 June (1983) and 30 September (1977). The former abundance of this and other wading birds is thought to be related to the availability of fish stranded each year by receding floodwaters of the untamed river.

Recent concentrations at Cibola NWR (20–50 birds) and in Palo Verde Valley suggest a current dispersal route via the Salton Sea in California, where large numbers still occur.

HABITAT This species prefers irrigated fields, canal banks, and marshy backwaters.

Fulvous Whistling-Duck (*Dendrocygna bicolor*)

STATUS In recent years a very rare and irregular visitor. Formerly more common, even in winter. All recent records are from the southern half of the valley north to Palo Verde. Max: 27; 16 November 1961, Martinez Lake (Fig. 19).

Since 1970, there are only four records: 1 found dead at Cibola NWR, 3 February 1974 (spec); 2 seen at Cibola, 1 July 1978; 1 there, 13 July 1978; and 17 at Martinez Lake during winter 1985–86. This species is currently an uncommon and declining summer resident at the Salton Sea.

HABITAT This duck has been found primarily in protected marshes, backwaters, and irrigated fields.

Tundra Swan (*Cygnus columbianus*)

STATUS In recent years a rare but regular winter visitor and transient in very small numbers from late November to late February; formerly more numerous. Max: 53; 15 December 1956–25 February 1957, Martinez Lake.

Most records since 1970 involve one to six birds wintering on Imperial NWR, near Parker, or near Topock Marsh. A few other migrants have been noted elsewhere, the largest group being 11 at Cibola NWR, 9 December 1978.

HABITAT Swans prefer protected river channels and marshes and rarely visit open agricultural fields.

Greater White-fronted Goose (*Anser albifrons*)

STATUS An uncommon transient in small flocks during late September (earliest, 10 September) and early October. Also a rare but probably regular winter resident. Usually occurs in small groups (two to seven birds) associated with other geese. Much more numerous until the 1950s, when it also occurred in flocks as an early spring transient. The few recent spring records have been in February and March. Max: 577; 28 September 1943, BWD.

The southward migration of this species is both earlier and much less protracted than in other geese. Since 1976, over three-quarters of all fall birds have passed through between 20 September and 10 October, and over a third have passed through during the last three days of September. Overwintering individuals normally depart in mid-February. From 1 to 6 individuals were noted in most recent springs, and 14 were at Cibola NWR in late February 1983. Because of its scarcity and early migrations, this species is usually not hunted in the LCRV.

HABITAT Wintering birds feed with other geese in open fields on or near national wildlife refuges. Migrants are typically seen flying over the river or lakes and occasionally resting on sandbars or in fields.

FIGURE 19. Fulvous Whistling-Ducks at Martinez Lake, 16 November 1961. (Photo by G. Monson.)

Snow Goose (*Chen caerulescens*)

STATUS An uncommon or fairly common and local winter resident and transient. Sometimes seen in large flocks from November to mid-March. Individuals may arrive as early as 5 October (33; BWD, 1952) and have lingered as late as 10 April. Most numerous at Topock Marsh, especially in late November. There are five records of the "Blue" morph from Topock Marsh, two from Martinez Lake, and one from Cibola. The Blue morph occurs regularly at the Salton Sea.

The largest flock winters near Topock on the Havasu NWR (up to 2,800 birds). These birds regularly commute to agricultural areas near Needles and Bullhead City to feed. At Cibola NWR usually only 25–50 birds are present, but up to 300–500 have gathered there in some years. A small flock of 10–25 birds consistently winters at Imperial NWR. Migrants are occasionally seen elsewhere along the river, primarily in November and late February.

HABITAT Wintering flocks occur in the vicinity of cultivated grain fields, primarily on or adjacent to national wildlife refuges. Migrant groups frequent protected marshes and sandbars and often fly very high over the river.

Ross' Goose (*Chen rossii*)

STATUS A rare but probably regular winter resident from November to March, usually associated with flocks of Snow Geese. Max: 20; 29 November 1981, Topock Farm.

The first record from the LCRV is a specimen from near Topock, 10 December 1948. There are about five other records before 1973, all from Imperial NWR, involving up to six individuals.

A few have wintered consistently since 1978 at Cibola and Havasu NWRs, and less regularly at Imperial NWR. Records away from these areas include six at Blythe, 9 December 1981; single birds south of Parker, 11 December 1981 and 17 November 1982; one at BWD, 25–26 January 1983; and two at Parker, 23 December 1984. This species winters commonly at the nearby Salton Sea and can be easily overlooked.

HABITAT Most sightings are from cultivated grain fields where other geese are feeding. Strays have also been found grazing on lawns or at ponds with domesticated waterfowl and coots. They also may occur on lakes, sewage ponds, or along the river.

Brant (*Branta bernicla*)

STATUS Recently a casual or very rare winter visitor and spring transient. About seven records including one record for two or three individuals with other geese on 16 February 1979 at Cibola NWR.

The other winter records, all of single individuals, are 23 December 1970 at Mittry Lake (photo), and 1 December 1978–16 February 1979 and 2 December 1981, both at Cibola NWR. Individual spring migrants were seen 28–29 May 1965, north of Yuma (photo); 14 April 1975, at Topock; and 14 April 1978, north of Blythe (photo). These probably were northbound migrants returning from the Gulf of California, where considerable numbers now winter. This overland migration route is further evidenced by concentrations in recent springs at the Salton Sea and other southern California lakes and reservoirs. All records pertain to the black form (*B. b. nigricans* Lawrence).

Canada Goose (*Branta canadensis*)

STATUS A locally common winter resident from mid-October to early March. Most numerous on national wildlife refuges where supplemental grain is provided; however, small flocks may be scattered throughout the valley. Summer records pertain to crippled birds. Max: 24,000; winter 1987–88, Cibola NWR.

This is the most common goose wintering in the LCRV. Recent increases are attributable to management practices on the three national wildlife refuges. Largest concentrations have been at Cibola (usually about 20,000–24,000 birds), where numbers have been steadily increasing. Other birds traditionally gather at Topock, BWD (500–700), and Imperial NWR (500–1,000).

An ongoing study, initiated in 1986, will hopefully lead to better understanding of the racial diversity and origins of LCRV birds. The most common race is a large bird known as *B. c. moffitti* Aldrich, which will change with publication of new data by Hanson (in press). Nearly half of the Canada Geese on the lower Colorado River have their origin at Bear Lake in northern Utah and breed in the Great Basin. Several specimens (and perhaps 5% of the wintering population) taken at Cibola NWR fall within the size range of the very large *B. c. maxima* Delacour from the northwestern prairies. Medium-sized individuals may be birds from the Teton Mountains in Montana, southeastern Idaho, and northwestern Wyoming. A much smaller and darker form, *B. c. minima* Ridgway, has occurred at least occasionally; e.g., 15 December 1949 and 14 November 1953 at Topock, and at Cibola NWR, 12 November 1977 and 20 October 1981 (2). This last race is considered to be a separate species by some authors, the Cackling Goose (*B. hutchinsii* (Richardson) (Monson and Phillips 1981) and Aleutian Goose (*B. leucopareia* (Brandt 1836)) (Hanson in press). At least two specimens of Richardson's Goose (*B. c. hutchinsii* (Richardson)) have been collected. This race is also considered by some (Phillips et al. 1964; Hanson in press) to be a separate species.

Goose hunting has become popular relatively recently in the LCRV and is associated with the dramatic winter population increases of this species, especially at Cibola, during the last decade. Wounded individuals are sometimes seen on the river channel or in alfalfa fields after the main flocks have headed north.

HABITAT During winter, this species roosts on large protected lakes, such as Cibola Lake and the Bill Williams arm of Lake Havasu. At dawn, flocks varying in size from 10 to >100 birds fly from their roost to protected alfalfa or other grain fields to feed. Thousands of geese passing overhead every winter morning at Cibola and BWD are an awesome sight, with their resonant honking echoing across the surrounding desert. Away from traditional roosts, transient geese may be found at other marshes, alfalfa fields, or flying high over the river.

Wood Duck (*Aix sponsa*)

STATUS A rare but regular visitor from late September to early March. Small flocks have wintered on ponds along the lower Bill Williams River. Recent summer records include a pair at Imperial NWR, 25 June

1977, and a pair seen 12 May 1982 among flooded cottonwoods at BWD, where breeding is at least possible. Max: 14; 24 December 1977, Bill Williams Delta CBC.

They are sparse late fall transients in Arizona and southeastern California, with relatively few birds wintering away from the Colorado River. Aside from the few summer records, this species has been found from 16 September (1976) to 20 March (1977), both at BWD.

HABITAT This duck is attracted to wooded ponds and protected backwaters and is rarely seen on open river channels or agricultural canals. During the 1970s, beaver activity in the BWD created much suitable habitat for these ducks and other species. Raised water levels destroyed most beaver ponds during the late 1970s and early 1980s. However, new ponds were being formed after the water level dropped again in 1983.

Green-winged Teal (*Anas crecca*)

STATUS A common to abundant early fall transient in late August and September, often occurring in flocks of 40–80 birds. Common through fall and winter in most years, with greatest numbers wintering at Cibola NWR (up to 3,000) and south of Parker (200–400). A northbound migration is usually evident as numbers increase in late February. Nearly all depart by mid-April, with only a few records from May through July. Max: 4,121; 23 December 1951, Havasu NWR.

Recent summer records include single birds near Parker 16 June 1977, 9 June 1978, and 10 June 1979; one at Cibola NWR, 5 June 1979; and a flock of 18 in the Dome Valley, northeast of Yuma, 15 June 1979.

This is the only teal occurring in large numbers during fall and winter. Along with the Northern Pintail, it is among the first ducks to arrive in fall, with a few early migrants appearing by late July. This is the most frequently hunted puddle duck in the LCRV.

HABITAT Wintering birds are scattered along the river, with small groups occupying backwaters, flooded fields, marshes, and unchannelized stretches of river. A favorite wintering spot is west of Poston, where nutrients from a large wasteway canal promote growth of extensive beds of Sago pondweed.

FOOD HABITS Of 101 gizzards examined, 37 contained Sago pondweed and other aquatic vegetation. Thirty-nine gizzards contained mostly seeds; 14, insects; and 3, shelled invertebrates. The remaining gizzards contained roughly equal amounts of insects and vegetation.

Mallard (*Anas platyrhynchos*)

STATUS An uncommon to fairly common winter resident and transient, arriving in early August and mostly departing by mid-April. Rare in late

April and May, and casual in June and July. Has bred, at least in 1943, 1946, 1948, 1959, and from 1982 to 1984. Max: 1,013; 17 December 1972, Yuma CBC.

Although a few usually pass through in August, there is often a gap before the wintering birds arrive in late October or November. Normal winter concentrations rarely exceed 100–300 birds, and are often much smaller. Mallards are rarely taken by hunters in the LCRV, with only 37 recorded between 1974 and 1983.

HABITAT Mallards occur in small flocks along stretches of unchannelized river, in backwaters and marshes, and occasionally in flooded fields.

BREEDING Documented breeding in the valley includes eight young at California Swamp, 16 June 1943; three young near Needles, 20 June 1946; three young at Topock, 18 June 1948; a female with four young at West Pond, 23 June 1959; and several pairs at Cibola NWR in May 1982, 1983, and 1984.

FOOD HABITS Of 25 gizzards examined, 8 contained almost exclusively plant parts; 7, mainly seeds; 5, shelled invertebrates; 1, insects; and the rest, a mixture of seeds, molluscs, and vegetation.

Northern Pintail (*Anas acuta*)

STATUS A common to abundant early fall transient, primarily from mid-August through September, with the first birds arriving in late July. Fairly common to common in late fall and winter, with most concentrations at Cibola NWR (2,000+). Spring migration is primarily in February. Rare after March, but a few may linger into May and small flocks are occasionally seen in June. Max: 2,280; 3 September 1952, Topock.

The southbound passage of pintails in August and September is sometimes impressive. Their shimmering flocks are one of the first sure signs of fall along the river. For example, 1,200 were at Cibola on 1 September 1978 and a flock of 750 was noted near Poston on 5 September 1979. An exceptionally large northbound migration occurred 12 February 1976 when 1,800–2,000 birds passed over Blythe. Winter numbers away from Cibola are often small. However, up to 2,000 wintered at Imperial NWR (1972) and 650 were counted near Parker in December 1977.

This species was unusually common in June 1977, when groups of 1–50 birds were noted on at least 13 occasions, totaling over 200 individuals. Many were in southbound flocks, thought to be migrating early from the drought-stricken Great Basin. Normally, only a few are found between March and July; other notable concentrations include 74 near Cibola NWR, 17 June 1961, and 30 northeast of Yuma, 13 June 1979.

Although sometimes abundant, this wary bird ranked only third among puddle ducks taken by hunters between 1974 and 1983.

HABITAT This species has responded well to waterfowl management practices and is sometimes the most numerous puddle duck in the valley. In winter, large flocks usually congregate in cultivated grain fields, but they may be rare or absent along many stretches of river.

FOOD HABITS The pintail is exclusively a vegetarian in the LCRV. Among 29 gizzards examined, 16 contained only vegetation parts, 7 contained only seeds (wheat and milo), and 2 contained roughly equal amounts of vegetation parts and seeds.

Blue-winged Teal (*Anas discors*)

STATUS A rare to uncommon transient from February to mid-May; occurs in pairs or small flocks with other teal. Rare but regular through summer and early fall, and irregular from November through January. Exact fall status is uncertain due to difficulty in separating females and eclipse-plumaged males from Cinnamon Teal. Max: 10; 1 July 1977, south of Blythe.

Recent fall records include a male at BWD, 15 September 1976; pairs at Cibola NWR, 22 September and 10 November 1978; four at BWD, 28 September 1981; a male at Parker, 29 October 1981; and a female there, 16–27 November 1982. In winter, this species has been found in five years since 1971 at Yuma, and in three years since 1976 at Parker.

Although always the rarest teal species, it was unusually numerous in 1977, when small numbers were seen throughout the spring and summer. In other years they have been nearly absent. The presence of paired birds in summer has raised suspicions of local breeding, but this is unconfirmed.

HABITAT This species occurs throughout the year in marshes, unchannelized river, or small ponds. A regular stopover for this species is the sewage treatment ponds at Parker.

Cinnamon Teal (*Anas cyanoptera*)

STATUS A fairly common transient from early February through mid-May and from late July through September, in small flocks. Uncommon and local through summer, with breeding noted only in 1916, 1979, 1981, and 1983. The first migrants may arrive by early January, and fall birds regularly linger through early October. Rare but regular in late December, when up to 16 have been noted near Yuma (1942, 1973, 1981) and Parker (1979). Max: 820; 22 September 1953, Topock Marsh.

This is one of the few primarily transient waterfowl species in the LCRV. It is the only common duck after wintering species depart in early spring. The sudden appearance of drakes along the river in mid-winter is always a welcome first sign of spring migration, and a reminder that no season here is without subtle and constant change.

There is a gap in late fall when this species is virtually absent from the LCRV. This implies that birds seen in late December may be early spring arrivals and not overwintering. Our only records between 13 October and 15 December are of two at Parker, 25 November 1979, and three at BWD, 5 November 1981. This species is hunted only in October and therefore is not an important gamebird in the LCRV.

HABITAT Cinnamon Teal may frequent a variety of wetland habitats, including marshes, unchannelized river, ponds, irrigated fields, and canals. The few known nestings have been in marshy canals near Poston and Yuma.

BREEDING After the unusual floods of 1915–16, a nest was found near Palo Verde on 30 June 1916; this nest produced six fledged young by 2 July (Wiley 1916). The next evidence of breeding was a group of four adults with four young in an agricultural canal west of Poston on 13 May 1979. A female with two young was seen there again on 25 May 1979. On 13 June 1979, three broods were noted in Dome Valley, east of Yuma, when an unusually large number (100–200) of summering birds was present. This species has bred successfully at Cibola NWR since 1981, with large numbers (200) present in flooded areas during the summer of 1984.

FOOD HABITS Among 20 gizzards examined, most contained a diverse mixture of seeds, vegetation parts, insects, and shelled invertebrates.

Northern Shoveler (*Anas clypeata*)

STATUS An uncommon transient and winter resident from late August to early April. Rare in May and early August and there are several midsummer records. Max: 875; 10 January 1961 and 3 January 1962, Laguna Dam and Imperial NWR, respectively.

This is now one of the scarcest of the regularly wintering puddle ducks, with concentrations rarely exceeding 50–100 birds. It is sometimes absent from some sections of the river, such as at Parker in 1981 and 1982. This scarcity is surprising, considering the thousands that winter at the Salton Sea.

HABITAT Small groups inhabit most unchannelized river stretches and marshes, and migrants occasionally rest in flooded fields and on lakes.

Gadwall (*Anas strepera*)

STATUS An uncommon to fairly common winter resident and transient from mid-September (rarely August) to early April, rarely lingering through May. Casual during summer months. This species has bred locally at West Pond, July 1957 (2 females with 15 young), and at Topock Marsh in 1949, 1951, 1952, and 1970. Max: 650; 12 December 1953, Topock Marsh.

Winter numbers may vary considerably from year to year, with birds being more numerous in the northern half of the valley. A rather large group of 50 transients was at Cibola NWR on the early date of 25 August 1982. Aside from the breeding records cited above, there are about 15 sightings in summer. This species is rarely hunted in the LCRV.

HABITAT This duck prefers unchannelized river channels, marshes, and ponds, although migrant groups sometimes rest on larger lakes.

FOOD HABITS Among 22 gizzards, 16 contained almost exclusively Sago pondweed or duckweed, 5 contained about equal proportions of pond-weed and molluscs, and 1 contained only insects.

Eurasian Wigeon (*Anas penelope*)

STATUS A casual transient or winter visitor, usually occurs among flocks of American Wigeons. Four records: 18 December 1947 at To-pock, 28 September 1974 near Poston, 20–28 March 1981 at Parker (photo), and 19 December 1987 on Yuma CBC.

There is little doubt that a wild population of this Eurasian duck regularly migrates to the West Coast and that our records and other inland Southwest records are a valid extension of that pattern. It is possible, of course, that individuals that escaped captivity may occur here as well.

American Wigeon (*Anas americana*)

STATUS A fairly common to common winter resident from November through February. Uncommon transient in September, October, and March; rarely arriving in August or lingering to mid-May. Casual in June and July. Max: 520; 15 December 1960, Imperial NWR.

In recent years, largest flocks have wintered between Parker and Poston. They are normally scarce in the southern half of the valley. This was the second most-heavily hunted puddle duck in the LCRV between 1974 and 1983.

HABITAT This species is most numerous on unchannelized river, marshes, and ponds. Numbers vary locally from year to year depending on water levels and availability of pondweed beds.

FOOD HABITS The American Wigeon is the most nearly vegetarian species among the common ducks occurring on the lower Colorado River. Among 89 gizzards, 81 contained almost exclusively Sago pondweed. Three specimens contained mainly insects; three, seeds; one, shelled invertebrates; and one, a mixture of insects, seeds, and vegetation parts.

Canvasback (*Aythya valisineria*)

STATUS A rare to locally uncommon resident from November through April; extreme dates are 15 October (1952) to 15 May (1954). Areas of regular occurrence include Imperial Dam, Lake Havasu, and Topock Marsh. Max: 147; 27 December 1955, Martinez Lake.

The scarcity of this duck along the lower Colorado River is surprising, especially considering its abundance at the Salton Sea.

HABITAT Canvasbacks favor ponds and large reservoirs and are rarely seen along the main river channels.

Redhead (*Aythya americana*)

STATUS An uncommon or locally fairly common transient and winter resident from mid-October through March, usually in small flocks. Rare but regular in August, September, April, and May; irregular through June and July. Breeding documented 7–23 June 1943 at California Swamp near Laguna Dam, and in June 1949 at Topock. Max: 600; 23 November 1943, Lake Havasu.

Redheads are usually encountered as small transient groups of up to 10 individuals. Wintering concentrations occur only sporadically and primarily in the northern half of the valley. Variation in numbers and local distribution is related to the presence of Sago pondweed beds which provide most of this species' diet in the LCRV (see below). This species is not heavily hunted in this area because of its scarcity. Also, its diet strongly flavors its meat, making it undesirable.

HABITAT Like other diving ducks, Redheads prefer deep water such as open lakes and channelized river, but may also seek marshes and ponds, especially for breeding.

FOOD HABITS Gizzard contents of 18 birds revealed that 14 ate mostly Sago pondweed, 3 contained a mixture of algae and seeds, and 1 contained only algae.

Ring-necked Duck (*Aythya collaris*)

STATUS An uncommon to fairly common winter visitor from mid-October to mid-April, usually in small flocks. However, large concentrations have occurred near Imperial Dam. Very rare outside that period. Max: 984; 20 December 1977, Yuma CBC.

As in many other waterfowl species, winter numbers are extremely variable, even at Imperial Dam where they are usually most numerous. Away from that area, concentrations are rare. Up to 122 have been noted near Parker (CBC) and 50–60 birds were on Lake Havasu, October–December 1981. Our only summer records are of two pairs at Imperial NWR, 5 June 1960; a male at Martinez Lake, 8 July 1978; a female below Parker Dam, 26 May 1979 (possibly injured); and an eclipse-plumaged male at Andrade, 9 July 1979 (photo). This species is not normally hunted in the LCRV.

HABITAT This duck prefers ponds, protected river channels, and marshes, and often occurs on agricultural canals.

FOOD HABITS Of 21 gizzards, 13 contained primarily algae; 7 contained vegetation parts or seeds; and 8 contained at least some animal foods, including insects, gastropods, and molluscs. This varied diet relates to the species' use of ponds, canals, and backwaters, as well as the main river channel for foraging.

Greater Scaup (*Aythya marila*)

STATUS A rare but regular winter visitor from mid-November to mid-April, usually associated with Lesser Scaup. Occurs regularly at Imperial Dam and below Parker Dam. Max: 25; 16 February 1982, north end of Lake Havasu.

This scaup was first detected in the LCRV on 24 January 1976 at Imperial Dam (flock of 10), and has occurred virtually every winter since then (at least through 1988). Specimens are an immature male south of Parker, 3 January 1977; an immature male near Poston, 17 December 1977; and an adult female from Cibola NWR, 10 November 1984. Four were photographed at Lake Havasu City, 21 December 1984. Curiously, the last scaup to linger in the valley are frequently Greaters (latest record, 20 April 1976).

HABITAT Greater Scaup are found below spillways of dams and on lakes where other diving ducks gather.

Lesser Scaup (*Aythya affinis*)

STATUS An uncommon to common winter visitor from November to April, often in large local concentrations. A few arrive in early October and some individuals on rare occasions linger into May. There are few summer records, most of which probably involve injured birds. Max: 508; 18 December 1982, Yuma CBC.

This species is among the most numerous wintering ducks in most years. Areas of frequent abundance include Imperial and Parker Dams and the Bill Williams arm of Lake Havasu. Notable late concentrations include 27 on Imperial NWR, 7 May 1978, with 3 remaining on Martinez Lake until 15 June.

HABITAT Lesser Scaup are found primarily below spillways of dams, in coves of large lakes, and along the channelized river.

Oldsquaw (*Clangula hyemalis*)

STATUS A rare and irregular winter visitor to northern parts of the valley from late November through February. Most records are from Davis Dam and Lake Mohave (nearly annual since 1975). Recorded south to Imperial NWR on 7 November 1959 and 25 February 1961 (2). There is one summer report: a male at Davis Dam, 6 July 1975. Max: 8; 20 January 1977, Davis Dam.

Other records away from Davis Dam include: at Lake Havasu on 9 February 1951, 10 November 1953 (spec), and 16–26 February 1982 (photo); at Parker Dam on 23 January 1983 (adult male), 27 December 1985–1 February 1986 (photo), and 22 November 1986–6 February 1987; and at Parker on 17–20 December 1980. This species is casual elsewhere in Arizona and at the Salton Sea.

HABITAT This duck occurs primarily in deep, open water, such as on large reservoirs and near dams.

Black Scoter (*Melanitta nigra*)

STATUS A casual winter visitor. One record, an immature male below Parker Dam, 9 January–1 April 1981 (photo). During this time it had largely acquired alternate (nuptial) plumage.

This scoter is casual anywhere inland. There are only four records at the Salton Sea (all 1976–1977) and four additional records in Arizona (all since 1975).

Surf Scoter (*Melanitta perspicillata*)

STATUS A rare and irregular late fall visitor. Most records are from Parker Dam, Lake Havasu, and Davis Dam in late October and November. Casual through winter, with all records since 1981. Max: 4; late November 1975, Davis Dam.

This is the most frequent scoter to occur inland in the Southwest. Small numbers probably migrate regularly overland to the Gulf of California, using the lower Colorado River lakes as stopovers, especially during storms. In addition to our 11 fall records, nearly all other Arizona records have been associated with November storms, including several flocks. However, most records at the Salton Sea and inland California lakes were in spring, suggesting a more western overland route for migrants returning from the Gulf of California. Winter records at Parker Dam are of one that molted into first-spring male plumage between 3 December 1981 and 19 March 1982, and a female from 22 November 1986 to 6 February 1987. In addition, one to two birds were at Davis Dam, 21 January–8 February 1984. An immature male seen at Parker Dam, 10–19 March 1988, may have wintered locally or may represent the only spring migrant.

White-winged Scoter (*Melanitta fusca*)

STATUS Recently a very rare winter visitor. Singles were at Davis Dam in late November 1975, 15 January 1984, and November 1984; at Imperial Dam from November 1976 to 23 April 1977; and south of Parker Dam from 11 November to 10 December 1984. Two were on Lake Havasu from 9 January to 25 February 1981 (photo), and three were near Parker Dam on 27 December 1985 (photos).

This species has occurred sporadically at other Arizona lakes in late fall and winter. Most records at the Salton Sea, however, have been in spring and summer.

Common Goldeneye (*Bucephala clangula*)

STATUS A fairly common to common winter visitor from mid-November through February, often in large flocks in the northern half of the valley, less common in the southern half. Numbers decrease through spring, with a few individuals on rare occasions lingering into May. June and July records probably are of injured birds. Max: 1,860; 21 December 1979, Bill Williams Delta CBC.

This goldeneye has increased as a wintering bird since about 1960, with very few at or below Parker until after 1954. Of the common ducks, it is among the last to arrive in the valley in large numbers each fall. Single wintering flocks of 500–1,000 birds have been seen at Parker and Davis Dams. Farther south, around Imperial Dam and Senator Wash Reservoir, numbers rarely exceed 50–100 and are often much less.

Of 1,117 ducks taken by hunters between 1974 and 1983, 439 were Common Goldeneyes, making them the most frequently hunted species in the LCRV. Although few hunters seek these birds, the handful who do are enthusiastic.

HABITAT Goldeneyes prefer deep open lakes and channelized stretches of river, such as near dams.

FOOD HABITS Among 227 gizzards, 154 contained mainly aquatic insects (primarily caddis fly larvae), 27 contained about equal amounts of insects and molluscs (primarily the Asiatic clam), and 46 contained primarily molluscs.

Barrow's Goldeneye (*Bucephala islandica*)

STATUS Since 1974, a rare but regular winter visitor from mid-November to early April. Usually occurs with flocks of Common Goldeneyes below Davis and Parker Dams, with scattered records south to Imperial Dam. Max: 57; 17 December 1974, Davis Dam.

The flock at Davis Dam was present each winter from 1974 until 1979, but did not appear again until 1983–84 (52 birds). Smaller numbers (up to 15) have been found at Parker Dam each winter since 1977. Records farther south are mostly of single individuals at Cibola, 14 February 1978; at Parker in December 1977, January–February 1979, 2 March 1982, 22 December 1984, and December 1985–February 1986 (up to 8); at Senator Wash Dam, 20 February 1979 (3); and at Imperial Dam, 31 January 1980. Specimens are all from Parker Dam: an adult male, 23 December 1985, and immature males, 20 December 1986 and 15 December 1987. There are very few records away from the Colorado River in Arizona or interior southern California, and all of these have been since 1977.

Bufflehead (*Bucephala albeola*)

STATUS An uncommon to fairly common winter visitor from the end of October to March in the northern half of the valley, less numerous farther south. Most individuals depart by April, but a few sometimes linger into May. June records are probably injured birds. Max: 464; 23 December 1977, Parker CBC.

Largest flocks are consistently in the vicinity of Big River Resort below Parker. Near Imperial Dam, concentrations of up to 165 have occurred in some years. Very few have been found between these two areas of concentration. It is not a popular gamebird among hunters, probably because of its small size and low palatability. Among 1,117 ducks checked in the bags of 286 hunters from 1974 to 1983, Buffleheads totaled 159.

HABITAT Like goldeneyes, this species concentrates on deep channelized river reaches, lakes, and near the spillways of dams. A few may also be found on ponds and canals.

FOOD HABITS From 111 gizzards, 88 contained primarily insects (nearly all caddis fly larvae) and the rest primarily contained remains of the Asiatic clam.

Hooded Merganser (*Lophodytes cucullatus*)

STATUS A rare but regular winter visitor from November to mid-March. Max: 13; 12 January 1950, Topock Marsh and 17 December 1988, Yuma CBC.

Most records are of females or immatures. However, a few drakes have been seen. This species is less regular south of Parker Dam, but up to six were near Yuma (December 1982). Extreme dates of occurrence are from 29 October (1981) to 27 March (1954).

HABITAT This species occurs on small ponds, as well as on large lakes and around the spillways of dams.

Common Merganser (*Mergus merganser*)

STATUS An uncommon to common winter visitor from November through February, decreasing rapidly by April. Occasionally occurs in large flocks during colder winters. Small numbers recorded during summer months at Parker Dam were most likely injured birds. Max: 1,863; 21 December 1979, Bill Williams Delta CBC.

This is the most often encountered merganser in winter in the LCRV. Its abundance and southern extent of occurrence in the valley may be correlated with the amount of fresh water that freezes north of the region. There are relatively few present during mild years. The largest concentrations are consistently on Lake Havasu; e.g., a single flock of 1,150 was seen 2 January 1953. However, a flock of 300 flew south past Cibola NWR, 5 November 1974, and 250 were near Imperial Dam later that winter. They may be very scarce in the southern half of the LCRV in some years. This species is rarely hunted in the LCRV, probably because most hunters consider them unpalatable.

HABITAT These mergansers frequent deep river channels, large lakes, and spillways of dams.

FOOD HABITS Fish, primarily threadfin shad, were the exclusive content of 17 gizzards examined.

Red-breasted Merganser (*Mergus serrator*)

STATUS A fairly common but irregular transient from mid-February through May, and again in late October through November. Rare but regular in winter (December–January) in very small numbers and irregular through summer. Max: 880; 19 April 1978, Lake Havasu.

Large numbers of this species may pass through within a short period and undoubtedly have gone undetected in some years. The appearance of early migrating flocks (e.g., 170 at Martinez Lake, 22 February 1957) may give the erroneous impression that they are common wintering birds. The largest concentration in fall was 244 at Imperial NWR, 8 November 1958. Unprecedented during winter was a count of 376 on 19 December 1987 on the Yuma CBC; possibly a concentration of late migrants. Up to 8 have been seen together in summer (24 June 1949) at Topock and up to 31 have wintered near Yuma (December 1976).

Ruddy Duck (*Oxyura jamaicensis*)

STATUS An uncommon to rare transient and winter resident from mid-August to mid-May. Rare but regular through the summer months; occasionally breeds (e.g., 1941, 1943, 1946, 1978, 1979). Max: 376; 8 October 1955, Imperial NWR.

They are quite locally distributed, with concentrations only at Imperial Dam in winter and occasionally on Lake Havasu in fall (up to 50 birds). Notable summer concentrations (nonbreeding) include 209 at Mittry Lake, 23 June 1943; 16 at West Pond, 5 July 1958; 44 at West Pond, 16 July 1959; 20 at Parker Dam, 26 May 1979; and 17 at Cibola NWR, 5 June 1979. This species never occurs in the large numbers found at the Salton Sea and on lakes in central Arizona, and it seems to have decreased steadily since the 1950s.

HABITAT For nesting, they have used marshy ponds and canals. At other times, Ruddy Ducks may appear anywhere along the river, on lakes, and rarely in smaller canals.

BREEDING Definite breeding records for this species include: Mittry Lake, 19 April–11 May 1941 (≤10 young); California Swamp above Laguna Dam, 2 May–June 1943 (≤12 young); near Headgate Rock Dam, California side, in June–July 1946 (female with 6 young); at Imperial NWR, 14 May 1978 (female with 5 young); and at the All American Canal at Andrade, 19–20 July 1979 (adult with 7 young).

Turkey Vulture (*Cathartes aura*)

STATUS A fairly common summer resident, with large movements into and out of the valley primarily in March and early October. Uncommon in winter, when it is usually absent from areas north of Parker. Usually occurs singly or in small groups, except when in large migratory flocks. Max: 750; 28 September 1982, Blythe.

Spectacular among migrating birds in the Southwest are the concentrated kettles of Turkey Vultures that follow the Colorado River each spring and fall. Their arrivals and departures fully embrace the period of hot weather in the valley and are good indicators of changing seasons. Largest fall numbers consistently pass through during a short period between 28 September and 5 October. A late flight of 41 birds was noted, 2 November 1955. Spring migration may be more protracted, and northbound flocks have been noted as early as 30 January (185 in 1982). The largest spring flight was 350 crossing the river at Imperial NWR, 15 March 1977. An unusual winter roost of 130 birds formed south of Ehrenberg in 1978–79.

HABITAT Vultures typically soar over agricultural lands searching for carrion, but individuals may forage over any habitat. Groups sometimes rest on sandbars and we have seen birds drink from the river. Nighttime roosts are in tall trees or on cliffs.

BREEDING Actual nests have rarely been observed, but they undoubtedly breed in cliff crevices and in mountains along the edge of the valley.

California Condor (*Gymnogyps californianus*)

STATUS Formerly probably an occasional visitor. Apparently seen at Yuma by Coues in September 1865, and also in March 1881 upriver at Pierce's Ferry near the Grand Canyon, where also known from remains found in caves. (Not shown in bar graphs.)

There are a few other historical records from Arizona, from as late as 1924.

Osprey (*Pandion haliaetus*)

STATUS An uncommon transient from mid-August through mid-October and from early March to early May. Rare but regular, and increasing, winter visitor from November through February; rare and irregular in midsummer, except in the Laguna–Imperial Dam areas and Lake Havasu where individuals occasionally are present year round. Max: 18; 19 December 1987, Yuma CBC.

This is one of the few primarily transient raptors, usually migrating singly. As many as 22 have been noted in a season (fall 1960).

HABITAT Ospreys may linger to fish or roost at backwaters, lakes, or near dams. Although this species appears to be present year round at Imperial Dam, and occasionally at Topock, there is no hint of nesting anywhere in the valley.

Black-shouldered Kite (*Elanus caeruleus*)

STATUS Recently a rare visitor; increasing and perhaps becoming resident. There are five sightings to date: 22 February 1979 and 22 March 1980 at Cibola NWR; 13 November 1980 near Poston; 10 September 1984 at Yuma; and four near Hunter's Hole south of Yuma during the fall and early winter of 1984, with two remaining until December 1985. Also seen in the Colorado River Delta, Sonora, Mexico.

The most likely source of these birds is southern California, where populations have increased greatly and dispersing individuals have recently been noted elsewhere (now regular year round in small numbers at the south end of the Salton Sea). However, a Mexican source cannot be discounted as that population also is spreading rapidly north. This species has reached southeastern Arizona regularly since about 1978, and since 1983 has established itself as a breeding bird (Gatz et al.

1987). It is likely that this graceful raptor will eventually become an established resident in the LCRV.

HABITAT This kite typically hunts over agricultural land and brushy vegetation bordering fields.

Mississippi Kite (*Ictinia mississippiensis*)

STATUS A casual visitor. Three records: an immature seen at BWD, 17 July 1982; a subadult at Blythe, 24 July 1985; and an adult flying over Cibola NWR, 6 June 1986.

This species has expanded its breeding range along southeastern Arizona rivers during the past 15 years, favoring native cottonwood groves. It has occurred elsewhere in western Arizona in midsummer, and it appears almost annually as a casual late spring migrant in southern California. Its occurrence along the Colorado River is, therefore, not unexpected.

Bald Eagle (*Haliaeetus leucocephalus*)

STATUS A rare to uncommon winter visitor, increasing in some years after midwinter. The first birds may arrive by mid-October and virtually all are gone by mid-March. One pair unsuccessfully attempted to breed at Topock each spring from 1975 to 1977. An unseasonal immature at Cibola NWR, 19 May 1978, may have been a wanderer from the Arizona breeding population.

In most years there is an influx of Bald Eagles into the valley in January or February. Up to 15 individuals have been accounted for in a single winter (1977–78) between Topock Marsh and Imperial NWR. Most sightings are of lone immatures.

Two nesting pairs were found breeding along the Bill Williams River near Alamo Dam in 1987. It may be possible, therefore, to find dispersing young or wide-ranging foraging adults during spring and summer on the Colorado mainstem in the near future. In addition, at least some of the birds arriving in midwinter are from the Arizona breeding population. In 1988, a radio-tracked fledgling from the Verde River, Arizona, was followed to British Columbia and then reappeared in the LCRV at Martinez Lake during December of the same year (Granger Hunt, pers. comm. *fide* Tom Gatz).

HABITAT Appropriately this federally endangered species is most often found on the three national wildlife refuges where there are protected backwater lakes and marshes. Individuals may be seen, however, along any stretch of the river.

BREEDING In the Southwest these birds require mature cottonwood trees or cliffs for nest sites. Arizona birds initiate breeding by early

January, and young are usually fledged by early June. In addition to the Topock pair, the closest nesting pairs are along the Bill Williams and the Big Sandy Rivers, at least since 1985. Hopefully, we may soon see eagles attempting to breed again on the lower Colorado River proper.

Northern Harrier (*Circus cyaneus*)

STATUS A common transient and winter resident throughout the valley from September through March, with numbers increasing steadily in late August and decreasing in April. Migrants may arrive in early August and linger to mid-May; casual through June and July. Max: 107; 22 December 1982, Parker CBC.

The Northern Harrier, previously known as Marsh Hawk, has probably increased as a wintering species in the LCRV since the clearing of riparian vegetation for agriculture. This species has recently nested in western Arizona (in April) and rarely in southern California. Our few late July records probably represent very early fall migrants; however, one apparently summered at Topock (seen 12 June and 25 July 1950), and one was south of Blythe, 30 June 1977.

HABITAT This is an open-country species that seems to be adapting well to agricultural development in our region. It is seen most often hunting over alfalfa or grass fields in early morning or late afternoon. Individuals also forage over sparse riparian vegetation, marshes, or even over open desert.

Sharp-shinned Hawk (*Accipiter striatus*)

STATUS A fairly common transient and winter resident. Migration is mainly from late March to early April and in late September, but recorded from the end of August to 10 May (1977). Max: 26; 23 December 1977, Parker CBC.

HABITAT This small hawk inhabits any patch of moderately dense vegetation that provides open pathways for its swift, maneuverable flight. Thus, mesquite and willow groves are preferred haunts, as are brushy borders of agricultural fields and canals. Migrating individuals may soar over open country. This is the most likely hawk to be seen around human residences.

Cooper's Hawk (*Accipiter cooperii*)

STATUS A fairly common transient and winter resident from late August to late April. Former breeding noted near Picacho in 1910, at Parker in 1946, 1948, and 1950, and at BWD in 1953 and 1954. Presence of birds through summer at Topock, 1946–47 and 1949–51, also indicated nesting. There is one recent summer record, 24 July 1982 at BWD, but this species is not known to breed at present in the valley. Max: 24; 23 December 1977, Parker CBC.

Cooper's Hawks normally outnumber Sharp-shinned Hawks during most of the year. They often arrive earlier and stay later, with extreme dates of migrants being 7 August (1976) and 18 May (1978). Historically, breeding was under way in April and May, and it is possible that some recent attempts at nesting have been overlooked.

HABITAT This hawk prefers extensive patches of riparian vegetation, especially those with cottonwoods or other tall trees. It appears to be dependent on mature cottonwood–willow groves for nesting.

Northern Goshawk (*Accipiter gentilis*)

STATUS A casual fall transient and winter visitor. Records include an old specimen from Palo Verde, 2 November 1916; one, 29 November 1972, at Topock Marsh (photo); an adult seen 15 October 1978 at Ehrenberg; and another adult north of Ehrenberg, 10 December 1982–21 February 1983. An additional sighting of an adult, 6 June 1964, at Earp was highly unseasonal.

Goshawks are well known for their wide dispersal during major flight years. Documented irruptions in Arizona corresponded with our records in 1916 and 1972, but not in 1978 or 1982.

Common Black-Hawk (*Buteogallus anthracinus*)

STATUS A casual transient and recent summer visitor. One bird, seen repeatedly between 23 March and late June 1979 at BWD, exhibited courtship behavior. One seen there also on 3–25 August 1981, 8–16 September 1982, and 13 April–9 July 1983. Migrants elsewhere include 25 August 1946 south of Parker, 11 April 1978 at Ehrenberg, 28 September 1978 at Davis Dam, 2 April 1980 at Fort Mohave, and 25 March 1983 at Dome Valley.

The flurry of recent records coincides with the discovery of this species in the Virgin River drainage in southeastern Nevada and southwestern Utah. These areas, as well as the Big Sandy River in northwestern Arizona, are likely sources for our migrants. As yet, there are no records specifically for the California side of the river, although one was seen farther west at Thousand Palms Oasis, 13 April 1985. An immature seen near Ehrenberg on 5 May 1979 showed characteristics of a Great Black-Hawk (*B. urubitinga*).

HABITAT Because this hawk requires tall cottonwood–willow associations near water, BWD offers the only potential nesting habitat in the valley today. The migrant in Dome Valley was flushed from a concrete-lined canal.

Harris' Hawk (*Parabuteo unicinctus*)

STATUS Formerly a fairly common resident and breeder in many parts of the valley, although not noted by Grinnell (1914) in 1910. Extirpated due to unknown causes by 1961 except at Topock Marsh, where a small population persisted until 1964. May still occur as a casual visitor, but status has been clouded by accidental releases by falconers and by recent attempts to reestablish a breeding population. Max: 30; 27 December 1950, Havasu NWR.

The earliest records in the LCRV seem to be those of Stephens (1903), who shot two near Ehrenberg in 1902 and saw two others on the California side. Wilder (1916) reported this species "in numbers in the river bottom near Palo Verde" on 1–3 December 1902. He noted "10–20 in the air at a time," with two birds shot. The first evidence of breeding was on 25 July 1916 when four full-grown young, not able to fly, were found south of Palo Verde, and a nest was found there the following spring (Wiley 1916, 1917). Five were seen in the Potholes region near Laguna Dam in December 1924, and a nest was found there in April 1930 (Miller 1925, 1930). A small population was apparently localized in that portion of California, with birds seen around nest sites in both summer and winter. At least a few individuals (1–5) were seen each winter on Imperial NWR during the 1940s and 1950s. The last record there was of a failed nest at Adobe Lake in 1961.

The total extent of this species' historic distribution is unclear. At Topock Marsh, our earliest record is of seven in December 1946. Eighteen were counted there in 1948, increasing to a high of 30 in 1950, with more than 20 present on winter counts at least until 1953. This population declined to two birds by the spring of 1961 and disappeared shortly thereafter. Other locality records include one at BWD, 20 December 1943; two at Parker, 31 December 1953; and one north of Blythe, 28 November 1964. This latter record is the last certain sighting of a wild Harris' Hawk in the LCRV. Recent sightings of wary, unbanded birds that may have been wild include 2 October 1977 north of Ehrenberg, 3–5 December 1978 at the Clark Ranch near Blythe, and 20 July 1979 near the Mexican boundary at Andrade.

Since 1978, the U.S. Bureau of Land Management has undertaken a reintroduction program by releasing birds of all ages in the vicinity of Mittry Lake, Imperial NWR, and Cibola NWR. Stray individuals from these releases have been found as far north as Parker. Several pairs have formed, with one pair successfully raising young in 1983 and three pairs nesting successfully in 1986. The reestablishment of a self-sustaining and increasing population in the LCRV, however, is uncertain.

HABITAT This is a bird of desert-scrub and mesquite habitats where saguaro cacti or riparian trees provide nest sites. This species used mesquite and willow groves along the Colorado River, often adjacent to marshy backwaters, as well as drowned trees at Topock Marsh. As these situations are now extremely rare throughout the valley, the reintroduction of Harris' Hawks will not succeed unless an effort is made to revegetate large stands of riparian habitat.

BREEDING The location of the first nest found in 1917 was a mesquite thicket with arrowweed underbrush, near a large thicket of tall willows and a slough with many dead trees, cattails, and tules (Wiley 1917). Two eggs were in the nest on 17 April and these hatched on 27 April. Four full-grown but flightless young were noted on 25 July 1916. A nest at the Potholes was 12 m up in a cottonwood and contained a 5- to 6-day-old chick on 18 April 1930 (Miller 1930).

This hawk was a common nesting bird at Topock Marsh from at least 1946 through the late 1950s. Nests were invariably built in dead mesquites which had been drowned by rising water levels following the completion of Parker Dam. Eggs were laid as early as 11 February, with most nests having young by 1 April. Through these same years, and up to the 1960s, nests could be found throughout the valley from Topock south. The last known nest on Imperial NWR was one at Adobe Lake, 29 May 1961, which held two dead juveniles.

FOOD HABITS Two stomach samples from the Potholes in December 1924 contained remains of several Green-winged Teal and a flicker. This hawk fed mainly on ducks, according to local residents at Potholes (Miller 1925). One was observed to eat a Common Moorhen and a Sora in April 1930 (Miller 1930). Nesting birds at Topock preyed chiefly on cotton rats, muskrats, and American Coots.

Red-shouldered Hawk (*Buteo lineatus*)

STATUS A rare and irregular visitor from the west; recorded now in all seasons and probably increasing in numbers. Has bred in the valley at least once (1970) at Mittry Lake.

The nesting attempt, as described by Glinski (1982), was in a cottonwood grove at the north end of Mittry Lake. The adult pair had been shot, and the two stolen young were confiscated by the Arizona Department of Game and Fish, reared in captivity, and identified as the California race *B. l. elegans* Cassin. In 1973, the cottonwood grove containing the nest was cleared during a "waterfowl habitat management" program.

There are now at least 13 other records: an adult seen 16 February 1962 across from Yuma; a subadult at BWD from 29 October 1977 to 31 January 1978 (photo); another adult at Yuma, 12 March–23 April 1978; two seen near Needles, 21 January 1978, with one there again on 18 February and 26 July 1978; a subadult north of Palo Verde, 21 November

1981; one at Davis Dam, 27 February 1982; an adult at Blythe, 21 August–18 December 1983; and an adult and subadult together at Clark Ranch north of Blythe during fall and winter 1983–84. One of these birds remained at Clark Ranch at least until June 1984, and an adult (possibly the same) was seen there 3 June 1985. The latest reports were of one near Poston, 17 January 1986, and one at Parker, 24 November 1989 through at least January 1990.

These occurrences are associated with a population expansion and increased dispersal eastward in southern California and southern Nevada. Although this may suggest repeated attempts to establish a viable population along the Colorado River, the fate of the 1970 nesting effort emphasizes the odds against this elegant bird gracing the valley as a resident in the near future.

Broad-winged Hawk (*Buteo platypterus*)

STATUS A casual transient. An adult was studied at Cibola NWR, 15 April 1982. Another probable sighting was of two flying high over BWD on 5 August 1980.

This highly migratory species is recorded annually in the Southwest and a few have been discovered recently wintering in coastal southern and central California and western Mexico. Its presence again in the LCRV may be expected.

Swainson's Hawk (*Buteo swainsoni*)

STATUS An uncommon transient from mid-August to mid-October and from late March to mid-May, in very small numbers. Early fall migrants were at Cibola NWR, 19 July 1977, and north of Ehrenberg, 29 July 1976. An adult at Cibola NWR, 30 June 1983, was unseasonal and may not be classified as a normal migrant. There is also one careful winter sighting of a light-morph adult on 5 January 1975 at Bard.

In recent years, this species has decreased substantially in the Southwest, even as a migrant. Some early records of large flocks of "migrating Harris' Hawks" in agricultural fields are probably referable to flocks of immature or dark-morph Swainson's Hawks (see Bent 1964). Concentrations are still noted east to the Phoenix area, but today large concentrations occur very rarely in the southern California deserts. The only recent report of a flock in our area was of 42 birds just east of the LCRV near Roll, Arizona, on the early date of 14 July 1951. Although July records are early for fall migration in southern California, they are within the period of high concentration in central Arizona and New Mexico. Since 1975, no more than 11 individuals have been observed in a single season (spring 1978), and no more than 4 have occurred at a single location (6 September 1976, south of Yuma). About a third of the migrants have been dark-morph birds.

HABITAT This open-country hawk is seen primarily in agricultural areas, where it is prone to perch on haystacks or on the ground in fields. Individuals may also roost in riparian groves, orchards, or other tall trees.

Zone-tailed Hawk (*Buteo albonotatus*)

STATUS A rare and irregular summer resident at BWD, where nests were found in 1943, 1946, and 1957, and a pair was courting on 10 May 1977. Additional sightings there include two on 28 March 1978, with one seen regularly until 19 June, and a lone bird on 20 May 1983. Three records of migrants: 3 April 1949, south of Parker; 25 April 1953, along lower Lake Havasu; and 22 September 1983, northeast of Yuma. There is one old winter specimen from Yuma, 23 January 1902. Also, a mounted specimen, shot locally, was in the W.C. Store in Somerton in 1942.

This species nests regularly as close to the LCRV as the Hualapai Mountains in west-central Arizona. There are also several recent breeding records for southern California, and a few spring migrants have been noted there at desert oases in recent years.

HABITAT This hawk is primarily a bird of Upper Sonoran Desert streams and mountain canyons. Because tall trees are required for nesting, the BWD provides the only remaining habitat suitable for this species along the Colorado River.

Red-tailed Hawk (*Buteo jamaicensis*)

STATUS A common transient and winter resident from late September to early April. Uncommon summer resident and breeder, primarily along the edges of the valley. Max: 91; 22 December 1984, Parker CBC.

The LCRV birds are typically light-morph (*B. j. calurus* (Cassin)) with indistinct or no bellybands, approaching the race *B. j. fuertesii* Sutton and Van Tyne. However, in summer a few very dark individuals have been noted. At other times nearly all the Western races can occur, and dark-morph birds from northern populations are sometimes numerous. Erythristic individuals, with chestnut underparts, also may occur and we have seen very pale birds with nearly all white tails that approach the plumage of Krider's Red-tailed Hawks (*B. j. kriderii* Hoopes). The Harlan's Hawk (*B. j. harlani* (Audubon)) is probably also a casual visitor, with typical individuals carefully studied during the winters of 1980–81, 1982–83, 1984–85, and 1989–90 between Ehrenberg and Parker. All sight identifications of Krider's and Harlan's Red-tailed Hawks are nevertheless suspect. Various intergradations between other races or phases may result in characters similar to those found in these populations (Phillips et al. 1964; Lish and Voelker 1986).

In addition, immature birds predominate in fall and winter and they may be heavily streaked or blotched and have highly variable tail patterns. Thus, this common species presents an excellent example of geographical and individual plumage variation and offers a challenge to even the most experienced birder who visits the LCRV.

HABITAT This is the most ubiquitous raptor in the valley, especially in winter. This species is common in both riparian and agricultural areas, as well as in desert washes and the outskirts of towns. The breeding population in the immediate valley is likely much reduced from former times. However, winter numbers may have increased since the clearing of dense riparian groves for agricultural use.

BREEDING Local birds seek the few remaining cottonwood trees or retire to the peripheral deserts where saguaro cacti or rocky cliffs provide nest sites. Nests are also found on the towers of large power transmission lines. Juveniles have been noted by early June.

Ferruginous Hawk (*Buteo regalis*)

STATUS An uncommon transient and winter resident from mid-October to mid-March; most numerous in the northern half of the valley. A few migrants may arrive by late September and linger into early April. There are always a few dark-morph individuals among wintering birds. Max: 38; 22 December 1982, Parker CBC.

Numbers on the Parker CBC increased steadily from 5 birds in 1976 to a high count of 38 in 1982. Local concentrations have appeared near Blythe since 1981. About 20 or more birds roosted in a single grass field west of Blythe in December 1983. This species is only occasionally found in the Yuma area.

HABITAT In recent years this hawk may have increased as a winter resident because of its association with agricultural land. It perches on the ground more often than the Red-tailed Hawk. Both species are equally fond of telephone poles.

Rough-legged Hawk (*Buteo lagopus*)

STATUS A rare and irregular visitor from late November through February, primarily in the northern half of the valley.

The LCRV marks one of the southern limits of its winter range, even in flight years. During one winter (1977–78), seven birds were noted between Bullhead City and Martinez Lake. In other years the species has been scarce or even absent.

HABITAT This hawk prefers open terrain and has been found primarily in agricultural land where other buteos concentrate.

Golden Eagle (*Aquila chrysaetos*)

STATUS A rare and irregular visitor to the immediate river valley; primarily seen in fall and winter.

One or a few were noted each year since 1976 and it has been recorded nearly every winter near Yuma, on the CBC. As many as four wintered regularly on Havasu NWR during the 1940s. The few summer records probably pertain to wandering birds from nesting sites in nearby desert mountains.

HABITAT This wide-ranging raptor could occur almost anywhere, but most sightings have been from the desert periphery of the valley.

Crested Caracara (*Polyborus plancus*)

STATUS Now a casual visitor. Records include one at Mittry Lake in January 1964, a sighting near Tacna about 55 km east of Yuma on 5 December 1978, and one seen 8 December 1980 at Cibola NWR. Historically a resident at Yuma, with the last record 15 January 1905.

Nearest breeding populations are in extreme south-central Arizona, adjacent Sonora, Mexico, and to the southwest in northern Baja California. There are no accepted records for the California portions of the valley, although occurrence there before 1900 was likely.

American Kestrel (*Falco sparverius*)

STATUS A locally uncommon summer resident and breeder; common transient and winter resident from mid-September to March. Max: 187; 22 December 1980, Parker CBC.

Although ubiquitous in the valley during the nonbreeding seasons, kestrels have been severely reduced as a nesting species. Visitors from other regions will be struck by the paleness of resident kestrels compared with those from farther north. They are also very small, and most likely represent a northward extension of the cactus-nesting race (*F. s. peninsularis* (Mearns)) from Baja California. Birds from populations north of the LCRV, *F. s. sparverius* Linneaus, are widespread here in winter.

HABITAT Being the only cavity-nesting falcon in the United States, it persists only where enough cottonwood trees or snags remain to attract larger woodpeckers that excavate suitable nest sites. Widespread agricultural development, although providing winter habitat, has dramatically reduced available trees and, therefore, nesting kestrels. They also nest in saguaros at the edge of the valley. Grinnell (1914) found this species primarily in saguaros and not throughout the riparian belt.

Merlin (*Falco columbarius*)

STATUS A rare but regular transient and winter visitor from late September to early April. Extreme dates of migrants are 7 September (1951) and 29 April (1952). Formerly more common until the 1950s. Max: 4; 20 December 1979, Parker CBC.

Several races of this small falcon may be noted in the LCRV with careful inspection. Very pale Merlins (*F. c. richardsonii* Ridgway) are migrants from the Canadian Prairie regions. Darker birds (*F. c. bendirei* Swann) originate mostly in the northern Rocky Mountains northwest to Alaska. Blackish individuals (*F. c. suckleyi* Ridgway) from coastal British Columbia are also to be looked for.

HABITAT This small but powerful falcon prefers semi-open habitats such as agricultural land bordered by riparian woodland or broken by shrub-lined canals or scattered trees. However, they also hunt around residences, even in towns.

Peregrine Falcon (*Falco peregrinus*)

STATUS At present a rare and irregular transient, winter resident, and postbreeding visitor. Recently recorded most often in fall. Nested near Parker Dam at least until 1954 and probably also at Imperial NWR (1942) and Topock Gorge (1953).

Peregrine Falcons were more numerous in winter and during migration before 1954, which roughly marks the time of their sudden decline throughout North America. However, as this species continues to recover through much of its range, we anticipate that it will regain its former status and may eventually nest again on cliffs along the lower Colorado River. The Peregine Falcon is listed as a federally endangered species.

Recent summer records include immatures, 15 July 1978 northeast of Yuma and 20 June 1979 near Parker, and an adult, 10 July 1982 near Palo Verde. Three were seen in migration at Cibola NWR, 22 September 1978.

Prairie Falcon (*Falco mexicanus*)

STATUS An uncommon transient and winter visitor from September to late March. Also rare but regular from May through August, most likely as a postbreeding visitor. May breed occasionally at BWD. Max: 8; 21 December 1983, Parker CBC.

Of roughly 135 sightings between 1976 and 1983, most were in fall and early winter. About 25 occurred during summer (May–August), and only 1 sighting was in April during the height of the nesting season.

HABITAT Wintering birds are scattered through agricultural areas where doves, quail, and other prey are abundant. They breed in the arid mountains along the periphery of the valley, and a pair was present in spring 1980 at BWD where the rugged Buckskin Mountains meet the riparian floodplain. Grinnell (1914) found this species nesting downstream from the Bill Williams River mouth, apparently on the same cliff occupied by Peregrine Falcons during the 1950s.

Ring-necked Pheasant (*Phasianus colchicus*)

STATUS Periodically introduced for hunting, but only a few persist outside the vicinity of Yuma. Recent sightings come from Parker and Cibola NWR. Pheasants are established and locally abundant in citrus groves south of Yuma, with an additional small population in Bard Valley (present as long ago as 1942).

Gambel's Quail (*Callipepla gambelii*)

STATUS A common to abundant permanent resident, with population peaks in late summer. Often occurs in large coveys. Max: 1,206; 18 December 1979, Parker CBC.

This is one of the most characteristic desert birds, and represents perhaps the most important game resource in the LCRV. For example, some 7,000 hunters (averaging 5.2 trips/hunter) take up to 110,300 quail per year, as in 1984 (see Table 2, Chapter 2). The Gambel's Quail undergoes dramatic changes in population size throughout the year and from year to year. Quail numbers vary markedly according to habitat as well. Local abundance is determined, in part, by food availability, hunting pressure, and weather.

HABITAT Overall, quail prefer relatively sparse, shrubby vegetation throughout the year, avoiding tall dense riparian groves (e.g., cottonwoods) and pure saltcedar. Densities in winter and spring are highest in honey mesquite woodland and adjacent desert washes where various shrubs (i.e., saltbush and quail bush) form dense patches and annual plants are often abundant. As summer progresses, quail begin to move to denser habitats and by fall the largest coveys are in dense screwbean mesquite woods.

Agricultural areas bordered by riparian vegetation also support large numbers during summer and fall, but these habitats are less important in winter and spring. However, agricultural-riparian edges may harbor more quail than most riparian habitats year round. This is due largely to the combination of dense quail bush, which provides cover, and abundant food provided by cultivated crops. Very few quail persist, however, in agricultural land far from riparian vegetation.

BREEDING The reproductive cycle begins in February, when unmated males begin uttering "cow" calls. These are heard until August. Male reproductive activity peaks from 15 March to as late as 7 August (reproductive data from 401 male specs). Mature ovaries in females are first detected about 21 March and are found until early September (314 female specs). Egg laying occurs from late March until late July. In general, female reproductive activity lags several weeks behind that of the male.

Broods occur from late April through August; young less than one-quarter grown may be seen throughout that period. After August, family groups begin converging to form large coveys which persist through fall. This quail has the longest breeding period of any species along the Colorado River, except the Inca Dove. Somehow they successfully produce young during the hottest period of the year in areas with little foliage cover. Their physiological tolerance to desert environments is discussed by Goldstein (1984).

FOOD HABITS Major food items (from 900 specs) include seeds from cultivated crops, mesquite, annual plants and other plant material (including mistletoe), and animal matter. Screwbean mesquite seeds are the major food item in late summer and fall. Annuals and cultivated crops are most important in winter and spring. Animal material (mainly ants and rodent feces) is taken largely during summer.

Proportions of major food items change from year to year, mostly in response to weather changes. During the springs of 1974 and 1977, a higher percentage of annuals were in the diet of quail, corresponding with major fall rains. During the winter of 1975–76, following a very dry year, annuals were less abundant in the diet. Cultivated crops were the major food source through most of that year. Overall diet diversity was highest during summer and lowest during winter and this diversity was higher after wet years than after dry years.

COMMENT Our five years of data on habitat, reproduction, and diet provide some important insights into the ecology of this species in the LCRV. Quail populations consistently suffer heavy mortality from hunting, food shortages, physiological stress, and other natural processes in fall and early winter, suggesting that this is the time of population limitation. Choice of winter habitat may affect breeding season survival and reproduction, supporting the theoretical predictions of Fretwell (1972). Most spring breeding takes place in honey mesquite woodland and seems strongly related to winter production of annual plants. Quail that successfully winter in honey mesquite would seem to have a reproductive advantage over quail wintering in other habitats. The single most important factor governing quail productivity appears to be the presence of annuals in winter and early spring. Annuals provide the bulk of the diet at those critical periods when mesquite pods are absent, particularly at the beginning of the breeding cycle.

Precipitation and germination of annual plants are closely related. During times of low rainfall, agricultural areas become important elements for quail in the LCRV. The amount of annuals and cultivated crops in the diet were inversely related throughout the study period. The alternative food source provided by agricultural land may help to mitigate fall and winter losses from hunting and starvation. It also ensures an ample food supply to stimulate early spring reproduction in dry years.

Although agricultural areas provide important food sources through summer, quail are heavily dependent on native riparian habitats for breeding and winter foraging. A mosaic of riparian habitat (primarily honey mesquite) and agricultural lands must be maintained to support healthy quail populations. As this mosaic shifts toward pure agriculture, the enhancing effect of the riparian-agricultural edge will disappear and isolated areas of honey mesquite will become refuges for quail. If the entire valley is converted to agriculture, some quail may persist in pockets of nonarable land or in desert washes adjacent to the LCRV, but this species will be lost as a gamebird resource.

Black Rail (*Laterallus jamaicensis*)

STATUS A permanent resident in small numbers in the vicinity of Imperial Dam. Also present at BWD in 1978, and apparently a resident there since 1982. The total population was estimated to be 100–200 individuals from Mittry Lake north to Imperial NWR (Repking and Ohmart 1977).

The Black Rail was first found below Imperial Dam on 15 June 1969. Aside from a small number present around the Salton Sea, the LCRV birds represent the only stable inland population in western North America. The first record at BWD was 23 April 1978. They were found there again in 1982 and are now resident, with up to 10 detected at one time. The American Ornithologists' Union (1983) checklist incorrectly lists this species as only wintering along the lower Colorado River and being "casual or accidental" in Arizona.

Black Rails are rarely seen, but their calls betray their presence. Peak calling activity is from late winter through spring. At least a few individuals call year round. Recent radiotelemetry work indicates that LCRV birds are largely diurnal, being active from 0.5 hour before sunrise to 0.5 hour after sunset (Flores and Eddleman 1988). We know very little of this species' migratory habits (if any) in the LCRV. However, the recent population expansion farther north to BWD suggests that at least some birds move. Radiotelemetry also indicated increased wandering during the winter months, but as yet no long-distance movements have been documented (Flores and Eddleman 1988). Young birds accounted for all dispersal, with adults remaining extremely sedentary (Eddleman pers. comm.).

The LCRV population is most representative of *L. j. coturniculus* Ridgway, the California Black Rail, found in brackish and salt marshes along the California coast. However, the taxonomic and distributional status of our Black Rail population is still poorly understood. Presently, California Department of Fish and Game and Arizona Department of Game and Fish list the California Black Rail as a State Threatened and State Endangered Species, respectively. The U.S. Fish and Wildlife Service also lists the California Black Rail as warranting endangered or threatened status. Water management operations, principally dramatic and long-term fluctuations in water levels, are seen as serious threats to the Black Rail's future existence.

HABITAT This very secretive species occurs almost exclusively in marshes where water levels are stable at a few centimeters, covering about 10% of the ground. Large mats of three-square bulrush are often associated with these water levels, but Black Rails are most often found where cattails and California bulrush dominate.

BREEDING Five nests were found from 19 April to 23 July 1988. These were the first documented nests for the species in Arizona. One nest was located 19 April and contained five eggs. Nests were well-defined bowls with a ramp and canopy and were constructed of cattail, California bulrush, and three-square bulrush. Both adults incubated, with the clutch hatching by 4 May; the incubation period lasted 17 days (Flores and Eddleman 1988). Range of clutch size was three to seven eggs. Territory size is estimated to be 0.4 ha (Eddleman pers. comm.).

FOOD HABITS Preliminary analysis of fecal samples identified spiders, beetles, ants, leafhoppers, snails, and bulrush and cattail seeds. Dramatic drops in available invertebrate biomass and corresponding drops in Black Rail biomass were found in winter (Flores and Eddleman 1988).

Clapper Rail (*Rallus longirostris*)

STATUS An uncommon to fairly common summer resident and breeder between February and September, north to Topock Marsh. More secretive and, possibly, less numerous in winter.

Until recently, most of the population was thought to retreat to Mexico during winter. Radiotelemetry studies, however, indicate that

FIGURE 20. Yuma Clapper Rail, an endangered subspecies largely endemic to the lower Colorado River Valley. (Photo by R. E. Tomlinson.)

over 70% of the breeding population winters along the lower Colorado River (Eddleman 1989). Before radiotelemetry studies, Yuma Clapper Rails were known during winter only by vocal detections and were thought to be restricted largely to the southern part of the valley, especially between Imperial and Laguna Dams.

Breeding population centers are Mittry Lake, West Pond, Imperial NWR, BWD, Topock Gorge, and Topock Marsh. These centers reflect the distribution of relatively large marshes. Smaller populations occur where moderately extensive emergent vegetation is persistent, including backwaters. Elsewhere in the interior Southwest, they occur at the Salton Sea and along the Gila and Salt Rivers east to Picacho Reservoir and Blue Point in central Arizona. Range extensions should be monitored north to Las Vegas Wash in the Lake Mead National Recreation Area and along the Gila and Salt River drainages wherever suitable habitat and food resources occur.

The Colorado River race of this species is the Yuma Clapper Rail (*R. l. yumanensis* Dickey) (Fig. 20). It is the only federally listed endangered bird species largely restricted to the LCRV. Therefore, factors affecting Clapper Rail population levels have received a great deal of our attention. To understand the present status of the Yuma Clapper Rail, we must refer to historical accounts of naturalists working in the region during the late nineteenth and early twentieth centuries.

Early naturalists on the Colorado River, many of whom were familiar with Clapper Rails, did not record them north of the Gila–Colorado confluence. Grinnell (1914), in particular, found no evidence that the Yuma Clapper Rail existed along the river. He described a depauperate waterbird fauna compared with what we find today. His description of the river's plant associations indicates that potentially suitable marshes were very small and very local in distribution. These points have caused some controversy, with other workers contending that extensive marshes did exist throughout the river valley (Todd 1987). However, the existence of many large, long-standing marshes north of the Gila–Colorado confluence before 1920 is not supported by any detailed description of the region from that period (Ohmart et al. 1977).

The first Clapper Rail specimen from the Colorado River, identified as *R. l. levipes* Bang, was collected by Brown on 25 August 1902 "at Yuma" (Swarth 1914). This specimen (now lost) was undoubtedly *R. l. yumanensis*, and went undescribed at the time (Dickey 1923). *R. l. levipes* is not known to stray far from its California coastal range.

The type specimens of the Yuma Clapper Rail were taken near Laguna Dam in 1921 (Dickey 1923). Laguna Dam was completed in 1909, and marsh vegetation became established along canals near the dam 10–11 years later. Huey (*in* Bent 1963:275–277), with Canfield, collected the type specimens. He commented that before 1924 the center of rail abundance was in the Colorado River Delta. Reaches of the river immediately above the delta may have been unsuitable for nesting during periods of unusually high water in 1921 and 1924, thus forcing some rails up river. After these floods receded, extensive marshes formed above Laguna Dam, including along the canals where the first rails were found.

Yuma Clapper Rails appeared north of Laguna Dam a few years after the completion of Parker, Imperial, and Headgate Rock dams in 1938, 1939, and 1942, respectively. Early sightings by Monson included one rail near Headgate Rock Dam on 12 August 1946, an adult and three young on 9 July 1948, two adults and three young on 17 July 1948, and one adult on 18 May and another on 17 June 1949. The first sightings in the BWD, by Monson on 12 May 1954, occurred 16 years after the completion of Parker Dam. An immature was collected there 16 August 1954. Monson had visited the area regularly on an annual basis prior to his 1954 observations. The species was not located at Topock Marsh until 1966. Today this is the most northern population center. Our northernmost record is now from Laughlin Bay, 20 May 1986, on the Nevada side of the river. All evidence indicates that the population was localized in the Yuma area before ca. 1940 and has since become more widespread (Ohmart and Smith 1973; Monson and Phillips 1981). Recent expansions into the newly formed Salton Sea in California by the 1940s, and Picacho Reservoir in central Arizona by the 1970s support this view.

The Yuma Clapper Rail is listed as a Federal Endangered Species and as a State Threatened Species by both California and Arizona. The present population size in the LCRV is estimated to be between 400 and 750 individuals in the United States, with 450–970 in Mexico (Eddleman 1989). However, this population is limited by, and has come under threat of reduction from, river management activities such as dredging, channelization, and stabilization of banks by riprapping, all detrimental to marsh habitat formation. Recent flooding has resulted in more pressure on water management agencies to increase channelization and bank stabilization activities which will result in a large reduction of available marsh habitat.

Yet another threat is the possibility of environmental contaminants affecting the health of Yuma Clapper Rails. Selenium (a trace metalloid) concentration was determined from the livers of five adult birds and from two sets of eggs. The concentration found in livers equalled or surpassed those found in ducks at Kesterson NWR in California, an area of extreme selenium contamination (Ohlendorf et al. 1986b; Radtke et al. 1988; Kepner unpubl. data). Rail eggs contained concentrations that were found to result in a 20% chance of death or deformation in American Coot embryos at Kesterson NWR (Ohlendorf et al. 1986a; Kepner unpubl. data). Crayfish, a major food item, also had selenium concentrations that could cause toxic effects in their predators (Lemly and Smith 1987; Kepner unpubl. data). Selenium can cause extensive metabolic problems in birds and may affect reproductive success. The sources of selenium in the LCRV appear to be from upstream coal-fire plants, mining, natural weathering, and, perhaps, irrigation-based agriculture. Agricultural activities in the LCRV proper do not appear to be contributing (Radtke et al. 1988).

HABITAT Clapper Rails are associated primarily with dense marsh vegetation, but high densities also occur in some moderately dense cattail/bulrush marshes. They may occur in dense reed and even sparse cattail/bulrush, but in reduced numbers. Habitat edges between marshes and terrestrial vegetation are important, but the main factors determining habitat use are the annual range in water depth and the existence of residual mats of marsh vegetation (Eddleman 1989). Most individuals remaining through winter are found in tall, dense cattail/bulrush stands, however some occur in flooded saltcedar and willow stands. Home ranges of individuals or pairs may encompass up to 43 ha and may extensively overlap with home ranges of other birds, but year-round home ranges averaged 7.5 ha. For further details on Clapper Rail habitat use along the lower Colorado River see Anderson and Ohmart (1985a), Eddleman (1989), and Todd (1987).

BREEDING Nesting behavior commences by February, with most eggs hatching during the first week of June. There is no evidence of more than one brood per season, despite the long breeding period (Eddleman 1989). Both adults are responsible for care of the eggs and young. Clutch size is usually six to eight eggs.

Young are precocial and follow the adults through the marsh within 48 hours of hatching. Adults lead the young to productive feeding grounds where they quickly learn to feed on their own. As with many precocial birds with large clutch sizes, young Clapper Rails suffer high mortality from predators (usually within their first month of life). Other details on breeding may be found in Smith (1975), Bennett and Ohmart (1978), and Eddleman (1989).

FOOD HABITS Crayfish are the major food of this species in the LCRV. Crayfish, like the rails themselves, are apparently recent invaders (since 1900) to the northern portions of the LCRV (Ohmart and Tomlinson 1977). They were introduced for use as fish bait. Grinnell (1914) commented on the surprising lack of invertebrate species and numbers during his Colorado River trip in 1910.

Crayfish abundance may be a limiting factor determining rail occurrence today. Seasonal shifts in habitat use by crayfish may affect use of habitats by the rails. For example, a lack of crayfish during winter was thought to be responsible for the migratory habits of this species in the LCRV. Recently, however, crayfish have been found in densities high enough to support wintering Yuma Clapper Rails, especially in dense marshes (Eddleman 1989).

Other dietary items include isopods, aquatic and terrestrial beetles, damselfly nymphs, earwigs, grasshoppers, spiders, freshwater shrimp, and freshwater clams (Ohmart and Tomlinson 1977). In 10 rail specimens examined north of the Gila–Colorado confluence, 95% of the diet by volume consisted of crayfish. Two specimens from the Gila–Colorado River confluence had 49% isopods and 50% freshwater clams. From the Colorado River Delta, in brackish water, four specimens examined had broader diets consisting of water beetles, fish, leeches, plant seeds, dragonfly nymphs, and shrimp.

COMMENTS The Yuma Clapper Rail is unique among Clapper Rail populations for three reasons: (1) it readily disperses after breeding and may be partly migratory, factors that probably increase its ability to colonize new habitats; (2) it effectively uses freshwater situations that offer radically different (and probably more restricted) food resources; and (3) wintering Yuma Clapper Rails are almost completely silent. Other populations of Clapper Rails are resident in brackish or saltwater marshes and most are vocal all year. These traits suggest a greater similarity between *R. l. yumanensis* and *R. elegans*, the King Rail, a closely related species whose relationship to the Clapper Rail in eastern North America is poorly understood (see Avise and Zink 1988).

Virginia Rail (*Rallus limicola*)

STATUS A common to abundant winter resident, and recently a fairly common and local spring and summer breeder. Max: 90; 21 December 1979, Bill Williams Delta CBC.

As with most marsh- and waterbirds, the Virginia Rail apparently has increased substantially since the creation of stable marsh habitats in the LCRV. It was apparently unknown in summer until 15 June 1969, when several were found between Laguna and Imperial Dams, along with the first Black Rails in the LCRV. By 1976, they were common in summer, at least at BWD, and in the following spring many were found in marshes throughout the valley.

HABITAT Cattail/bulrush marshes are the primary habitats throughout the year. Small ponds, agricultural canals, and flooded riparian vegetation are also used when cattails or bulrushes are present.

BREEDING Breeding was first verified in 1978 at BWD (downy young seen 29 March); young were also found there in 1982 and 1983. The Virginia Rail's "kicker" call, which has been attributed to several other rail species and remained an ornithological mystery for decades, has been heard in spring and summer along the Bill Williams River and near Imperial Dam. Breeding also occurs elsewhere in the valley with nests and young found from 1985 to 1988 at Mittry Lake. At least some breeding birds are permanent residents as evidenced from radiotelemetry (Eddleman pers. comm.).

Sora (*Porzana carolina*)

STATUS A common winter resident and transient between late July and early May. Summer records exist for Topock Marsh, Bard, Palo Verde, and BWD, but there is no evidence of breeding. Max: 110; 19 December 1981, Yuma CBC.

Soras, fitted with radios for tracking their movements, were all migratory (Eddleman pers. comm.). From 1985 to 1988, wintering individuals stayed no later than 7 May (1988) and arrived no earlier than 22 July (1987, 1988) the subsequent fall. Complete and simultaneous wing and tail molts were found to occur soon after adults arrived in fall (Eddleman pers. comm.). This apparently represents the first documentation for simultaneous wing molt on the wintering grounds for rallids.

HABITAT Cattail/bulrush marshes, small ponds, and agricultural canals with marsh vegetation are the most often used habitats. A few may be found at nearly any marshy location, especially during migration.

Common Moorhen (*Gallinula chloropus*)

STATUS A fairly common and local permanent resident and breeder throughout the valley. Numbers decrease somewhat during winter north of Parker. Max: 76; 20 December 1983, Yuma CBC.

HABITAT Marshes and agricultural canals serve as both breeding and wintering habitats. Individuals occasionally may venture out into open water, but remain close to cover. When concealed, they are easily located by their loud, bugle-like calls.

BREEDING Nests of reeds are placed on marsh edges, near open water. Ten to 12 eggs are laid from March through July. Recently hatched young are noted most often from mid-April through June.

American Coot (*Fulica americana*)

STATUS A locally common breeder and abundant winter resident throughout the valley. Large concentrations occur in some years on the larger reservoirs. Max: 50,000; December 1953, Havasu NWR.

The coot is probably the most ubiquitous waterbird in the valley year round. It is gregarious except in late spring and summer when pairs or family groups are frequently encountered. Nonbreeding flocks begin to form in August and increase in size through the fall. Size and distribution of winter concentrations vary with water level fluctuations and probably with the extent of unfrozen open water north of our region. Very few coots were noted by Grinnell (1914) in spring 1910, suggesting that changes in riverine habitats since construction of dams have benefited this species.

HABITAT Marshes, small ponds, and agricultural canals are used for nesting. During winter, large flocks congregate in coves of major reservoirs (i.e., Lake Mohave and Lake Havasu). Large numbers are also strewn along the river channel, especially near dams and backwaters. Riprapped sections, distant from dams, are the least-used portions by coots.

BREEDING Nests usually are concealed in bulrushes and cattails. Eight to 12 eggs are laid from April to August. Recently hatched young have been noted in late May and June and, exceptionally, as early as 6 April (1941; Arnold 1942).

FOOD HABITS Sago pondweed and cattail were the most heavily used food items in fall and winter (Eley and Harris 1976). Invertebrates are taken during the breeding season in addition to seeds and aquatic plants.

Sandhill Crane (*Grus canadensis*)

STATUS A fairly common and local winter resident between early October and late February. A few individuals linger into early March. Flocks of 300–600 have traditionally wintered near Poston, Bullhead City, and at Cibola NWR in recent years. Max: 778; 23 December 1977, Parker CBC.

This species has increased as a wintering bird during the last two decades in response to agricultural practices. Extreme dates of occurrence in recent years are from 26 September (1977, 1978) to 27 March (1979; flock of 57 flying north over Ehrenberg). An old account describes cranes as "abundant" below Yuma on 9 April 1862. Grinnell (1914) saw them migrating daily from 1 to 9 March 1910. Evidently, this species was formerly more common as a winter visitor, at least around Yuma and into Mexico. We know of no recent records south of Cibola.

Both the Alaskan (Little Brown; *G. c. canadensis* (Linneaus)) and the Canadian Prairie Province (Greater; *G. c. tabida* (Peters)) populations of the Sandhill Crane are represented, with the latter being more common. Few natural events along the lower Colorado River rival a flock of trumpeting and whistling Sandhill Cranes silhouetted against the desert mountains on a brisk winter morning. They may be observed "dancing" on warmer days in late winter and early spring.

HABITAT Cranes feed in agricultural fields such as alfalfa and milo. Protected sandbars or shallow ponds are required for roosts. Management for feeding and roosting sites at Cibola and Havasu (Topock Division) NWRs has led to recent increases in these areas. In response, some local farmers use sound cannons to discourage cranes from feeding in alfalfa fields.

Black-bellied Plover (*Pluvialis squatarola*)

STATUS An uncommon transient in very small numbers from late March to late May (mainly in April) and from late July through mid-October. Occasional birds may linger into late fall (e.g., 26 October–17 November 1979 at Imperial Dam). There are two winter records: 30 December 1940 (2) at Yuma, and 10 December 1989 at Topock Marsh. Max: 9; 2 April 1979, Blythe.

Nearly all occurrences are of one or two birds, in contrast to the large flocks along the coast. This plover is fairly common at the Salton Sea, where some winter regularly. It is uncommon or rare elsewhere in interior California and Arizona.

HABITAT Like other plovers, they prefer plowed agricultural fields, with only a few records from mudflats or lakeshores.

Lesser Golden-Plover (*Pluvialis dominica*)

STATUS A very rare or casual transient. Four definite records: 12 October 1977 south of Blythe; 6 May 1978 (photo) at Parker; 22 September 1978 near Palo Verde; and 18 October 1979 near Poston.

They have been noted very rarely throughout the inland Southwest, primarily in fall. All records probably pertain to the nominate race, *P. d. dominica* (Muller), which migrates from Arctic North America to South American wintering grounds.

HABITAT This plover has been found in plowed agricultural fields and on riverbanks.

Snowy Plover (*Charadrius alexandrinus*)

STATUS A rare spring transient from late March to early June, and uncommon from July to early September. Rare and irregular in late fall and winter. Max: 27; 22 July 1954, Topock Marsh.

Dates of apparent spring migrants are from 24 March to 6 June (1951, Topock); most records are for early April. One at West Pond, 27 June 1960, was probably a very early fall migrant, and 10 at Topock, 6 October 1988, were exceptionally late. The only recent winter record is one at Lake Havasu, 20 January 1982. This species is fairly common for much of the year at the Salton Sea.

HABITAT This species is most frequently observed in plowed agricultural fields and occasionally on exposed mud flats or shorelines. They breed to the east and west of the Colorado River and might nest in the valley if suitable shoreline habitat were protected from human disturbance.

Semipalmated Plover (*Charadrius semipalmatus*)

STATUS An uncommon transient in small numbers from mid-April to mid-May and from mid-July to mid-October. One seen 22 June 1968 at West Pond was likely a very early southbound migrant. One at Lake Havasu, 10 December 1981, may have been wintering. Max: 7; 14 May 1978, Imperial NWR.

Our records of this Arctic-nesting species are about equally divided between spring and fall. As many as 32 individuals were seen during spring 1978, but normally fewer than 10 were noted each season. As with many other shorebird species, large numbers occur at the nearby Salton Sea during migration (up to 2,000 in April), and a few regularly winter there as well.

HABITAT Semipalmated Plovers generally occur with other migrating shorebirds on mud flats, riverbanks, pond edges, or in plowed and flooded agricultural fields.

Killdeer (*Charadrius vociferus*)

STATUS A fairly common summer breeder throughout the valley. Also a common to abundant transient and winter resident from August through May; often seen in large flocks. Max: 1,248; 23 December 1977, Parker CBC.

This is the only shorebird that regularly nests in the LCRV. It is highly gregarious when not nesting. Largest flocks (up to 300 birds) have been found in October and November on exposed gravel bars in the river or in plowed fields. Concentrations of up to 90 birds have been seen as early as 4 August. Winter numbers are highest in agricultural valleys, but a few may be found at nearly every point along the river.

HABITAT This is the most numerous and adaptable shorebird in our region, occurring in almost any open situation close to water. Nesting pairs may be found along agricultural canals, at the edges of plowed fields, on gravel dredge spoil, or on natural shorelines.

BREEDING Breeding activity has been noted from March to July.

Mountain Plover (*Charadrius montanus*)

STATUS An uncommon transient and irregular winter resident in small flocks between mid-October and late March in agricultural areas near Parker, Blythe, and Yuma. Max: 205; 1 December 1981, Blythe.

In some years small flocks appear in February in areas where none wintered, and these remain through March. These dates correspond with their departure from Imperial Valley, California. It is possible that these birds regularly stop along the Colorado River before migrating to their breeding grounds. Extreme dates of occurrence are from 4 September (1951) at Topock to 29 March (1980) near Parker.

HABITAT This plover favors bare plowed fields that have not been irrigated. They blend perfectly with the gray-brown soil and are detected only by their short running movements and sudden stops.

FOOD HABITS A sample of six stomachs from one field near Parker contained weevils, other beetles and larvae, maggots, and earwigs.

Black-necked Stilt (*Himantopus mexicanus*)

STATUS A common transient in small flocks from mid-March to mid-May and from early July to mid-September. Fairly common throughout late May and June and in early October. Also a rare and local breeder in some years (e.g., 1916, 1950, 1954, 1956, 1959, 1977, 1980, 1983). There are several recent winter records including two seen at Parker, 20 December 1979, and one at Imperial NWR, 18 December 1982. Max: 200; 31 July 1976, Ripley.

This is certainly the most numerous migrant shorebird to visit the LCRV. Still, its abundance pales beside the thousands that both breed and winter at the Salton Sea. Presumably, a lack of suitable nesting habitat (diked ponds or marshes with stable water levels) prevents a breeding population from establishing itself in the valley. Nearly all of our breeding-season records (late May–June) are of small groups moving north or south along the river.

Spring and fall migrations of stilts merge imperceptibly, or even overlap in June, making the exact determination of each migration impossible (see Fig. 13, p. 81). Peak fall passage is in late July and lasts throughout August, when concentrations of 100–200 birds are not unusual. Spring flocks rarely reach 40–60 birds, with numbers rather evenly distributed throughout the season.

HABITAT Black-necked Stilts are particularly fond of irrigated fields. However, flocks may visit nearly any area with shallow water.

BREEDING The few records include young seen in 1916 near Palo Verde; 14 August 1956 at Headgate Dam above Parker; a chick present at Martinez Lake on 20 July 1977; two pairs at the Poston sewage ponds, June 1980; and two pairs at Cibola NWR in June 1983. In this latter case, one nest contained four eggs on 19 June but failed because of rising water levels.

American Avocet (*Recurvirostra americana*)

STATUS A fairly common spring transient from mid-March to early May, and common fall transient from late July through mid-October. Fall migration is often extremely protracted, with records from late June (exceptionally, 9 June 1977) to late November. Rare and somewhat irregular through winter. Max: 2,500; 21 September 1959, Topock Marsh.

The peak of fall migration is late for a Great Basin-nesting species (see Fig. 13, p. 81). Small flocks trickle through during the entire season, but the last birds in October are often concentrated in a single large flock; e.g., 650 at Havasu NWR on 1 October 1959, 320 at Imperial NWR on 11 October 1977, and 295 near Palo Verde on 15 October 1981. A very late concentration of 1,400 was at Topock, 2 November 1952. Peak spring passage is throughout April, with flocks of up to 80 birds. Small groups in mid-June are probably early fall migrants as there are virtually no records after mid-May. One individual seen at Imperial NWR on 25 May and 4 June 1977 was probably summering. Winter records are from 11 different years, primarily in the Yuma area and at BWD, where up to six were present in 1948–49. One was at Topock, 2 February 1988. This species winters and summers abundantly at the nearby Salton Sea and has bred regularly in southern Arizona.

HABITAT Flocks rest on sandbars, islands, and marshy shorelines, and commonly feed in irrigated fields.

Northern Jacana (*Jacana spinosa*)

STATUS A casual visitor. An adult at Mittry Lake from 6 to 30 June 1986 (photo) was found by U.S. Fish and Wildlife Service personnel studying Yuma Clapper Rails.

This primarily tropical species normally ranges as far north as Sinaloa in western Mexico and irregularly to south Texas. The only other record in the Southwest is of one present near Nogales, Arizona, from June 1985 to January 1986.

Greater Yellowlegs (*Tringa melanoleuca*)

STATUS An uncommon winter resident and fairly common transient from early July (earliest, 22 June 1953) to early September and from mid-March to mid-April; rarely encountered into early May. Max: 43; 1 October 1960, near Somerton.

Winter numbers are usually very small and increase from north to south. There are few winter records north of Parker. Local influxes into nonwintering localities indicate that migration is under way by early March and probably continues into November. Peak passage appears to be during April and August. During migration, small groups of 2–8 birds are typically found, with concentrations of up to 20 or more birds noted on only a few occasions. In winter, it is usually encountered singly, with the largest flock being eight at a Cibola backwater in 1982.

HABITAT This species frequents various water-edge situations, including canal banks, irrigated fields, shallow ponds, marshes, and lakeshores.

Lesser Yellowlegs (*Tringa flavipes*)

STATUS A rare to uncommon spring transient in April and early May, and uncommon fall transient from early July (exceptionally, 30 June 1953) to early October. Max: 20; 17 August 1977, near Parker.

This species occurs typically in groups of 1–10 birds, usually associated with other migratory shorebirds. Numbers are usually three to four times higher in fall than in spring, with peak passage in August.

HABITAT Lesser Yellowlegs are most frequently observed in plowed and irrigated agricultural fields during August. A few have been noted at small ponds or backwaters.

Solitary Sandpiper (*Tringa solitaria*)

STATUS An uncommon fall transient from mid-July (exceptionally, 8 July 1953) to late September (exceptionally, 14 October 1982); rare in spring from early April to early May (≈20 records). One winter record, 26 December 1950, near Yuma. Max: 9; 25 August 1982, Poston.

During spring migration they become increasingly scarce from east to west in the interior Southwest; there are only two records from the Salton Sea. Even in fall this species is more frequent along the lower Colorado River than at the Salton Sea, although there are two additional winter records from the latter region.

HABITAT These sandpipers are fond of secluded wet spots, frequently close to some vegetative cover. They will occasionally join other shorebirds in irrigated fields, but about two-thirds of all records are of solitary individuals.

Willet (*Catoptrophorus semipalmatus*)

STATUS A fairly common transient, usually in small flocks, from early April to mid-May and again from mid-June (earliest, 3 June 1946) through August. Rare in early September; the only later records were 10 at Cibola on 10 October 1981, and 4 on 26 October 1979 near Yuma. A few lone birds have been noted in early spring as early as 6 March (1979). Max: 260; 30 July 1953, Topock Marsh.

As with a few other Great Basin-nesting waterbirds, the southbound migration is extremely early, often peaking before most other migrants appear (see Fig. 13, p. 81). The spring migration is compressed primarily into a 2-week period at the end of April, with the largest concentration being 116 at Martinez Lake, 25 April 1978. This species winters in large numbers at the Salton Sea but is virtually never seen along the lower Colorado River after the end of August (only three records since 1975).

HABITAT Willets prefer open marshes, sandbars, and lakeshores, and only rarely visit irrigated fields.

Spotted Sandpiper (*Actitis macularia*)

STATUS A common transient and winter resident between August and April. Rare but regular through summer, increasing steadily in fall and decreasing in spring. Max: 84; 23 December 1977, Parker CBC.

This sandpiper is not gregarious, with most records of single birds or pairs. However, local concentrations may occur, such as 50 around Imperial Dam on 7 September 1981. A few summer records are of pairs in alternate plumage; however, these have not shown evidence of nesting. This species has nested farther north in the Grand Canyon.

HABITAT They are encountered along the entire riverbank and along canal banks and lakeshores. Occasionally, they are seen at sewage ponds but rarely in irrigated plowed fields.

Upland Sandpiper (*Bartramia longicauda*)

STATUS A casual fall transient. Two sight records: 11 September 1952 at Lake Havasu and 9 September 1978 at Imperial NWR.

These dates match those of two other fall records in southern California. One recent Arizona record was in late September and two were in October. This species has also been found six times in southern California during mid- and late May.

Whimbrel (*Numenius phaeopus*)

STATUS An uncommon transient from late March to late May, and from July (exceptionally, 30 June 1982) to mid-September; occasionally in small flocks. One winter sight record, a flock of nine near Somerton, 12 January 1940 (Arnold 1942). Most regularly seen near Blythe and Poston in early August. Max: 96; 28 April 1984, near Yuma.

This species was formerly considered a very rare migrant, with all but one of the few records before 1970 being in mid-September. Since then, there have been about 40 records involving >200 individuals. Only one of these has been in September (5 September 1981). This change in known status likely reflects the increased fieldwork in agricultural habitats during midsummer. The relative rarity of this species is still inexplicable in light of its abundance at the nearby Salton Sea, where up to 10,000 have been noted during spring migration.

HABITAT Whimbrels almost exclusively visit plowed and irrigated fields with other shorebirds. They are rarely found on sandbars along the river.

Long-billed Curlew (*Numenius americanus*)

STATUS An uncommon to fairly common transient between early March and early October, with peak migration in April and again from mid-July through early September. Occasionally seen in small flocks at those times; early summer records are usually of singles or pairs. A notably late influx occurred in October 1980 (43 records), and in 1981 up to 14 lingered near Palo Verde until late November, with several seen throughout that winter; 15 were also near Parker, 9 January 1982. Other records as early as 12 February (1979, 1982) were probably of early spring migrants. Max: 190; 28 September 1974, San Luis, Arizona.

Although records now span the entire year, occurrence of this species is highly erratic, especially outside the peak migration periods (see Fig. 13, p. 81). This is surprising, in light of the large flocks that winter at the nearby Salton Sea.

HABITAT Curlews frequent plowed or grassy agricultural fields and occasionally rest with other shorebirds on sandbars or lakeshores.

Marbled Godwit (*Limosa fedoa*)

STATUS An uncommon to rare spring transient from late March to early May; fairly common fall transient in small flocks from early July to late August. Rare outside the main migration period, as early as June (exceptionally, 9 June 1977) and as late as mid-November (exceptionally, 9 December 1977). Earliest spring date is 10 March 1988 at Topock Marsh. One midwinter record of two at Yuma, 31 January 1981. Max: 180; 29 April 1952, Topock Marsh.

This species was apparently more numerous during the 1950s when, in addition to the maximum count given above, 164 were at Topock Marsh on 30 July 1953. The only large flock noted in recent years is of 50 birds at BWD on 15 July 1979. Only one to seven individuals have been found annually in spring since 1976 (see Fig. 13, p. 81).

HABITAT Godwits frequent plowed or grassy fields that have been flooded. They often roost on sandbars, islands, or the spillways of dams.

Ruddy Turnstone (*Arenaria interpres*)

STATUS A casual transient. Three fall records from Lake Havasu on 16 September 1952 (two birds, spec), 21 August 1953, and 3 September 1977. Two additional spring sightings are of one at Havasu NWR, April 1966, and two at Cibola NWR, 16 May 1979.

At the Salton Sea, small flocks are regular in spring. This species is casual elsewhere in the Southwest.

Black Turnstone (*Arenaria melanocephala*)

STATUS A casual visitor. One seen 21 May 1948 on the California shore of Lake Havasu.

Wintering birds in the Gulf of California are a potential source of such inland strays. This is supported by at least seven other spring records from the Salton Sea.

Red Knot (*Calidris canutus*)

STATUS A casual or very rare transient. Five fall records: 9 August 1950 at Lake Havasu, 23 July 1952 at Topock (spec), 2 October 1959 near Imperial Dam, 28 November 1975 near Davis Dam, and 6 October 1979 east of Yuma. Also one spring sighting, 1 May 1975, near Davis Dam.

This species is extremely rare throughout the inland Southwest, with most records in fall. Surprisingly, it is a fairly common transient at the Salton Sea, where 200–300 birds are noted occasionally in spring.

Sanderling (*Calidris alba*)

STATUS A rare but regular fall transient from late August to late September. Two later records: 28 November 1975 near Davis Dam, and 21 November 1978 at Imperial Dam. Rare and irregular in spring from mid-April to late May (about eight records).

In addition to the above records is a probable sighting of 60 Sanderlings in a single flock flying low, downstream over the river on 22 September 1980 south of Blythe. Also in mid-September 1980, 15–20 Sanderlings were seen at Cibola NWR. Although such numbers are unprecedented in Arizona and most of interior California, this species does concentrate regularly at the Salton Sea. Nearly all other records are of 1 to 4 birds, with the largest spring count being seven at West Pond, 24 May 1959.

HABITAT They occur primarily on exposed sandbars and river channels below dams, and rarely in flooded fields.

Semipalmated Sandpiper (*Calidris pusilla*)

STATUS A casual transient. One record, a juvenile on 31 August 1980, at the north end of Lake Havasu (Fig. 21).

As familiarity with juvenile plumages of *Calidris* sandpipers has improved, this species is demonstrated to be a regular migrant elsewhere in Arizona and southern California. Future records are to be expected, and it should be looked for in spring as well.

FIGURE 21. Juvenile Semipalmated Sandpiper at the north end of Lake Havasu, 31 August 1980. (Photo by K. V. Rosenberg.)

Western Sandpiper (*Calidris mauri*)

STATUS A fairly common spring transient from April (exceptionally, 29 March 1983) to mid-May. Also fairly common during fall migration which is more protracted, with numbers increasing throughout July, peaking in August, and decreasing through September to early October (exceptionally, 31 October 1978). A few substantiated records in December and January indicate that this species may winter irregularly in very small numbers. Max: 172; 19 August 1953, Topock.

This species is outnumbered by the Least Sandpiper in all months except May. Although Western Sandpipers winter commonly at the Salton Sea, they are casual or very rare elsewhere in the interior Southwest. All winter sightings from the LCRV require documentation. Those that we accept are 1 at Draper Lake, 29 January 1957; 23 December 1960 at West Pond (2); 18 December 1978 at Parker; 2 December 1979 at Cibola NWR; 16 December 1981 at Lake Havasu; 19 December 1981 at Imperial Dam; and 31 December 1987 (2) at Imperial Dam. An additional sighting of 80 at Mittry Lake, 11 November 1942, may have included wintering individuals.

HABITAT Western Sandpipers concentrate primarily with other shorebirds in flooded fields, on sandbars, and lakeshores.

Least Sandpiper (*Calidris minutilla*)

STATUS A common to abundant transient and winter resident from mid-July (exceptionally, 18 June 1977) through April; very scarce in early May (latest record, 9 May 1978). Max: 650; 6 October 1960, Imperial NWR.

It is impossible to separate migrants from wintering individuals in this species, although numbers are always highest in fall. Single flocks of 100–200 birds have been noted as early as 18 July and are common through November. Winter concentrations are smaller, although 248 were at BWD on 9 January 1953; 240 were at Lake Havasu, 16 February 1982; and up to 552 have been found on the Parker CBC (1977).

HABITAT They may be found in nearly any wet area including flooded fields, canals, muddy banks, and marshes. However, largest numbers concentrate on exposed sandbars in the river channel.

Baird's Sandpiper (*Calidris bairdii*)

STATUS An uncommon fall transient in August and September; rarely to mid-October (exceptionally, 6 November 1978). Two seen, 8 July 1953, at Topock Marsh and one near Yuma, 20 July 1979, were very early and were likely adults. One reliable spring sight record on 2 May 1979 at Blythe. Max: 5; 20 August 1980, Blythe.

From 5 to 15 individuals are noted each fall. This is in contrast with the large concentrations that occasionally visit desert lakes and ponds north and east of the LCRV.

HABITAT Virtually all of our sightings are from irrigated fields.

Pectoral Sandpiper (*Calidris melanotos*)

STATUS An uncommon fall transient from mid-September to early November, rarely appearing as early as late August. Two reliable spring sight records: 9 April 1975 (2) near Davis Dam and 15 April 1979 (2) at Blythe. One winter sighting 30 December 1957 at Martinez Lake. Max: 24; 14 September 1947, Lake Havasu.

Since 1976, fall totals have ranged from 2 to 40 individuals. It is a regular fall migrant across southern Arizona but is very rare at the Salton Sea.

HABITAT Nearly all records are from irrigated fields.

Dunlin (*Calidris alpina*)

STATUS A rare but regular late fall transient and winter visitor from mid-October to early March. Only one earlier record, 16 September

1955, at Imperial NWR. Also seven spring records in late April and early May. Max: 25; 20 December 1980, Yuma CBC.

Dunlins are the latest shorebirds to arrive in fall, and birds seen in December may still be migrating. They most often appear in small groups (2–10 birds). Notable wintering flocks include 21 at BWD in January 1978, and 17–19 all winter in 1981–82 at the north end of Lake Havasu. Three of the spring records involve small flocks—10 at Icehouse Bend, 11 May 1950; 10 at Martinez Lake, 27 April 1956; and 7 east of Yuma, 9 May 1979. Specimens examined from the LCRV are *C. a. pacifica* (Coues).

HABITAT Dunlins are most often seen on exposed mudflats or gravel bars where other shorebirds are concentrated.

Stilt Sandpiper (*Calidris himantopus*)

STATUS A casual transient. The only record is of two seen 17 August 1979 near Bard.

The regularity of this species throughout Arizona and at the Salton Sea makes its rarity along the Colorado River an enigma.

Short-billed Dowitcher (*Limnodromus griseus*)

STATUS A rare but regular fall transient from late July (exceptionally, 1 July 1973) to early October (latest, 6 October 1988). Max: 13; 18 September 1977, Parker to Lake Havasu.

This species has occurred regularly in early fall in recent years, as it has elsewhere in southern Arizona. However, at the nearby Salton Sea it is common during both fall and spring migrations. This disparity in status between these two inland regions is shared by other long-distance migrant shorebirds with primarily coastal distributions (e.g., Black-bellied Plover, Whimbrel, Red Knot). Most records are of calling juveniles. Therefore, careful attention to molt and plumage patterns should aid in separating this species from the more numerous Long-billed Dowitchers (see Prater et al. 1977:116–118).

HABITAT Most records have been of one to four birds in irrigated fields or shallow marshes.

Long-billed Dowitcher (*Limnodromus scolopaceus*)

STATUS An uncommon winter resident and fairly common transient from mid-July to mid-May. The only records outside that period are from 1978—one at Imperial NWR on 21 May, and three on 15 June and one on 8 July at Martinez Lake. Max: 350; 20 October 1942, Mittry Lake.

Peaks of migration are in April and October, the largest flocks being 75 on 29 April 1979 at BWD, 195 on 16 October 1954, and the above maximum at Mittry Lake. However, local concentrations in nonwintering areas during November and late March indicate that migration is under way at these times.

HABITAT Favored wintering sites are exposed mudflats, such as at BWD, the north end of Lake Havasu, and the Cibola backwater channel. Migrants visit flooded fields, shallow marshes, sandbars, and canals.

Common Snipe (*Gallinago gallinago*)

STATUS A fairly common transient and winter resident throughout the valley from September to late April. First birds arrive in late August, and late transients are recorded into early May. One bird on 18 July 1979 (photo) at Poston was highly unseasonal. Max: 100; 17 December 1957, BWD.

HABITAT Snipes may be found in any habitat that provides wet, muddy ground. These include plowed and irrigated agricultural fields, dirt-lined canals, marshes, and partly flooded willows. Its true abundance may escape those who choose to stay on firm, dry ground.

Wilson's Phalarope (*Phalaropus tricolor*)

STATUS An uncommon spring transient in late April and early May (exceptionally, to 17 May 1979), with one very early record, 27 March 1959 (2), at West Pond. Fall migration is more protracted; fairly common in August and uncommon in July and early September, with a few later records 7 October 1982 and 19 October 1981. A few seen in June are probably also southbound migrants; these included 2 at West Pond, 2 June 1960, and as many as 140 at Walker Lake, Imperial NWR, 27 June 1961. There is one sighting in winter, 19 December 1971 at Martinez Lake. Max: 550; 31 August 1979, near Yuma.

The bulk of the fall passage usually occurs in one or more large concentrations during August (see Fig. 13, p. 81). Flocks of more than 200 birds have been noted several times between 7 August and 2 September. Outside of that period, birds trickle through in small groups of up to 20 individuals. Spring numbers are usually very low, but 115 were at West Pond on 3 May 1959. As with many other shorebird species, they are much more common (even abundant at times) at the Salton Sea, where a few have also wintered.

HABITAT This phalarope is found primarily in irrigated fields and shallow ponds. They almost never occur on large reservoirs or on the main river channels where the water is too deep for the birds to stand.

Red-necked Phalarope (*Phalaropus lobatus*)

STATUS An uncommon fall transient, but sometimes in large flocks, from August to mid-October. Rare and irregular in spring with five records: 16 May 1952 at BWD (24); 8 May 1961 at Imperial NWR (4); and 19 May 1952, 26 May 1979, and 17 May 1982 (7) at Lake Havasu. Max: 644; 18 August 1954, Topock Marsh.

Migrations of this species are primarily coastal. However, large numbers regularly concentrate at large inland lakes in the western United States, especially in fall. At the Salton Sea, they may be abundant at times in early fall and concentrations have been noted in spring. Extreme dates of fall migration in the LCRV are 21 July (1978) to 2 November (1979). Other large flocks were 200 on 14 September 1977 and 425 on 4 October 1982, both at Lake Havasu.

HABITAT In contrast with the previous species, the Red-necked Phalarope is usually encountered in tight flocks, far from shore on large reservoirs or in the main river channel.

Red Phalarope (*Phalaropus fulicaria*)

STATUS A rare and irregular fall transient from mid-September to late November (about ten records). Also one early fall record, 4 August 1946 at Parker, and one spring record, 1 April 1977 at BWD.

All records are of single birds, except for two together on Lake Havasu, 20 October 1988. This phalarope is almost exclusively pelagic in distribution. However, a few are noted annually at inland bodies of water in the Southwest.

HABITAT This species has been recorded from large lakes and irrigation canals in the LCRV.

Pomarine Jaeger (*Stercorarius pomarinus*)

STATUS A casual transient. Two records from Lake Havasu: 26 September 1950 (spec) and 3–5 September 1977 (photo).

There were three other inland reports of this species during September 1977 at the Salton Sea.

Parasitic Jaeger (*Stercorarius parasiticus*)

STATUS A rare but possibly regular fall transient from late August to early October. Recorded on Lake Havasu in at least 1947, 1953, 1977, 1980, and 1981, and at Davis Dam, 17–19 September 1976.

This is the only jaeger to be found with some regularity at inland bodies of water, although its occurrence at any single locality is unpre-

FIGURE 22. Immature Long-tailed Jaeger at Lake Havasu, 3 September 1977. Note the blackish secondaries contrasting with the rest of the wing, and blunt-tipped central tail feathers. (Photo by K. V. Rosenberg.)

dictable. Nearly all jaegers on Lake Havasu were seen from boats on the widest portion of the lake and were not visible from shore. Thus even a regular passage of this species could be easily missed. Interestingly, they have been observed in association with migrating Common Terns on Lake Havasu, much as they are in coastal waters.

All but one record of this species involve dark, immature birds. Although immatures are difficult to identify with certainty, our records include specimens on 13 October 1947 and 19 September 1953, and individuals photographed on 31 August 1980 and 14–24 September 1981. The only adult occurred on 14 September 1977 (photo). The fall of 1977 may have been an exceptional one for inland-migrating jaegers. In addition to all three species being sighted on Lake Havasu in September, all three were found elsewhere in interior southern California. At least 7 other Long-tailed Jaegers and 16 Parasitic Jaegers were reported from other inland locations in western North America.

Long-tailed Jaeger (*Stercorarius longicaudus*)

STATUS A casual transient. Records from Lake Havasu in 1977 are an immature on 3 September (Fig. 22); two adults on 4 September (photo), with one present on 5 September; and another immature carefully identified on 14 September (Fig. 23).

This is the rarest jaeger inland, with most individuals migrating well offshore in the Pacific Ocean. However, there is a scattering of other interior Southwest reports, all in late August or early September.

The appearance of so many jaegers on Lake Havasu in September 1977 has resulted in some confusion in the published literature. The immature photographed on 3 September was originally identified and published as a Parasitic Jaeger (Witzeman et al. 1978). The identification was recently corrected by J. L. Dunn. Monson and Phillips (1981) do not cite a record of either species on that date, but list an immature Parasitic from 4 September that may have been this bird. Whether an immature "Parasitic" on 27 August was also this bird remains unclear. Monson and Phillips (1981) also incorrectly cite the date of the two adult Long-tailed Jaegers as *14 September*. It is possible that with the "reidentification" of these and several other recent Arizona records of immature jaegers, Long-tailed Jaegers may prove to be the most regular of the three species in our area.

Laughing Gull (*Larus atricilla*)

STATUS A casual transient. Three records: one on 3 September 1960 near Imperial Dam (spec), an immature on 17–20 March 1979 at Davis Dam, and an adult seen 16 June 1981 at Cibola NWR.

Although this species is common in summer at the Salton Sea and formerly bred there as well, there are virtually no other records from the inland Southwest.

Franklin's Gull (*Larus pipixcan*)

STATUS A rare and irregular fall transient from late August through November, and in spring from mid-March (earliest, 10 March 1988) to early June.

Of roughly 25 records to date involving about 40 individuals (see Fig. 13, p. 81), 3 were in late March, 3 in late April–early May (10 individuals), 4 in early June, 3 in late August–early September (5 individuals), and 12 in October–November (20 + individuals). Most occurrences are of single birds with other gulls; however, up to 6 were together on 13 November 1979 at Parker, and on 26 April 1982 at Lake Havasu. Nearly all fall birds have been immatures, whereas spring birds are often adults with completely black heads and rose-tinged underparts.

Franklin's Gulls exhibit a similarly complex pattern of occurrence throughout the Southwest. They are somewhat more numerous in fall (October) at the Salton Sea, and in spring throughout Arizona. Directly north of our region, this species is abundant at Great Salt Lake in Utah.

However, unlike other Great Basin- and prairie-nesting waterbirds, the Franklin's Gull migrates primarily toward the east, through Texas and eastern Mexico.

HABITAT This gull frequents small ponds as well as large reservoirs and, in general, wherever other gulls are concentrated.

Bonaparte's Gull (*Larus philadelphia*)

STATUS A rare transient and winter visitor. Records for every month of the year, but regular only in November. A large flight occurred in late spring 1977 with over 30 seen between 22 April and 1 July. Max: 12; 27 October 1949, BWD.

Outside of the 1977 influx, transients have been noted sporadically between late February and late May and from August through November. Only 1–10 individuals have been seen in a season. However, eight were seen below Martinez Lake, 4 November 1960, and up to six birds were found together on 11 November 1978 at Lake Havasu. In contrast, the hundreds often noted in spring at the Salton Sea and other California inland lakes indicate that the northward passage of this species from the Gulf of California occurs mostly west of our region. Eight

FIGURE 23. Immature Long-tailed Jaeger, Lake Havasu, 14 September 1977. Note blunt-tipped tail feathers; this individual was much whiter below than the one seen 3 September. (Photo by S. W. Cardiff.)

winter records are mostly of single birds: 4 January 1955 at Imperial NWR, 13 January–3 February 1976 at Davis Dam, mid-December–25 January 1980 at Parker Dam, 4 December–12 January 1981 at Lake Havasu, and 6 February 1987 at Parker Dam. Seven were at Topock on 15 December 1952, two were near Imperial Dam on 17 December 1988, and eight were on Lake Havasu, 10 December 1989.

HABITAT Nearly all sightings have occurred with other gulls or terns over open water, particularly near dams and large reservoirs.

Heermann's Gull (*Larus heermanni*)

STATUS A very rare but possibly regular fall transient in recent years. Recorded nine times (10 individuals) between 17 September and 13 November in seven different years since 1975. Over half were in the brief period 9–13 November in four different years. Records outside this period were 19 April (the only adult) and 9–13 June 1983 at Cibola NWR, and 28 December 1985 at Parker.

This primarily coastal gull migrates between its breeding grounds in the Gulf of California and beaches along the Pacific Coast north to British Columbia. An overland route between these two regions, by at least some individuals, would account for the cluster of fall records from the Colorado River and across southern Arizona. Curiously, this species is virtually unreported during fall at the Salton Sea, where it occurs instead as a rare postbreeding visitor in summer. Our June record fits this latter pattern. However, the April sighting of an adult was when this species should be breeding at the Gulf of California and when other such inland records are virtually nonexistent.

Mew Gull (*Larus canus*)

STATUS A casual winter visitor or transient. One adult and two first-winter immatures (photo) present 19–20 March 1979 at Davis Dam. The two immatures remained until 1–3 April, and a second-winter individual was present also 31 March–17 April (Fig. 24). Also, a probable immature at Taylor Lake, 25 February 1960, and a probable adult, seen 25 January 1978, above Davis Dam.

This species is somewhat regular in winter and spring at the Salton Sea and possibly along the Nevada shore of Lake Mead. However, the photographed birds mentioned above represent the first confirmed records for Arizona.

Ring-billed Gull (*Larus delawarensis*)

STATUS A fairly common to locally common winter resident throughout the valley from November through March. Uncommon fall tran-

FIGURE 24. Second-winter Mew Gull at Davis Dam, 1–3 April 1979. One of four individuals present; first confirmed record for Arizona. (Photo by B. M. Whitney.)

sient in small flocks beginning in late July (exceptionally, 26 June 1978). Rare between mid-April and July, although this is the most likely gull to occur during that period. Max: 1,600; 29 December 1954, Imperial Dam and 1,500; Ferguson Lake, 21 March 1961.

This is the only gull normally encountered, except in April when it is outnumbered by transient California Gulls. Flocks of 40–80 birds are regular by early August, and winter flocks may number several hundred in some years.

An interesting behavior of this species was observed at the Lake Havasu City Golf Course in October 1976. A flock was observed circling cultivated date palms, hovering in midair to pluck the fruits. This behavior also has been noted at the Salton Sea.

HABITAT Wintering concentrations often form at dams and large reservoirs, although small numbers may appear near any water including ponds, backwaters, and agricultural canals. Migrant flocks or individuals are more partial to flooded fields, especially in late summer. These gulls are primarily seen flying in loose lines above the river or resting on sandbars.

California Gull (*Larus californicus*)

STATUS An uncommon fall transient from late July through October, and uncommon to fairly common spring transient from mid-March through April. Rare but regular in winter, except at Davis Dam where a sizeable flock occurs in most years. Only one record outside these periods, 4 seen 23 May 1983 near Poston. Max: 275; 1 April 1948, Blankenship Bend.

Away from Davis Dam, this gull usually occurs as singles or in small groups of two to five immatures, except in April when flocks of migrating adults stop along the river (i.e., up to 130, 14 April 1979 at Blythe, and see Max above). Our only fall concentrations are of 13 on 23 July 1977 near Bullhead City; 20 on 24 July 1978 and 24 on 4 August 1982 near Palo Verde; and 29 on 5 August 1954 at Havasu Lake. Numbers at Davis Dam vary greatly from year to year, and peaks often do not occur there until late winter, perhaps augmented by early spring migrants.

HABITAT In winter they frequent dams and lakeshores where other gulls concentrate. Migrants, however, are most often seen in flooded, plowed fields.

Herring Gull (*Larus argentatus*)

STATUS A rare but regular winter visitor between mid-October and mid-April. The only earlier records are 2 August–2 September 1961 (second year) at Martinez Lake and in 1982 near Blythe on 30 July (immature), 10 August (second year), 25 September (adult), and 1 October (adult).

This is the most likely large gull to occur here, being rather common in winter at the Salton Sea. About two-thirds of all records have been from north of Parker Dam. Immatures are renowned for their extreme variability in plumage and bill color, making the certain identification of other similar species very difficult. Birds of all ages have been seen, with up to four individuals together.

HABITAT They primarily occur with other gulls near dams, marinas, or on large lakes. Apparent migrants have also visited flooded agricultural fields and rested on sandbars in the river.

Thayer's Gull (*Larus thayeri*)

STATUS A casual winter visitor. One specimen record 13 December 1946 from Parker Dam. Three careful sightings: 12 December 1974–18 February 1975 (one or two immatures) at Davis Dam, 17 January 1976 (three immatures) at Davis Dam, and 26 November 1978 at Imperial Dam.

These records fit a pattern of consistent wandering to inland waters such as the Salton Sea. However, because much confusion surrounds the identification of immatures, all records must be well documented.

Yellow-footed Gull (*Larus livens*)

STATUS A casual postbreeding visitor; one adult seen at Senator Wash Dam, 27 July 1989. (Not shown in bar graphs.)

Until recently this species was considered conspecific with the Western Gull. It has increased since 1975 as a postbreeding wanderer to the Salton Sea from breeding areas in the Gulf of California. This gull's appearance in the LCRV was therefore to be expected. An adult, dark-backed gull seen 18 July 1979 at Painted Rock Reservoir on the lower Gila River (Monson and Phillips 1981) was also probably this species.

Western Gull (*Larus occidentalis*)

STATUS A casual winter visitor. One immature collected on 12 December 1946 at Parker Dam.

This specimen was identified as the more northern nominate race *L. o. occidentalis* Audubon (Monson and Phillips 1981). This constitutes one of very few acceptable records of this gull away from the immediate Pacific Coast.

Glaucous-winged Gull (*Larus glaucescens*)

STATUS A casual winter visitor. Two specimen records: 4–24 February 1954 on Lake Havasu and 17 November 1956 at Imperial NWR. Sight records of immatures include all winter 1975–76 at Davis Dam (second-winter), 19 December 1977 at Imperial Dam, 30 October 1981 at Lake Havasu City, and 19 November 1982 at Cibola NWR.

This species is found regularly in small numbers in winter at the Salton Sea. A few may regularly move between the Pacific Coast and the Gulf of California where they also winter in small numbers. All but the bird at Davis Dam were in pale first-winter plumage. Hybrids between this species and the Western Gull are frequently noted in the Pacific Northwest, and pure Glaucous-winged Gulls may be impossible to detect in the field. However, there is little doubt that birds seen along the lower Colorado River originated along the Pacific Coast, and thus contribute to the pattern of occurrence of this species in the Southwest. Caution must be used in separating Glaucous-winged Gulls from immature Thayer's Gulls, as well as from worn, faded, or leucistic Herring Gulls.

Black-legged Kittiwake (*Rissa tridactyla*)

STATUS A casual late fall and winter visitor. Eight records involved 13 individuals since 1973.

FIGURE 25. Immature Black-legged Kittiwake below Parker Dam, 21 December 1980. This bird was part of a record invasion of this species into the Southwest that brought at least four to the lower Colorado River Valley. (Photo by K. V. Rosenberg.)

This normally oceanic species was first recorded 24 November 1973 at Martinez Lake. One was at Davis Dam, 17–20 February 1975 (photo); two immatures were there in early December 1975; and an adult and immature were there 25 February 1976. Another adult was at Parker Dam, 11–19 November 1978 (photo). In late fall 1980, the inland Southwest was invaded by Black-legged Kittiwakes. The Colorado River attracted two immatures at Lake Havasu, 28 November–5 December (one spec), and two others were below Parker Dam, 21 December–6 February 1981 (Fig. 25). The single specimen proved to be of the Pacific race *R. t. pollicaris* Ridgway. More kittiwakes occurred to the north on Lake Mead during 1980–81, and also at the Salton Sea. There are scattered records from these locations in other winters as well.

Sabine's Gull (*Xema sabini*)

STATUS A rare but regular fall transient from mid-September to mid-October. At least 19 individuals have been recorded in 12 different falls between 1948 and 1983. Also a flock of seven was at Martinez Lake on 13 April 1956, with one lingering to 27 April, and one was at Imperial Dam 22–23 June 1974.

The consistent, annual occurrence of this pelagic gull at inland bodies of water throughout western North America suggests a regular overland migration route from its arctic breeding grounds to the tropical oceans where it winters. All fall records are of lone juveniles, except three together on Lake Havasu, 17 October 1957 and 11 October 1980, and a single adult at Cibola on 6 October 1983.

The more unusual spring sightings are much earlier than the normal northward passage of this gull along the Pacific Coast. However, the lone June bird corresponds with other late spring–summer records at the Salton Sea.

HABITAT Most Sabine's Gulls have been seen far from shore on Lake Havasu. The remainder were scattered along most stretches of the main river channel between Davis and Imperial Dams.

Gull-billed Tern (*Sterna nilotica*)

STATUS A casual visitor. One of two collected on 24 May 1959 at Imperial NWR (Arizona side, above Draper Rock).

This species is a common breeder in summer at the Salton Sea; thus future records are to be expected. There is only one additional record from Arizona.

Caspian Tern (*Sterna caspia*)

STATUS A rare to uncommon spring transient from late March to mid-May and uncommon to fairly common fall transient from early June through early October. A few appear again in late December and remain through winter (about 10 records). Max: 34; 11 September 1953, BWD.

Most spring migrants have occurred in April and early May. Six to eight individuals have been noted per season in recent years. Apparent southbound birds pass through in two distinct periods, June–early July and late August–September, with fewer noted between 15 July and 15 August. Groups of 5–10 individuals are noted frequently during times of peak passage (see Fig. 13, p. 81).

Curiously, there are no records for late October or November. Winter visitors suddenly appear in mid to late December (on CBCs) or even later. These records are mostly from between Laguna Dam and Martinez Lake and include one on 30 December 1961, four on 23 December 1974, one to two from 16 December 1978 to 27 January 1979, one on 24 January 1980, seven on 20 December 1981, three on 20 December 1986, two on 23 December 1987, and seven on 17 December 1988. Winter records north of Martinez Lake are of singles in early February 1981 and from 17 December 1981 to 17 February 1982 at Cibola NWR, and on 16 December 1981 at Lake Havasu. These persistent occurrences may indicate a recent change in status.

This species is very common from April through September at the Salton Sea, where they also winter more regularly. Wandering birds from this population may account for part of the early summer influx in the LCRV. However, individuals flying south over the river at that time suggest true southbound migration, as in other Great Basin-nesting species. In contrast, this species is exceedingly rare east of the Colorado River across Arizona.

HABITAT These terns are observed most frequently foraging on larger bodies of water or resting on sandbars with other terns and gulls. Migrants normally fly fairly high above the main river channel and a few have been found in flooded fields.

Common Tern (*Sterna hirundo*)

STATUS An uncommon to fairly common fall transient in August and September, rarely until mid-October. Also, an immature carefully identified, 7 March 1983, northeast of Yuma. There are five additional spring and summer sight records; however, its certain occurrence at these seasons requires confirmation. Max: 100+; 3 September 1977, Lake Havasu.

The fall passage is more concentrated and localized than that of the Forster's Tern and, in general, occurs later. In recent years, they have far outnumbered Forster's Terns on Lake Havasu in late August and early September. Elsewhere in the Southwest they are numerous only at the Salton Sea, indicating their general preference for larger bodies of water. They also occur regularly in late spring and summer at the Salton Sea, lending support to our sightings during that period.

Arctic Tern (*Sterna paradisaea*)

STATUS Probably a casual transient; 1 adult carefully observed at Lake Havasu, 4 September 1981, and another possible sighting of an adult with 65 Common Terns at BWD, 27 August 1978.

There are two additional fall records of this primarily oceanic species for Arizona. Nearly all interior California records, however, have been of adults in late spring (June).

Forster's Tern (*Sterna forsteri*)

STATUS An uncommon spring transient, primarily in late April and May but exceptionally from 4 March 1978. A fairly common fall migrant from June through mid-October, with peak passage from mid-July to mid-August. Prior to 1985, small numbers rarely lingered into late fall and winter. More recently, this species has been found regularly in winter, sometimes in large concentrations. Max: 62; 17 December 1988, Yuma CBC.

Migrations are typical of Great Basin-nesting species, with birds appearing continuously through the summer months (see Fig. 13, p. 81). We have not seen individuals lingering at specific locations through summer, and all such birds should be treated as transients. Indeed, most June records are of one to eight adults flying south along the river channel. As with several other species (e.g., Willet, Black-necked Stilt), it is impossible to delineate exactly the north- and southbound migrations. It is a common summer resident and sporadic breeding species at the nearby Salton Sea, where it is also fairly common in winter.

This species usually occurs in small groups (2–10 birds), although concentrations of 20–40 have been seen on occasion in late summer. In 1981, a flock of about 30 birds lingered at Lake Havasu through fall to early November, then decreased steadily until the last sighting of 1 on 16 December. More recently, winter records have proliferated. These records include large concentrations: 2 at BWD from November 1985 until 17 January 1986 (photo); up to 20 around Imperial Dam on 20 December 1986 and 39 there again, 19 December 1987; 3 at Topock Marsh, 2 February 1988; 20–38 at Lake Havasu, 25 November 1988–30 January 1989; and 62 around Imperial Dam, 17 December 1988, reduced to 17 there 11–12 February 1989. These may signal a recent change in status of this species, similar to what has occurred in the Caspian Tern. It is fairly common in winter at the Salton Sea.

HABITAT Most Forster's Terns are observed flying over open water or resting on sandbars. Occasionally, tarrying individuals will feed over shallow marshes, backwaters, ponds, or flooded fields.

Least Tern (*Sterna antillarum*)

STATUS A casual transient or postbreeding summer visitor. Five records: 18 June 1953 at Lake Havasu (spec), 30 July 1959 and 1 July 1973 at Imperial Dam, and 19–20 June 1980 and 5–6 June 1986 at Cibola NWR.

These dates correspond well with other sporadic records of this species across southern Arizona and at the Salton Sea.

Black Tern (*Chlidonias niger*)

STATUS An uncommon to fairly common transient from late April through May and again from mid-June through mid-September. Rare in early June (about nine records). Max: 66; 11 September 1959, near Yuma.

Fall numbers are usually far larger than those in spring, and records span a longer period. As with several other Great Basin-nesting species,

FIGURE 26. Black Skimmer, with Common Terns, at the north end of Lake Havasu, 3 September 1977. (Photo by K. V. Rosenberg.)

there is virtually no gap separating the appearance of north- and south-bound migrants in early June (see Fig. 13, p. 81). The only spring concentrations were associated with severe storms in 1977 (e.g., 59 on 13 May near Parker). Occasionally, flocks of 15–25 have been noted in fall, but most records are of 1–5 individuals. These numbers pale beside the thousands that gather in late summer at the nearby Salton Sea.

HABITAT Black Terns occur at marshy shores of lakes and ponds, as well as in flooded fields. They are frequently seen flying low over the river or resting on sandbars during migration.

Black Skimmer (*Rynchops niger*)

STATUS A casual transient or postbreeding visitor. Five records: 12 June 1977 near Martinez Lake (photo), 1–3 September 1977 at Lake Havasu (Fig. 26), 30 August 1979 at Cibola NWR, 30 July 1982 near Palo Verde (2 adults in flooded field; photo), and 16 May 1989 at Martinez Lake.

Recently, this species has colonized and bred at the Salton Sea (since 1972). Therefore, more records along the lower Colorado River are to be expected.

Rock Dove (*Columba livia*)

STATUS A locally common resident and breeder wherever residential areas occur.

HABITAT Most individuals remain near cities, trailer parks, farm houses, and feedlots.

Band-tailed Pigeon (*Columba fasciata*)

STATUS A casual visitor or transient. Seven records: an old specimen from east of Somerton, 20 November 1941; one seen at BWD, 28 October 1977; individuals at Clark Ranch north of Blythe on 10 July 1978, August 1981 (weakened bird captured and rehabilitated), and 10–11 July 1983; one at Blythe, 22–31 May 1979 (Fig. 27); and one seen at Parker, 7–15 November 1981.

In Arizona, they are migratory, only rarely remaining in the southern mountains after October. Migrant individuals are encountered somewhat regularly along Upper Sonoran Desert streams or at other lowland localities. The nearest breeding population is in the Hualapai Mountains. In coastal California mountains this species is considered to be

FIGURE 27. Band-tailed Pigeon at Blythe, California, 31 May 1979. (Photo by K. V. Rosenberg.)

resident, with seasonal altitudinal movements. A few have been noted in late spring and fall at desert oases and in summer in desert mountains of southeastern California.

The likelihood that the lower Colorado River birds originated in Arizona is supported by the Yuma specimen that was identified as *C. f. fasciata* Say. It is also likely that birds found recently in southeastern California mountains are part of this migratory population and are a potential source of transients in the LCRV and at other desert oases.

HABITAT Far from their normal haunts of oak or mixed pine-oak woodlands, in the LCRV they have been attracted to any tall trees available, especially cottonwoods.

White-winged Dove (*Zenaida asiatica*)

STATUS A locally abundant summer resident and breeder from April to mid-August. The first spring migrants may arrive in mid-March (exceptionally, 3 March 1978) and most have departed before the hunting season begins in early September. A few are seen in October in protected areas, and individuals occasionally winter in the Yuma area. Max: 2,200; 20 July 1979, Imperial Dam to Yuma.

White-winged Doves have expanded their range in Arizona since the late 1800s in association with the expansion of agriculture in major river valleys. They were described as common at Yuma in 1881 but were not found north to Fort Mohave at that time. Stephens (1903) saw and heard this species at Needles from June to August 1902. In 1910, Grinnell (1914) recorded the species as fairly common in willow–cottonwood habitat, but Wiley (1916) found no nests at Palo Verde in 1916. In recent years, the White-winged Dove has become one of the most abundant species throughout the valley where dense riparian woodlands remain. Numbers now appear to be declining with agricultural shifts from grain to cotton, and with decreases in preferred nesting vegetation.

This is an important gamebird in Arizona, which draws many out-of-state hunters. Most of the breeding population leaves the LCRV before hunting season, which traditionally begins on Labor Day. Young of the year, especially those raised late in the season, are heavily harvested. By the end of the first week of dove season, most doves either have migrated or have been killed. About 30,000 White-winged Doves are taken annually during this short period.

HABITAT Before agriculture, this dove was primarily a bird of saguaro-dominated deserts. Its breeding cycle closely corresponded with the flowering and fruiting of the saguaro (Alcorn et al. 1961; Haughey 1986). When these doves began feeding extensively in cultivated grain fields, use of adjacent riparian woodlands for nesting increased.

At present, largest numbers breed in tall stands of screwbean mesquite mixed with saltcedar, pure saltcedar stands that are allowed to mature, and in mature citrus orchards. Lesser numbers breed in cottonwood–willow, desert washes, and urban habitats. They feed almost exclusively in agricultural fields, and large flocks move daily between these feeding areas and nesting groves. The continued clearing and burning of native screwbean mesquite and saltcedar stands, such as north of Blythe, threatens some of the largest breeding colonies in the LCRV.

BREEDING Breeding commences by early May and continues through July. Nests are loosely constructed platforms usually placed 3–6 m high in screwbean mesquite or saltcedar trees. Clutch size is invariably two eggs and at least two broods are raised per season. This species may attain breeding densities of up to 1,400 pairs/40 ha in suitable habitat, and their nests may be placed as close as 1 m apart (Butler 1977).

Combined fledging success from all habitats in the LCRV is from 44 to 65% (Butler 1977). The major cause of nest failures is predation, primarily by Gila Woodpeckers, Cactus Wrens, and Crissal Thrashers, as well as by coachwhips, common kingsnakes, and gopher snakes. The predation rate was found to be positively related to openness of the habitat, with nests in pure, short-statured mesquite incurring the highest predation.

This is among the few species that breeds during midsummer in saltcedar habitats along the lower Colorado River. Among the stresses they face are extreme temperatures that exceed the physiological limits of their exposed eggs. Russell (1969) suggested that behavioral-physiological modifications in adult White-winged Doves helped to alleviate extreme temperatures. The probable process, investigated in Mourning Doves, involves the adult dropping its own body temperature to act as a heat sink for the eggs (Walsberg and Voss-Roberts 1983). This process is described in more detail under Mourning Dove.

FOOD HABITS The diet (from 138 stomachs) consists primarily of cultivated grain, with wheat, barley, milo, and corn among the most common items. Average seed size taken is substantially larger than those taken by Mourning Doves. Wild seeds of annuals, perennials, and leguminous trees are also eaten.

Mourning Dove (*Zenaida macroura*)

STATUS An abundant spring and summer breeder; less numerous during fall and winter but still common throughout the valley. Max: 3,118; 23 December 1977, Parker CBC.

This dove is an extremely popular gamebird, and the opening of dove season on Labor Day weekend turns the usually serene valley into a shooting gallery. Hunters come from as far as San Diego and Los Angeles to hunt both Mourning and White-winged Doves. About 200,000 Mourning Doves are taken each year.

HABITAT This species breeds in most riparian habitats as well as in citrus orchards. It is less partial to taller trees than the White-winged Dove, and some will nest in sparse saltcedar or open desert habitats. They feed primarily in agricultural areas. Outside the breeding season, Mourning Doves leave riparian areas and are found primarily in agricultural, desert, and urban habitats.

BREEDING Breeding begins in April (but sometimes as early as mid-February) and continues through September, although some nests are found throughout the year. Two eggs are laid in the loosely constructed nest, which may be on the ground or to 9 m above ground. Nest-site selection is more generalized than in White-winged Doves. Nests are commonly placed in saltcedar, screwbean and honey mesquite, or citrus trees but can be found in just about any species of tree or shrub and even on the ground. Multiple broods are raised, depending on when nesting commenced.

This is among the few open-nesting species breeding in midsummer in very open riparian habitats. One of the major problems these birds face is how to protect their eggs from extremely high midday air temperatures. How the birds accomplish this has been the focus of recent physiological research.

Russell (1969) suggested that adult White-winged Doves may maintain lowered body temperatures to draw heat away from the eggs and into the brood patch. Walsberg and Voss-Roberts (1983) investigated this possibility in Mourning and Ringed Turtle (*Streptopelia risoria*) Doves. They found depressed body temperatures for Mourning Doves during midday incubation averaged 38°C, compared with nonincubating body temperatures of 42–44°C for birds exposed to air temperatures of 39°C. Depressed body temperatures during incubation apparently allowed egg temperatures to average 40°C when air temperatures averaged 44–45°C. The adult dove can maintain lower temperatures by evaporative cooling through panting, gular flutter, and esophageal heat exchange, but this is not done without causing stress from dehydration. It is the male that attends the eggs from 0900 until about 1700, when the female relieves him. Leaving the nest during the middle of the day, even for a short time, could cause the eggs to overheat. Thus, the male Mourning Dove must regulate his own hydric stress as well as temperature of the eggs to breed successfully. The availability of persistent surface water is critical, therefore, to these desert-nesting doves.

FOOD HABITS The diet, from 446 stomach samples, was composed primarily of weed and agricultural seeds. Smaller seeds were selected than those eaten by White-winged Doves, and weed seeds constituted a higher proportion of the diet.

Inca Dove (*Columbina inca*)

STATUS Recently a fairly common local resident around Blythe, Parker, Needles, and Bullhead City; rare in the Yuma area. Max: 55; 22 December 1982, Parker CBC.

This is a relatively recent natural invader into Arizona, with no records before 1872. This species is almost totally dependent on human dwellings and apparently spread northward along with permanent ranches and settlements. Monson (1949) discovered the first resident colony at Parker in 1946. However, the species had been reported as a stray since 1942, when a small population probably was established at Yuma. The species has since spread north to Katherine Landing, just above Davis Dam, and to the southern tip of Nevada (to Boulder City). It is much more local on the California side of the river but is reliably found at Blythe, Clark Ranch, Parker Dam residential area (first found in 1948), and Needles.

Although they have been found periodically in the southern parts of the valley (e.g., Laguna Dam, Bard, Yuma), they have not become established in these areas. Instead, the Common Ground-Dove is the small dove inhabiting agricultural and suburban habitats around Yuma.

HABITAT This dove remains restricted to residential areas, although some may be seen away from heavily populated areas such as at ranches in agricultural areas.

BREEDING This species has the longest and most productive breeding season of any Arizona bird, with up to five broods being raised in one season. Nests are found from January to November. These are flattened platforms of twigs, grass, and rootlets placed in almost any native or cultivated trees.

Common Ground-Dove (*Columbina passerina*)

STATUS A fairly common but local resident, tending to be more common in the southern half of the valley. Max: 55; 18 December 1982, Yuma CBC.

HABITAT Common Ground-Doves prefer agricultural edges, orchards, and sparse riparian vegetation. This species occurs in suburban habitats at Yuma, replacing the Inca Dove which is the small suburban dove elsewhere in the LCRV and throughout most of central and southern Arizona.

BREEDING Nests are found from May through October and are well built with small twigs and grass. Any tree may be used, with willows and mesquites near water preferred.

Ruddy Ground-Dove (*Columbina talpacoti*)

STATUS A casual winter visitor. One male seen at Bard, 17 December 1988 (Yuma CBC), with a flock of sparrows in a citrus orchard. Two were in the same area in December 1989, and one was at Parker on 10 December 1989. (Not shown in bar graphs.)

These records fit the recent history of fall and winter vagrancy by this species into the American Southwest. Since 1981, Ruddy Ground-Doves have been reported in Arizona at least 17 times with many of these reports including two individuals. More than 10 birds have been reported from southern California since 1984, with records close to the LCRV during the falls of 1984 and 1985 at Iron Mountain Pump Station. Additional records since 1984 are scattered throughout southern New Mexico and west Texas. The origin of these birds is still being debated; however, the regularity of fall–winter dispersal north of breeding areas (similar to Common Ground-Dove) and recent northward expansion in Sonora argues strongly for naturally occurring Ruddy-Ground Doves.

Yellow-billed Cuckoo (*Coccyzus americanus*)

STATUS Until very recently, a fairly common but local summer breeder from June through August. A few arrive in late May and linger through early September (extreme date, 3 October 1959). The largest remaining population is at BWD; dramatic declines continue to occur elsewhere.

This cuckoo is the last migratory summer breeder to arrive in the LCRV, with some individuals still arriving in mid- to late June. Territories are quickly established, young are raised, and they depart by the end of August. Thus, the cuckoo spends only about one-quarter of its annual cycle in the valley, the shortest of the LCRV breeding species.

The plight of this species exemplifies the history of habitat change along the lower Colorado River, as well as the struggle of all riparian-dependent species to persist in the face of tremendous odds. The cuckoo needs strong federal protection, at least in California where it is listed as State Endangered, and has received much attention at its western-most population stronghold along the lower Colorado River. This species is also listed as Threatened by the Arizona Department of Game and Fish. The success of our own revegetation efforts was gauged, in part, by our ability to attract this species.

Historical accounts indicate that this species was fairly common along the main rivers of southern Arizona, including the Gila and lower Colorado River drainages (Swarth 1914). In the late 1800s and early 1900s it was also widespread in California, including along the southern Pacific Coast and in the Central Valley. Unfortunately, Grinnell (1914) and his party completed their journey a month before "spring" arrival of the cuckoo, so we do not have an historical account of its natural history along the lower Colorado River.

There was little information on this species in the valley until our detailed censusing began in 1975. Cuckoo densities were determined for each habitat from 1975 to 1979. Based on our inventory of available riparian habitat, 242 cuckoos were estimated to occur along the lower Colorado River mainstream in 1976 with an additional 208 at BWD. Presently this species is declining very rapidly throughout the valley and has disappeared from many areas of previous occurrence since water levels were raised in 1983.

Estimates based on 1975–79 densities and 1986 vegetation data showed a population decline of 42% since 1976. Actual field surveys in 1986, however, indicate a much steeper decline. No more than 18 cuckoos were found along the lower Colorado mainstream in 1986, constituting a 93% drop from 1976 levels. At BWD, only 50–60 cuckoos were estimated to occur in 1986, constituting a 71–75% decline. Similar dramatic declines have been noted throughout California since the early 1960s. The species has also declined along much of the lower Gila River within the past decade. Declines may be attributed almost solely to habitat loss. Documentation of this continuing decline may be found in Gaines and Laymon (1984), Hunter et al. (1987a, b), and Laymon and Halterman (1987), and we urge the continued monitoring of remaining populations at the BWD.

HABITAT Mature cottonwood–willow stands provide the primary habitat in the LCRV. Willows or isolated cottonwoods mixed with tall mesquites (both screwbean and honey) are also used but to a lesser extent. Foraging birds may be found in stands of smaller mesquite trees and even saltcedar, but no nesting is evident in these habitats. Therefore, the decline of cottonwood habitats (see Chapter 2) has had an extremely negative effect on this species.

The restriction of cuckoos to cottonwood–willow groves in the LCRV is interesting in light of their more extensive use of saltcedar and honey mesquite habitats farther east (Hunter 1988, Hunter et al. 1988). In our region the cuckoo, as a midsummer breeder, must face the extremely high midsummer temperatures that would kill unprotected eggs. Therefore, the cuckoo must be a nest-site specialist or develop mechanisms for cooling eggs behaviorally or physiologically (see Mourning Dove account). Mature cottonwoods, with willows forming a subcanopy layer, provide the best shading of any riparian habitat; saltcedar and open mesquite bosques are inadequate in buffering lethal temperatures. In addition, standing water in many cottonwood–willow groves may help to lower the air temperature by evaporative cooling. Thus, the declines in cuckoo populations may be attributed largely to the removal of necessary thermal cover.

One of the goals of our revegetation efforts was to attract declining bird species such as the Yellow-billed Cuckoo (Chapter 4). Revegetated stands were predicted to be mature enough to harbor cuckoos by 1980. Indeed, cuckoos were found nesting successfully on the revegetation site at Cibola NWR during 1981 and 1982when cottonwood trees were about 15 m tall; two nests were found in willows. In 1983, six were detected on the site but no nesting was documented. Cuckoos have been absent from the site since 1984 and it appears that their absence is a reflection of an overall decline. Laymon and Halterman (1987) suggest that even this 30-ha revegetation site was too small and, therefore, was not capable of supporting a viable population center. The return of healthy cuckoo populations to the LCRV is completely dependent on the return of extensive, mature stands of cottonwood and willow trees, either through natural regeneration or revegetation.

BREEDING   Nesting commences by early July and continues through August. One brood of two to three young is raised per season. Nests are platforms of sticks usually placed on horizontal branches in dense foliage from 4.5 to 14 m above ground. Willows near water seem to be preferred. A nest with two eggs was in a small screwbean mesquite 1.8 m above ground at BWD, 20 August 1950. Also, nests were found occasionally in saltcedar trees within screwbean mesquite- or cottonwood–willow-dominated habitats; brood success for these nests was not determined. The restriction of this species' breeding to the midsummer period is thought to be a response to the seasonal peak in large insect abundance, most notably cicadas (see Rosenberg et al. 1982).

Cuckoos near their nests are highly secretive. During intensive observation in 1977 and 1978 at BWD, only two nests of this species were found. On several occasions, as birds approached a nest site with food, they would sit motionless for long periods or move slowly from branch to branch for up to 30 minutes. They would then finally swallow the insect and fly off. On 15 July 1977, two copulations by a pair were observed. The first lasted 20–30 seconds, during which time the female exchanged her green katydid for the grasshopper in the male's bill. During the second copulation, which lasted 36 seconds, the female slowly removed another large insect from the male's bill.

FOOD HABITS Large insects made up the largest proportion of the diet of six birds; cicadas (40%) were the most important item. Mantids (30%), grasshoppers (25%), and caterpillars (5%) were also taken. Occasionally, cuckoos will catch lizards and tree frogs.

Cuckoos appear to forage in slow motion and are reminiscent of several tropical forest species. A bird will move along the smaller branches in the outer portions of a tree, often pausing on a perch for several minutes while slowly cocking its head at all possible angles to search the foliage. Once prey is spotted the bird will either move forward on the branch to glean it from a leaf or, about half the time, will swoop from its perch, pluck the insect from the leaf in midair, and continue to a new perch. Of 48 observed foraging attempts, about two-thirds were in willows, and the remainder in cottonwood trees. Nearly all attempts were higher than 6 m above ground.

Greater Roadrunner (*Geococcyx californianus*)

STATUS A common resident throughout the valley. Max: 72; 20 December 1979, Parker CBC.

This familiar bird is a characteristic sight along the lower Colorado River. It lives up to its name by traveling the many farm and levee roads that traverse the valley. At any time of the year, the loud bill-clacking may be heard at some distance. The relatively high population levels of this species in the valley are evidenced by national high counts on the Parker CBC from 1976 through 1980, with the 1979 tally constituting an all-time national high.

HABITAT Roadrunners use a wide variety of terrestrial habitats. Riparian habitats, especially honey mesquite, screwbean mesquite, and sometimes saltcedar, as well as desert washes, are important for breeding. During fall and winter, roadrunners are frequently found foraging along agricultural-riparian edges, but they are rarely seen far from protective cover.

BREEDING Nesting occurs from April to June. The distinctive, soft, descending series of male coos can be heard as early as February. The nest is compact and about 0.3 m in diameter, and is usually placed 1–4.5 m high in a large tree. Three to five eggs usually are laid.

FOOD HABITS Roadrunners are famous for eating large reptiles, including rattlesnakes. They also take rodents and birds, but their primary food source consists of large insects such as grasshoppers, cicadas, and beetles. One individual at the Lost Lake Resort, north of Blythe, dined repeatedly on Gambel's Quail chicks that fed along the edge of the park.

Barn Owl (*Tyto alba*)

STATUS A fairly common resident throughout the valley. Max: 18; 20 December 1980, Yuma CBC.

This is among the most strictly nocturnal of the resident owls, rarely venturing out to forage before total darkness. It is detected most often as a ghost-like form passing in front of car headlights, or as a screeching hiss breaking the silence of the desert night. It also is found regularly during the day as it flushes readily when disturbed from the roost.

HABITAT Barn Owls roost and nest in tall trees, buildings, and cliffs. During the summer months, they are often found in thickets of tall saltcedar and screwbean mesquite. This owl feeds primarily over agricultural land, but adjacent desert habitats are also probably used. This is the most likely owl to be found in towns.

BREEDING The season extends from January to June. Most nest sites are at small caves in cliffs above the river or lakes, such as on the California side below Needles and Parker and the Arizona side below Ehrenberg. Cavities in trees or old buildings may also be used. Five to seven eggs are laid each season.

Flammulated Owl (*Otus flammeolus*)

STATUS A casual visitor with one record of an individual calling repeatedly from 9 March to 18 April 1979 at BWD.

This owl is highly migratory, breeding throughout the coniferous-forested mountains of the western United States and wintering in Mexico. Early spring migrants are known to linger in the lowlands elsewhere in Arizona and California. In 1984, individuals were found north of our region at Las Vegas and Beaver Dam Wash.

Western Screech-Owl (*Otus kennicottii*)

STATUS A fairly common resident throughout the valley. Max: 17; 4 June 1978, BWD.

This owl may be surprisingly numerous in areas where much suitable habitats remain. For example, a density of 60 birds/40 ha was estimated in the BWD cottonwoods in June 1978. Grinnell (1914) noted this species very frequently in 1910 along the entire length of the river. It is likely that the population is severely reduced at present and will probably continue to decline as more tall native trees are lost.

Marshall (1967; also *in* Phillips et al. 1964) accepts as valid the designation of the lower Colorado River population as *O. k. yumanensis* (Miller and Miller). This subspecies is very pale gray dorsally, suffused with a pinkish-sandy color. The crossbars of the underparts are less conspicuous than in any other Western Screech-Owl population.

HABITAT This owl is most common in native riparian habitats, especially in stands of cottonwood–willow, screwbean mesquite, and honey mesquite where cavities are present. In winter they are also found in saltcedar habitats. Desert washes and urban areas are used throughout the year.

BREEDING Nests are in cavities of native riparian trees and in the few saguaros within the confines of the valley. Recently fledged juveniles have been noted at BWD throughout June to 20 July. Grinnell (1914) noted newly hatched young in a saguaro nest on 23 April 1910.

Great Horned Owl (*Bubo virginianus*)

STATUS A fairly common resident throughout the valley. Max: 13; 18 December 1976 and 22 December 1980, Parker CBC.

This owl is most frequently seen at dawn or dusk, when it perches on trees or poles along roadsides. This is also the most often-heard owl in most parts of the valley. These birds are the palest of all Great Horned Owls. They are sandy colored and their indistinctly marked plumage blends perfectly with the desert surroundings.

HABITAT This species is a habitat generalist. It uses riparian woodlands, desert washes, agricultural areas (for feeding), and sometimes even towns.

BREEDING Great Horned Owls often use old Red-tailed Hawk or raven nests in small caves, cottonwoods, or saguaros. Nesting begins in December with recently fledged young found into early June. Few breed within the valley proper; most occur in desert washes or cliffs on the edge of the valley.

Elf Owl (*Micrathene whitneyi*)

STATUS A rare and very local summer breeder between early March and early September. Most individuals are detected from April to July. Small populations remain at BWD, near Needles, north of Blythe, in the Fort Mohave area, and near Cibola NWR. Formerly present south of Yuma but apparently never more widespread or common.

This small owl is characteristic of the Sonoran Desert and is at the western edge of its range along the lower Colorado River. Still unclear is whether this species was ever very numerous or even whether populations have been continually present. Therefore, it is ironic that the Elf Owl was first discovered and described to science by Cooper (1861)

FIGURE 28. Last cottonwood tree to be bulldozed north of Needles in February 1978. This grove supported one of the last nesting populations of Elf Owls in California. Its destruction symbolizes the plight of many riparian forest-dependent species in the lower Colorado River Valley. (Photo by K. V. Rosenberg.)

from Fort Mohave. Brown (1903) and Grinnell (1914) described a population breeding in saguaros on the California side near Laguna Dam (Potholes region). These saguaros have long since vanished along with the owls. There seem to be no other historical records.

Elf Owls were recently found in the Fort Mohave–Needles area, where birds were first noted in California in 1972 and the southern tip of Nevada in 1975. In 1976, a few were found on the Arizona side near Fort Mohave, and in 1977 a population was discovered at BWD. During this period, breeding was noted at an isolated cottonwood–willow grove north of Needles that was being cleared for agriculture. The trees were completely bulldozed by 1978 (Fig. 28). In that year, 10 breeding pairs were found at the nearby Soto Ranch (see Chapter 6), where only 5 pairs were found in 1987 after further habitat loss (Halterman et al. 1988). The only other specific records from the Arizona side are from the Yuma Proving Grounds on 4 June 1980 (2), an area with saguaro cacti. However, populations on the Arizona side have never been adequately surveyed.

The Elf Owl's future in the LCRV is precarious. The species is currently listed as State Endangered by the California Department of Fish and Game, and it is critical that what little habitat remains for this owl be protected. Presently, no more than 17 pairs are estimated to occur in California (Halterman et al. 1989). Reintroduction efforts are being planned by the California Department of Fish and Game.

HABITAT Along the lower Colorado River, Elf Owls have been found primarily in cottonwood–willow habitats and where saguaros occur at the edge of the valley. On the California side, where saguaros and cottonwood groves have been virtually eliminated, the few remaining birds persist in tall mesquite groves containing cottonwood or willow snags.

BREEDING Historic breeding in the saguaros north of Yuma was documented by two sets of four eggs collected on 17 May 1903 by Brown. Grinnell (1914) collected two females with developing eggs on 22–23 April 1910. More recently, fledged young have been noted near Needles as early as 10 April (1976) and as late as 26 June (1982). Natural or woodpecker-made cavities in tall trees (primarily cottonwoods but including mesquite), snags, or saguaros are required for nest sites.

Burrowing Owl (*Athene cunicularia*)

STATUS A fairly common resident throughout the valley, becoming less common in northern areas in winter. Max: 33; 20 December 1979, Parker CBC.

The Burrowing Owl may have expanded its distribution in the LCRV with the expansion of agriculture, as none were recorded by Grinnell (1914). Their diurnal habits and preference for open habitats make this species easier to observe than other owls. They are not very vocal, however, and are most frequently detected when flushed from the edge of a road or canal bank. A completely albino Burrowing Owl was seen on 11 March 1978 near Poston.

HABITAT This owl uses primarily agricultural lands, especially where dirt embankments are extensive. They are usually terrestrial, but will often perch on fence posts and sometimes on telephone poles or wires. The species is occasionally found in sandy, sparsely vegetated riparian woodlands and along desert washes.

BREEDING Family groups have been noted frequently in midsummer, indicating widespread breeding. However, the breeding habits of this species in the LCRV have not been studied.

Long-eared Owl (*Asio otus*)

STATUS A rare but probably regular transient or visitor, with records from throughout the year. Possibly resident in some years, with breeding occasionally noted.

Of roughly 20 specific records, half are from winter; 5 were in early spring (March–April); 1 was heard south of Parker Dam, 17 August 1982; and 1 was seen at BWD, 25 September 1978. Summer records that suggest local breeding are 9 July 1948 at Lake Havasu (adult found freshly dead in water); 7 May 1952 north of Blythe; 2 May and 6 June 1960 at Imperial NWR; 29 May 1978 at Topock Marsh (fledgling; Fig. 29); 17 May 1979 (photo) and 30 June 1979 near Needles; and 11 June 1981 north of Blythe (remains of a juvenile found). Large roosts and breeding have been noted in desert areas east and west of the river valley.

HABITAT Most reports are from dense stands of tall riparian trees, such as cottonwoods or athel tamarisks, and from densely vegetated desert washes.

BREEDING Little is known about the length of breeding season or the nesting habits of this species in the LCRV. Based upon the two records of juveniles, breeding probably commences sometime in April.

Short-eared Owl (*Asio flammeus*)

STATUS A rare to locally uncommon winter resident from November through March. Usually encountered singly, but occasionally small concentrations are found. Max: 9; 1 January 1959 and 27 November 1976, near Yuma.

This owl is often thought of as a nocturnal counterpart of the Northern Harrier, although it is much less common than the harrier. Both species have probably benefited from agricultural development in our region. At least one or two owls were found each winter between 1976 and 1982. A roost of nine birds near Yuma on 27 November 1976 still contained five owls on 8 March 1977, indicating roughly the overwintering period in the LCRV.

HABITAT Short-eared Owls are most often found feeding at twilight or flushed from agricultural fields (primarily tall alfalfa), grassy margins, or marshes.

Northern Saw-whet Owl (*Aegolius acadicus*)

STATUS A casual winter visitor. Two records of individuals found dead, 21 December 1979 at BWD and 14 December 1985 east of Yuma.

This owl is highly migratory and is notorious for straying far from its usual range. It has been recorded elsewhere in western Arizona deserts and even at Puerto Peñasco, Sonora, Mexico.

Lesser Nighthawk (*Chordeiles acutipennis*)

STATUS A common summer resident and breeder between mid-April and mid-September. Numbers increase gradually after the first migrants

FIGURE 29. Fledgling Long-eared Owl at Topock, 29 May 1978. (Photo by R. J. Dummer.)

arrive in early March (extreme date, 25 February 1982), and decrease sharply through October. Individuals linger rarely into December and there are a few midwinter records. Max: 550; 14 May 1977, north of Needles.

Spring arrival is about a month earlier than the first arrivals farther east toward Phoenix and Tucson. However, our early dates correspond well with the period of occurrence in southern California deserts. Lesser Nighthawks tend to remain farther north during mild winters. For example, in 1977–78 one was at Yuma on 20 December, one was at Parker on 23 December, two were at Martinez Lake from 10 December to 7 January, and two were at Blythe on 31 January and 1 February. We have records from at least five other winters, including five reported on the Yuma CBC, 17 December 1988.

HABITAT They forage aerially over almost any habitat and often concentrate at dusk above the river channel. For roosting and breeding, this nighthawk is restricted to sparsely vegetated or burned riparian areas, desert habitats, and, occasionally, flat roofs in towns.

BREEDING Courtship displays consist of lazy chases with both adults uttering eerie whining calls. These displays are most often seen on June mornings. Nests are little more than scrapes on the ground, usually in gravelly situations but also on sandy or clay soils away from sparse vegetation. Eggs have been found as early as 8 May (1910), and a fledgling was found as late as 14 August (1976).

Lesser Nighthawks breed primarily during hot midsummer months and their eggs may face potentially lethal air temperatures. There is some evidence that they can lower their body temperature to draw heat away from the eggs during incubation, as some doves apparently do (Grant 1982; also see our Mourning Dove account). The two to three young are semiprecocial and remain hidden under nearby brush until they fledge.

FOOD HABITS Lesser Nighthawks are aerial foragers and take a wide variety of flying insects. The 28 stomachs examined contained 22 insect taxa including moths, ants, various beetles, cicadas, and leafhoppers. They feed more by day than do other nightjars, with birds commonly flying about until midmorning. Largest foraging concentrations are at dusk, however, and individuals may be seen feeding at lights throughout the night.

Common Nighthawk (*Chordeiles minor*)

STATUS A casual transient, with one, 16 October 1924 (spec), north of Bard, and one heard at BWD on 13 June 1979.

This species is a very late spring migrant in the Southwest, with most arriving from South America in early June. Only a handful of lowland desert records exist elsewhere in Arizona and southern California.

Common Poorwill (*Phalaenoptilus nuttallii*)

STATUS A fairly common summer resident along the edge of the river valley between mid-March and mid-October (exceptionally, 11 November 1978). The occasional presence of individuals in February and one north of Yuma, 15 December 1979, raises speculation of an overwintering population, but this remains to be demonstrated. Max: 15; 11 April 1987, near Lost Lake.

The high concentration of poorwills noted at Lost Lake consisted of calling individuals found at night during owl surveys in and around a large recent burn in the riparian vegetation. These may have been visitors from the adjacent desert. Most other records in the valley proper are of single birds flushed in the day during migration. At BWD, however, several may be heard on spring and summer nights along the surrounding desert cliffs.

Individuals have been documented to hibernate in some areas, including the Riverside Mountains that form part of the western boundary of the LCRV (Jaeger 1948, 1949; Marshall 1955; Rea 1983). It is possible that some overwinter along the periphery of the LCRV and at BWD. Examples of probable migrants include two flushed from screwbean mesquite south of Parker on 7 October 1976, one found dead the same day in farmland near Poston, and one at Cibola NWR from 14 October to 10 November 1982.

HABITAT Upland desert habitats and cliffs are used for roosting and nesting. At least some individuals feed in adjacent valley habitats, where they may sit on roads or riverbanks or even skim the surface of the water.

Black Swift (*Cypseloides niger*)

STATUS A casual transient. At least one reliable sighting, 9 May 1977, south of Ehrenberg.

This species has been seen both in spring and fall elsewhere in Arizona, but no fully documented record exists as yet. There are only a few records of migrants from the California desert region, but these include one from the Salton Sea in early May. The individual seen was in direct comparison with migrating Vaux's Swifts and swallows.

Chimney Swift (*Chaetura pelagica*)

STATUS A casual transient, with one secured on 6 May 1930 (spec) at Bard and one seen on 23 June 1974 at Laguna Dam.

This species has been recorded regularly in recent years throughout the Southwest. In addition, four birds were seen, west of the valley, at the Iron Mountain Pump Station on 2 June 1984.

Vaux's Swift (*Chaetura vauxi*)

STATUS A fairly common but irregular spring migrant from mid-April through mid-May. Uncommon fall migrant from the end of August (exceptionally, 15 August 1976) through late September, rarely to mid-October. Max: 220; 9 May 1977, Parker to Ehrenberg.

Migration of this species in the LCRV is very light, with usually only one or a few individuals seen among flocks of swallows. The large numbers seen in May 1977 were associated with a series of severe storms. Early May appears to be the time of peak passage in most years, including 4–5 May 1910 ("very many" seen by Grinnell) and 10 May 1957 (100 seen at Blythe). The only fall flocks noted were in early September 1976 during tropical storm Kathleen.

HABITAT The species can be found foraging aerially over rivers, lakes, marshes, and riparian vegetation. An exposed nocturnal roosting site was found on 8 May 1964, 3 km south of Davis Dam on the Arizona side. This flock was packed into a 1.2-m × 36-cm space, 3 birds deep, on an athel tamarisk (photo; Stager 1965).

White-throated Swift (*Aeronautes saxatalis*)

STATUS A fairly common to abundant resident in the northern half of the valley; rare in the southern half. Occasionally, very large flocks are seen in winter. Max: 1,210; 20 December 1979, Parker CBC.

Largest winter concentrations are consistently at Davis Dam, BWD, and south of Parker. CBCs in the latter two areas frequently have recorded national high counts for this species, illustrating the importance of the LCRV as a wintering area. In other seasons, and south of the Parker Valley, sightings are irregular and of few individuals. Some are present year round at BWD and Parker Dam where they undoubtedly nest, and small numbers may be resident where desert hills meet the river near Imperial Dam. The highest count in the Yuma area was only 28 on the CBC in 1973. Large nonbreeding flocks have formed at Parker as early as 23 August 1978 (350 birds), but these large flocks are not seen after February.

HABITAT These swifts breed on desert cliffs, such as at BWD and Lake Havasu. Foraging flocks range widely over riverine, agricultural, and riparian habitats. They may swarm close to the ground, darting by at dizzying speeds, or rise up to beyond the limit of sight, their presence belied only by their shrill, twittering calls.

BREEDING Nests are placed far inside crevices along cliffs. Courtship displays are spectacular, with the pair of birds copulating in flight, then swirling and falling towards the ground like a child's "whirlybird." Swifts have been observed at their nesting cliffs beginning in March and April.

Broad-billed Hummingbird (*Cynanthus latirostris*)

STATUS A casual fall and winter visitor. Five records: a male, 6–8 September 1979 east of Yuma; a female, 30 September 1979 at BWD (photo); an immature male, 28 September–November 1979 north of Blythe; and adult males, December 1975–February 1976 and 4 October 1981–13 February 1982 at Blythe.

These records are part of a recently well-documented pattern of dispersal northwestward from its normal fall and winter range in southeastern Arizona. This hummingbird has been found somewhat regularly in central Arizona and westward to the southern Pacific Coast in California. It has also reached southern Nevada and Utah.

HABITAT This species is especially attracted to blooming tree tobacco in fall in southern Arizona. The BWD bird was among riparian willows and tree tobacco; however, all other records were from feeders. Both of the Blythe birds were in the same suburban yard.

Black-chinned Hummingbird (*Archilochus alexandri*)

STATUS A common summer resident and breeder from mid-March to mid-September. The first migrants (usually adult males) arrive in late February (earliest, 20 February 1979) and late birds linger at feeders through October (latest, 25 October 1981). Max: 20; 8 June 1981, Parker to Ehrenberg.

Arrival in the LCRV is a full month earlier than along the Pacific Coast, but corresponds well with records elsewhere across southern Arizona. Winter records in the LCRV or anywhere else in Arizona are not substantiated. Reports (primarily from CBCs) are invariably of females and in all likelihood refer to female Costa's Hummingbirds, which do winter in small numbers. Any future winter reports need to be confirmed by careful descriptions of vocalizations, plumage, and relative bill length, or preferably by specimens.

HABITAT They are widespread in riparian vegetation and around human residences. Dense willows are preferred and this is one of few species found to breed commonly in pure dense saltcedar. This species is also attracted to flowering tree tobacco.

BREEDING Nesting begins by early April, with most young leaving the nest in mid-May. Nests are composed of willow fluff, down, and spider webs and are usually placed on a diagonal twig 1–2 m above ground. Phillips et al. (1964) state there is considerable evidence for the raising of two broods in southern Arizona. We have not noted this along the lower Colorado River. Adult males are rarely seen after early summer.

FOOD HABITS Food habits of this species have not been studied in detail along the Colorado River. However, individuals have been observed at a variety of flowering shrubs and trees including ironwood, palo verde, wolfberry, and tree tobacco. Small insects are also picked from flowers, foliage, or spider webs. Of course, this species is also attracted to sugar-water feeders.

Anna's Hummingbird (*Calypte anna*)

STATUS Presently a fairly common winter resident and breeder; uncommon from late April to August. Records occur throughout the year, primarily from feeders. Definite breeding has been noted at Yuma, Blythe, Parker, and Parker Dam. Max: 25; 20 December 1986, Yuma CBC.

Anna's Hummingbird has been spreading and increasing throughout the Southwest since the 1940s, occurring primarily, until recently, as a fall and winter visitor (Zimmerman 1973). Breeding in Arizona was not definitely known until the 1960s, with the first report of young fledging on 15 March 1962 at Yuma. By 1977, breeding was established as far north as BWD (photo). Residence throughout the summer was noted at Yuma in 1978 and at Blythe in 1979. Up to 20 were in a single yard in Blythe on 12 August 1982. Our northernmost record is of four at Needles, 10 May 1981. Anna's Hummingbird will probably continue to increase and may eventually be a common year-round resident throughout the valley.

HABITAT This species is found primarily in residential areas, especially where feeders are permanently maintained, or where flowering eucalyptus trees are planted. Occasionally it can be found in riparian areas with tall willows, cottonwoods, or athel tamarisks. Most records away from feeders are in fall.

BREEDING This species is among the earliest to begin breeding, initiating courtship and nesting by December. Most breeding is completed by April, but some individuals continue through May (e.g., two recently fledged young seen 26 May 1979 at Parker Dam).

FOOD HABITS Because of this species' preference for urban habitats, it feeds primarily on various cultivated flowering trees and shrubs, including eucalyptus.

Costa's Hummingbird (*Calypte costae*)

STATUS A fairly common but local breeder in early spring along the periphery of the valley. First arrives in early January and usually departs by mid-May. A few individuals linger through summer, fall, and early winter, usually at feeders.

This is, with the possible exception of the Violet-green Swallow, the earliest terrestrial breeding bird to arrive in spring. The seasonal occurrence of this species is complex throughout its range in the Southwest. An east-to-west migration has been postulated, as many appear on the California coast in May and June, the time of departure from hot desert regions (Phillips et al. 1964). This, however, remains to be proven. It is also possible that the species' status has changed in recent years, perhaps due to the availability of feeders. Individuals are being found with increasing regularity in December. At least some of these are displaying males, such as two near Parker on 22 December 1984. This may suggest a shift toward an even earlier spring arrival.

HABITAT This hummingbird is most often found along desert washes and only occasionally enters adjacent riparian vegetation. A few also breed and occasionally linger around houses, especially where feeders are permanently maintained. Overall, they prefer drier and sparser habitats than either Black-chinned or Anna's Hummingbirds.

BREEDING As with Anna's Hummingbird, breeding for this species starts earlier than that of any other lower Colorado River species. Two eggs are laid usually in March and young are out of the nest before May.

Breeding individuals are frequently detected by the very high and wiry call given by the male during his circular courtship display. This call was once thought to be produced through the primaries as the bird displayed; however, the male occasionally gives the same call with the same intensity while perched.

FOOD HABITS Costa's Hummingbirds feed at various desert annuals, shrubs, and flowering trees, although their specific food habits in the LCRV have not been studied.

Calliope Hummingbird (*Stellula calliope*)

STATUS A rare and irregular spring migrant from mid-April to early May (latest, 9 May 1972). One fall record of a male seen 2 September 1976 at Ehrenberg.

Curiously, we have not noted this species in spring since 1976, during our intensive fieldwork. Most records are from cultivated plantings on the California side of the river.

Broad-tailed Hummingbird (*Selasphorus platycercus*)

STATUS A casual spring migrant. Three records: one at Earp on 12 April 1975, a male seen and heard on 29 April 1979 at BWD, and a female seen on 15 April 1982 at Cibola NWR.

This species is rarely noted in the lowlands anywhere in the Southwest, although nearly all other such records are in April. One was west of the Colorado River at the Iron Mountain Pump Station on 29 April 1984.

Rufous Hummingbird (*Selasphorus rufus*)

STATUS An uncommon spring migrant from early March (earliest, 26 February 1982) through mid-April (exceptionally, 3 May 1977). Also a fairly common fall migrant in August and September, with the first birds (usually adult males) arriving in late July and individuals lingering occasionally (at feeders) to mid-October. A female *Selasphorus* seen 27 December 1977 at Parker Dam, and 11 February 1979 may also have been this species. Max: 30+; 11 September 1976 and 14 August 1981, Blythe.

Spring migration of this species is primarily along the Pacific Coast, with the lower Colorado River marking the eastern boundary of the flight path. Conversely, fall migration covers the entire state of Arizona. Most individuals move through mountainous areas north and east of the lower Colorado River, with fewer birds along the Pacific Coast and through Southwestern deserts. Although the LCRV is on the fringe of both spring and fall routes, the species may be quite numerous at times.

Exceptional numbers (over 80 records) were present throughout September 1976 in association with tropical storm Kathleen, both at feeders and in the "wild." Up to 10–15 individuals may be seen at feeders during peak fall passage (mid-August to mid-September). However, only two to five individuals are usually noted each spring. Much of our information on this and other hummingbird species comes from the observations and records of Mrs. Ione Arnold at Blythe.

HABITAT This hummingbird is found primarily among flowering plants and at feeders in residential areas and only occasionally in riparian woodland.

Allen's Hummingbird (*Selasphorus sasin*)

STATUS Probably a very rare early spring and early fall transient. Sightings of apparent adult males include 27 July 1953 at Parker, 28 August–9 September 1960 at Yuma, 17 February 1975 at Somerton, 12 July 1978 north of Blythe, and 6–10 July 1979 at Blythe.

Some records may be suspect, even of adult males, as a small fraction of adult male Rufous Hummingbirds are said to have partially green backs. However, Allen's Hummingbirds are undoubtedly more frequent in the LCRV than the few records indicate, as many females and immatures must pass through unnoticed.

Belted Kingfisher (*Ceryle alcyon*)

STATUS A fairly common winter resident from September through April, with migrants arriving throughout August (exceptionally, 24 July 1977). Some individuals linger through May (latest, 3 June 1977). One seen along the flooded Gila River east of Yuma, 19 June 1979, may have been summering, but breeding is not suspected. Max: 63; 20 December 1986, Yuma CBC.

The desert portion of the southwestern United States represents one of the few areas of North America that lacks a breeding kingfisher species. Whether this species is absent for physiological reasons or, as Grinnell (1914) suggested, because the Colorado River and its tributaries were too muddy in their natural state to allow profitable fishing is a matter of speculation.

HABITAT This is a characteristic species of riverbanks, ponds, and agricultural canals.

Green Kingfisher (*Chloroceryle americana*)

STATUS A casual visitor; one male seen at Topock Marsh, 25 March 1988. Also, an old report (Coues 1866; unsubstantiated) of several seen in autumn 1865. (Not shown in bar graphs.)

This small kingfisher is primarily a tropical species found north to Sonora, Mexico, and somewhat regularly to ponds and streams in southeastern Arizona. Our recent record coincides with an increase in sightings in the latter area, including the first confirmed breeding in Arizona (July 1988). It also represents the northwesternmost occurrence of the species. The older reports by Coues (1866) are questioned by most authors.

Lewis' Woodpecker (*Melanerpes lewis*)

STATUS A rare and irregular fall and winter visitor from higher elevations during "flight years." A few other fall and spring records occur during nonflight years. Recorded from 20 September (1950 at Topock) to 13 April (1979).

This species is famous for its nomadic wanderings and movements into the lowlands, sometimes involving hundreds of individuals. Only small numbers have ever reached the lower Colorado River, however. These flights were documented at least in fall–winter 1946–47, 1950–51, 1957–58, 1960–61, 1978–79 (Fig. 30), 1983–84, 1987–88, and 1989–90, but may have occurred more frequently. Up to 10 birds were present at Parker during the last flight, and there are several records south to the Yuma area.

HABITAT This woodpecker should be looked for in any area with tall trees, although this species is especially fond of pecan orchards.

Acorn Woodpecker (*Melanerpes formicivorus*)

STATUS A casual visitor. Four records: 23 October 1950 at BWD, 24 October 1971 at Picacho, 23 November 1977 at Ehrenberg, and 25 June 1977 near Poston.

This woodpecker is known to stray from its normal mountainous haunts during all seasons and, as with the Lewis' Woodpecker, there are "flights" in certain falls but in much smaller numbers. Several similar records exist across Arizona and California deserts, including from the Salton Sea.

FIGURE 30. Lewis' Woodpecker in a pecan orchard below Parker in February 1979. (Photo by K. V. Rosenberg.)

Gila Woodpecker (*Melanerpes uropygialis*)

STATUS A locally common resident and breeder throughout the valley where tall trees occur. Max: 78; 19 December 1978, Bill Williams Delta CBC.

This species was formerly more common and widespread. Coues (1866) reported them as abundant. Grinnell (1914) called them common, as they were found at every station along both sides of the river. Today they are restricted to the relatively few areas where native habitats retain some tall trees. This species is listed as State Endangered by California Department of Fish and Game. About 200 breeding individuals were estimated to occur in California in 1983, with about half of these on private ranches, residences, or parks (Hunter 1984). On the Arizona side the Gila Woodpecker is probably more common. The total population in riparian habitats was estimated to be about 650 birds in 1976, with a decline to about 600 in 1983 and another decline to 561 in 1986. Overall, the lower Colorado River population is probably just over 1,000 individuals.

Van Rossem (1942) assigned subspecific status to the population in the LCRV. Phillips et al. (1964) and Short (1982) argue *M. u. albescens* (Van Rossem) to be invalid, as it does not share plumage characters found in subspecies of other species along the lower Colorado River (e.g., the overall paleness as exhibited in Western Screech-Owls and Song Sparrows).

HABITAT Gila Woodpeckers can be found where large nest trees or saguaros occur. They are most common in the few remaining native cottonwood–willow-dominated woodlands, but also use very tall culti-vated trees such as eucalyptus and athel tamarisk. A few individuals use honey or screwbean mesquite stands, especially those with very tall old trees or snags.

Use of riparian habitats broadens during postbreeding dispersal, when individuals may occasionally be found in saltcedar, sparse honey mesquite, and other suboptimal habitats. By winter, birds in subopti-mal habitats disappear and the remaining birds occur almost exclu-sively in cottonwood–willow and tall, old mesquite stands, where they later breed.

BREEDING Nests are in cavities excavated in tall trees or saguaros. Cot-tonwoods and willows are apparently preferred because Gila Wood-peckers appear to have difficulty excavating cavities in the harder wood of mesquite (Brush et al. 1983). In residential areas they use very tall eucalyptus, athel tamarisk, date palms, orchard trees, and other orna-mental species.

Of 29 nests observed at BWD during 1977–78, 6 were in saguaros, 5 in cottonwoods, and the remainder in willows. All tree nests were 8–14 m above ground in either dead limbs or the dead tops of broken-off trunks. Most cavities faced to the north or northwest, and none faced east, presumably to avoid direct insolation.

Gila Woodpeckers are noisy and conspicuous when breeding. This begins in February when pairing and much territorial chasing is evi-dent. At BWD in 1977 and 1978, nesting was under way in March and the first sets of eggs hatched in mid-April. Young were out of the nests in mid- or late May and the adults wasted no time in beginning a second brood. Some birds began excavating new nests even while still feeding young in the first nests!

Family groups remained together as the adults attended their second nests. When the second broods fledged at the end of June, local density was extremely high. During July, aggression was observed again, and both adults and juveniles were molting. By the end of the month, very few young of the year remained in the cottonwood–willow habitat. This rather frantic progression of events resulted in extremely high productivity, as virtually every nest produced three to five young.

This high productivity is not apparent, however, away from extensive riparian habitats. In other areas, this species is heavily victimized by European Starlings, which evict the woodpeckers from their nesting cavities. At the Clark Ranch north of Blythe, for example, observations were taken on the effects of starling pressure on nesting pairs (Hunter 1984). The Clarks removed approximately 150 starlings from the vicinity of an active nesting territory in late April to late June 1983. Despite this removal, three to four pairs of starlings were always present in the nesting territory, while hundreds of starlings foraged in nearby fields. These woodpeckers abandoned three different nest cavities from mid-April to early June, and fledged no young. On the last expulsion, the starlings removed the woodpecker eggs from the cavity. Two woodpecker nests under study at the Parker Dam residences suffered a similar fate without starling control.

This species is facing two problems in the LCRV, especially on the California side of the river. First, the reduction of suitable native habitat severely restricts the viability of local populations. Isolated, mature cottonwood–willow groves <20 ha are devoid of woodpeckers and, in general, the smaller the habitat, the less likely that birds will occur. Although the species seems to be common at private residences, resorts, and parks, it is here that starlings interfere with nesting success. Though some young are produced, possibly enough to keep the population levels stable, little habitat exists to which young woodpeckers may disperse. Thus the lack of native habitat coupled with reduced productivity at suboptimal sites may limit any expansion of the population into unused habitats, and may signal continued population declines.

FOOD HABITS Based on 15 stomachs from riparian habitats in summer, the diet consists of cicadas (>50%), ants, termites, true bugs, beetles, and insect larvae. A few cactus fruits were also eaten. The preponderance of cicadas taken is surprising, considering the typical woodpecker habits of this species. We rarely saw them sally out in pursuit of these prey. It is likely that most cicadas were captured from tree trunks, perhaps as they first emerged.

Foraging at BWD was primarily above 6 m on the trunk or large limbs of willows or cottonwoods. Birds often probed under the coarse bark of the willows or gleaned from the smooth surface of the cottonwoods, but rarely hammered forcibly at wood. Food habits of this species are less well known in winter, but they are occasionally noted feeding on mistletoe berries in honey mesquite woods and in desert washes.

Yellow-bellied Sapsucker (*Sphyrapicus varius*)

STATUS A rare but possibly regular winter visitor from mid-October to late April. About nine records since 1974, mostly of immatures.

This is the familiar sapsucker of eastern North America. Before the 1970s the Yellow-bellied Sapsucker was considered to consist of four fairly well-defined subspecies. Recent evidence supports the separation of the sapsuckers into three full species: the Yellow-bellied Sapsucker, Red-naped Sapsucker (*S. nuchalis*), and Red-breasted Sapsucker (*S. ruber* including *S. r. ruber* (Gmelin) and *S. r. daggetti* Grinnell). All three species were once found together at Ehrenberg on 25 October 1981.

The Yellow-bellied Sapsucker is very rarely reported in Arizona or California. However, its status may be clouded by its similarity to the more common Red-naped Sapsucker. Immature-plumaged (i.e., brownish) sapsuckers in the lowlands during fall and winter are almost certainly Yellow-bellied Sapsuckers, as immature Red-naped Sapsuckers attain adult-like plumage before they leave their mountain breeding grounds. More care given to identifying adult sapsuckers may better define the status of the Yellow-bellied Sapsucker in the LCRV. For example, a small number of adult female *S. varius* acquire a solid black instead of red crown, a trait not known in *S. nuchalis*.

HABITAT Yellow-bellied Sapsuckers have been found in tall riparian trees such as cottonwoods, willows, and athel tamarisk, as well as among ornamental trees in suburban areas.

Red-naped Sapsucker (*Sphyrapicus nuchalis*)

STATUS A fairly common but local winter resident and transient from mid-October to mid-March. The first migrants may appear by mid-September (exceptionally, 8 September 1982) and a few linger to early April. Max: 15; 24 December 1977, Bill Williams Delta CBC.

This sapsucker winters commonly in our region, arriving from breeding grounds in the Rocky Mountains and Great Basin ranges. It is a quiet woodpecker, detected most often by its gentle tapping. See comments under Yellow-bellied Sapsucker.

HABITAT Sapsuckers prefer tall riparian trees, especially cottonwoods, willows, and athel tamarisks. They will also use ornamental trees in urban areas. Continued loss of mature cottonwood–willow stands may negatively affect this species in the LCRV.

FOOD HABITS They typically forage by drilling rows of holes around tree trunks to draw sap. The sap attracts various insects which become entrapped and are then easy prey for the sapsuckers. Various small fruits are also eaten, especially at ornamental plantings.

Red-breasted Sapsucker (*Sphyrapicus ruber*)

STATUS A rare but regular visitor from late October to early March. About 20 records since 1976.

There are two subspecies of this sapsucker, both of which have occurred in Arizona and southeastern California. Nearly all of our records are of birds that *look like S. r. daggetti* Grinnell of the southern Pacific

Coast. Although the occurrence of this subspecies is supported by specimens (e.g., BWD, 23 January 1953), Phillips et al. (1964) point out that
hybrids between *S. r. ruber* (Gmelin) (of the northern Pacific Coast) and
either *S. varius* or *S. nuchalis* may resemble *S. r. daggetti*. They concluded that the probable origin of many Arizona *S. ruber* is where the
ranges of *S. r. ruber* and *S. varius* overlap in the Rocky Mountains of
British Columbia. Intergrades between *S. r. daggetti* and *S. r. ruber* also
may occur. At least one bird seen 28 January 1979 at Big River Resort
appeared typical of *S. r. ruber*, and others were clearly intermediate
toward *S. nuchalis*.

The LCRV, lying south of these zones of hybridization and intergradation, attracts sapsuckers exhibiting a wide array of characters. Therefore, a detailed series of specimens from the valley would help clarify
the status of each form.

HABITAT This species is found primarily in tall riparian trees; especially
cottonwoods and willows are used. This species has also been found in
clumps of athel tamarisk and in ornamental trees in urban areas.

Williamson's Sapsucker (*Sphyrapicus thyroideus*)

STATUS A casual visitor from higher elevations. Three sight records: 12
March 1861 at Fort Mohave; a female, 7 December 1946 at BWD; and a
male, 7 December 1978 at Dome Valley, northeast of Yuma.

The latter individual was in athel tamarisks and was coaxed into
view by tapping gently on a wooden board, imitating the bird's own
tapping sounds. There are other lowland records in Arizona and southeastern California, including near the Salton Sea.

Ladder-backed Woodpecker (*Picoides scalaris*)

STATUS A common resident and breeder throughout the valley. Max:
99; 24 December 1977, Bill Williams Delta CBC.

This is the common small woodpecker throughout the desert Southwest, occurring in most wooded areas. The largest continuous populations, however, are in extensive riparian floodplains such as the LCRV.

HABITAT This woodpecker occurs in most terrestrial habitats including
riparian woodlands, desert washes and uplands, and agricultural-riparian edges. Highest densities (10–20 birds/40 ha) are in remaining cottonwood–willow habitats. However, this species occurs in low densities (1–2 birds/40 ha) even in saltcedar.

BREEDING Unlike nearly all other permanent resident breeding birds in
the LCRV, this species raises only one brood of three to four young each
year. Nest excavation and pairing are first observed in February. At
BWD in 1977 and 1978, nesting behavior was observed through March

and April, with young not fledging until mid-May. In contrast to nearby nesting Gila Woodpeckers, both adult and young Ladder-backed Woodpeckers were extremely quiet at the nest. Family groups remained together through June and July. By the end of July only adults were present on their territories after the juveniles had dispersed. Thus these birds devote nearly half of the year to producing a single brood.

Most nests were found in either willow or cottonwood snags, 6 to 12 m above ground. In mesquite and desert wash habitats, nests were placed much lower, with the diameter of the limb or trunk (usually 10–13 cm) appearing to play a more important role than height above ground. Overall, this species is much more general about the location of its nest cavities than are larger woodpeckers. Their smaller size allows them to use shorter trees including honey mesquite, saltcedar, and various desert species. They are also better able to drill holes in very hard wood such as mesquite. This species is, therefore, not restricted by the lack of cottonwoods or willows for nesting sites as are other cavity-nesting species in the LCRV.

FOOD HABITS This woodpecker feeds on bark-dwelling insects and larvae. Four stomachs contained mostly termites and beetle larvae.

They forage about equally on trunks and smaller lateral branches and from near the ground to the highest tree tops, depending on the habitat. They tend to hammer more vigorously and loudly than the larger woodpeckers, often drilling deep into wood to extract their prey.

Northern Flicker (*Colaptes auratus*)

STATUS A rare to locally uncommon permanent resident and breeder (Gilded race); fairly common overall from October to early April. Max: 279; 18 December 1978, Parker CBC.

The flicker's status can be divided among three well-marked subspecific groups, which can all occur simultaneously in the LCRV. The Gilded subspecies (*C. a. mearnsi* Ridgway) is the only one that breeds along the lower Colorado River. This desert form is the most distinct of the flicker races and has often been considered a separate species. At BWD the Gilded Flicker is still fairly common, primarily in association with saguaros in the adjacent desert uplands. Away from BWD the Gilded Flicker is now rare, with only scattered pairs persisting at Fort Mohave (at least formerly), the Colorado River Indian Reservation, Cibola and Imperial NWRs, and between Imperial and Laguna Dams. Historical records indicate that Gilded Flickers were more common and widespread throughout the river valley and were associated with both saguaros and cottonwood forests (Grinnell 1914; Swarth 1914).

The Gilded subspecies is listed as State Endangered by California Department of Fish and Game. No more than 40 individuals are estimated to occur in the California portion of the LCRV (Hunter 1984). As with the Gila Woodpecker, more individuals occur on the Arizona side. The estimated total population for the valley (including BWD), based on 1983 vegetation type maps, is ≈272 individuals. However, surveys would suggest the total number outside of BWD to be <100 individuals (Hunter 1984). Despite the rarity of the Gilded Flicker in the LCRV, it is still a characteristic bird of the Arizona Sonoran Desert, only 30 km east of the river.

The Red-shafted Flicker (*C. a. collaris* Vigors) is the common wintering form throughout the valley from early October (exceptionally, 16 September) to early April. Numbers may vary considerably from year to year. These are often supplemented by a few Yellow-shafted Flickers (*C. a. luteus* Bangs), with about 15 records since 1976. In addition, a few Red-shafted × Yellow-shafted intergrades have been noted.

There exists a red variant unique to the Gilded Flicker population of the lower Colorado River. These were first believed by Grinnell (1914) to be derived from some local physiological condition and not by intergradation with the Red-shafted Flicker, a view shared by Phillips et al. (1964). Short (1965), however, stated that specimens from the lower Colorado River were representative of a "hybrid swarm." Individuals along the Bill Williams have purer Gilded characteristics than do those along the main river channel. This situation is still partly unresolved and would require additional specimens, especially in the northern half of the LCRV. Unfortunately, the rarity of the species at present would almost certainly preclude such a study.

HABITAT Gilded Flickers are restricted to riparian woodlands containing cottonwood and willow trees and, occasionally, mesquite habitats with tall snags. Where saguaros occur, desert washes and adjacent uplands also are used. During winter, Red-shafted and Yellow-shafted Flickers may be found in virtually any habitat with trees.

BREEDING Breeding commences in February, with two broods raised per season. Young are fledged in late May and early July. Nest cavities are constructed 3–8 m high in saguaros, cottonwoods, willows, or occasionally tall honey mesquites. Trees with relatively soft wood or cacti are preferred.

The Gilded Flicker apparently prefers to nest in saguaros rather than in riparian trees. Grinnell (1914) found them commonly only where saguaros were present near Laguna Dam. At BWD they nested almost exclusively in saguaros in 1977–78, although they commonly foraged in the riparian forest. However, the few remaining flickers within the valley proper nest exclusively in riparian trees because saguaros are virtually gone. Flickers rarely nest near human dwellings (exceptions are Willow Valley Estates and Clark Ranch). Therefore, the decline in native trees has led to an even greater restriction of nest sites than incurred by Gila Woodpeckers.

FOOD HABITS The summer diet is almost exclusively ants and termites, taken primarily from the ground or downed logs. The winter diet is similar, but mistletoe berries are eaten as well. Flickers often flush from the ground into trees, and will climb trunks, but they rarely forage there. They seldom, if ever, hammer into wood, but rather probe the bill into rotting trees or soft soil. These behaviors are consistent with the species' feeding habits throughout its range.

Olive-sided Flycatcher (*Contopus borealis*)

STATUS An uncommon migrant from late April to early June and from late August through mid-September (exceptionally, 8 October 1981). Max: 6; 9 May 1977, at Parker and BWD.

This species normally occurs singly but may be quite conspicuous because of its habit of perching in the open. They are about twice as numerous in spring than in fall, with only one to five birds noted during fall in recent years.

HABITAT This flycatcher may be found in any situation with tall trees in or adjacent to riparian habitats. There the birds perch as high as possible, often on exposed snags, sallying out to capture flying insects.

Greater Pewee (*Contopus pertinax*)

STATUS A very rare winter visitor but almost regular in recent years. Six records during five separate winters: 13 January 1972 at Yuma; 15 December 1977 at Cibola NWR; 24 December 1977–9 March 1978, and what was probably the same individual, December 1978 through mid-January 1979 at Parker Dam (photo); 28–29 November 1979 at Blythe; and 22 December 1983 at Parker Dam. One seen 17 April 1983 at Dome Valley was most likely a wayward spring migrant. The pattern of recent winter occurrence extends across extreme southern California and southern Arizona, where a few lowland migrants have been noted as well.

Western Wood-Pewee (*Contopus sordidulus*)

STATUS A common spring migrant from late April through early June, and an uncommon to fairly common fall migrant from early August through late September. A few individuals are regularly noted outside these periods: in mid-April (earliest, 11 April 1950), mid to late June (latest, 18 June 1951 and 23 June 1974 [3 birds]), late July (earliest, 19 July 1977), and mid-October (latest, 22 October 1980). A very late pewee was seen at Ehrenberg on 23 November 1983. Individuals noted on Imperial NWR, 4 July 1961, and at Parker, 1 July 1952, are not classifiable as spring or fall migrants. Max: 68; 18 May 1977, Parker area.

This is among the most numerous and conspicuous landbird migrants in the valley (see Fig. 12, p. 75). During late May 1977, when large numbers of migrants were grounded by rain, several hundred pewees were counted. Normally, however, from 2 to 10 birds may be seen daily in spring, with only 1–5 individuals encountered daily in August and September.

HABITAT Migrant pewees may abound in any riparian woodland, where they sally from exposed perches at the edge of vegetation. They may frequent other trees or exotic plantings.

Willow Flycatcher (*Empidonax traillii*)

STATUS At present an uncommon late spring migrant from late May through mid-June, with most recent records after 1 June (formerly earlier). An uncommon to fairly common fall transient from early August to mid-October with a peak in early September (see Fig. 12, p. 75). Formerly bred.

Grinnell (1914) frequently observed them from 28 April–15 May 1910, and although he found no nests he suspected breeding. Thirty-four nests were collected in the Yuma area by Herbert Brown (Unitt 1987), indicating that this species was a fairly common breeder in the early 1900s. Historical records indicate that it was also formerly more common as a transient.

At least three subspecies of Willow Flycatcher may occur throughout the Southwest. Specimens that have been examined suggest that low-elevation breeding populations constitute the subspecies *E. t. extimus* Phillips from southern California through Arizona and New Mexico to west Texas (Phillips 1948; Oberholser 1975; Unitt 1987). *E. t. brewsteri* Oberholser of the Sierra Nevada and *E. t. adastus* Oberholser of the Great Basin are both restricted to higher elevations. Low-elevation birds, potentially all *E. t. extimus*, arrive on their breeding grounds in late April and early May. Montane-breeding populations arrive in the lowlands in mid-May and continue to pass through into mid-June. Unfortunately, the late spring migration of montane populations has led to errors in understanding the actual breeding status of the species in the lowland Southwest. Territorial individuals observed from late April through mid-May and from late June through late July should be investigated closely for actual nesting. Recent records in the LCRV that fit these criteria include singing birds on 21 April 1978 and 7 May 1979 on Imperial NWR, 9 May–20 June 1977 (2) at BWD, and 9 May 1979 near Yuma. All were found in isolated willow groves.

This is one of several midsummer-breeding, cottonwood–willow obligate species that have declined dramatically in the Southwest, especially at lower elevations (Hunter et al. 1987b; Unitt 1987). This species requires immediate attention from federal and state agencies to preserve its populations in the Southwest. Interestingly, the healthiest population in Arizona is at the bottom of the Grand Canyon, where up to 11 pairs nest primarily in saltcedar thickets (Brown 1988; Brown and Trosset 1989).

HABITAT Migrants may appear in any riparian woodland, but especially in willows near water. Breeding was apparently restricted to willow habitats.

BREEDING Although this species no longer breeds in the LCRV, nest data from the early 1900s shed light on some nesting characteristics (Unitt 1987). Clutch size in the LCRV was three eggs for 82% of the nests, with the remaining nests containing two eggs. Nests were almost always placed in willows at heights averaging 2 m, often near marshes. Only 2 of 34 nests collected by Brown in the early 1900s contained cowbird eggs. Although cowbirds were common during this period, it is not known what role they played in the extirpation of breeding Willow Flycatchers in the LCRV.

Hammond's Flycatcher (*Empidonax hammondii*)

STATUS A fairly common spring migrant from early April to mid-May, uncommon in late May. A rare but regular fall migrant (1–5 individuals each fall) from early September to mid- or even late October (3 specs). Also a rare and irregular winter resident with several sight records and one specimen, 12 February–21 March 1977 at BWD. There are four sightings of Hammond's/Dusky Flycatchers in mid-late August and several late fall sightings (identified as Hammond's), including 8 November 1981 at Ehrenberg and 15 November 1976 at BWD. Max: 12; 9 May 1977, Parker and BWD.

Although much progress has been made regarding field identification within this difficult genus (see Whitney and Kaufman 1985), the certain identification of many individuals remains beyond the capability of most observers. Nonetheless, specimens and careful attention to calling birds have established the status of this species as a migrant through the LCRV (see Fig. 12, p. 75). In 1977, we noted nearly 80 individuals during April and May, with 2–10 seen daily around Parker and BWD. In winter, it is the least common (or least-often documented) Western *Empidonax* in the southwestern United States, except perhaps in south-central Arizona where it is regular. This species may be identified by its high "peet" call, reminiscent of the Pygmy Nuthatch.

HABITAT Unlike their propensity to remain fairly high in tall conifers during the breeding season, migrating birds perch close to the ground, where they occasionally pounce on small insects from low, shaded branches. Individuals may occur in any area with dense trees or shrubs, including those planted around human residences. The wintering bird at BWD appeared to be territorial along a small stretch of the willow-lined Bill Williams River.

Dusky Flycatcher (*Empidonax oberholseri*)

STATUS Apparently a casual or very rare migrant. Specimens taken 25 September 1925 and 25 April 1930 near Bard, and 9 April 1977 at BWD. Three careful sight records of migrants: 13 April and 25 September 1978 at BWD, and 21 April 1978 at Yuma. Also a rare and irregular winter resident, with specimens on 19 February 1910 near Needles, 6 December 1953 at Parker, 16 December 1978 at Cibola NWR, and 16 January 1981 north of Ehrenberg. Also two careful sightings of winter birds on 24 December 1977 and 22 December 1983 at BWD.

This flycatcher is probably the rarest of all birds breeding in the western United States that migrate through our region, even though it is a fairly common breeder at higher elevations. This species may be a more regular migrant through the valley than the few records indicate, but difficulty in identification may allow some individuals to pass through undetected. They are also rather rare transients through most of California, but become more common eastward across Arizona and New Mexico. Surprisingly, the Dusky Flycatcher is more regular in winter in the LCRV than Hammond's Flycatcher; the latter is the more regular migrant. This is also true along rivers in southern Arizona.

Although certain identification of many *Empidonax* flycatchers is exceedingly difficult, careful attention to calls and behavior may help to establish additional records of this and similar species in the LCRV. The Dusky Flycatcher gives a dry "wit" call, as does the Gray Flycatcher. This is very different from the high "peet" call of the Hammond's. The Dusky characteristically flicks its tail upward like all *Empidonax* except the Gray Flycatcher.

Gray Flycatcher (*Empidonax wrightii*)

STATUS An uncommon transient from late March to early May and mid-September to mid-October; also a locally uncommon winter resident in very small numbers.

This is the only expected wintering *Empidonax*. Widely scattered individuals may be territorial at that season, often singing at dawn from high, exposed perches. These wintering birds, which remain at least until the end of March, cloud the status of pure transients in many locations. Probable migrants at closely monitored sites have occurred as early as 20 March and as late as 19 October.

Interestingly, Grinnell (1914) noted that all his specimens were males. He speculated that male Gray Flycatchers may winter farther north than females. This phenomenon has been acknowledged in other regions but has not been adequately demonstrated for many species. Field identification is facilitated by this species' habit of dipping the tail downward rather than flicking it up, as in other *Empidonax*. This species gives a characteristic dry "wit" call note.

HABITAT This flycatcher is most consistently found in honey mesquite woodlands, especially near Poston and Ehrenberg, and in sandy openings and riparian edges along the Bill Williams River. Other individuals and migrants may occur in other sparse riparian woodlands and desert washes. Most foraging is done from low perches from which Gray Flycatchers frequently sally to the ground to capture small insects.

Pacific-slope Flycatcher (*Empidonax difficilis*)

STATUS An uncommon to fairly common migrant from late March to early June, and from late July through mid-October. Multiple records in late June and July probably represent both late spring and early fall migrants. Also a rare but regular winter resident in very small numbers, especially at BWD. Max: 32; 9 May 1977, at Parker and BWD.

Pacific-slope Flycatchers have been found along the lower Colorado River in every month of the year, illustrating the complexity of their migratory status (see Fig. 12, p. 75). They are most numerous throughout May, when two to five individuals can usually be found daily in suitable habitats. An unusually large influx (>100 individuals) occurred in mid-May 1977, when storms grounded large numbers of landbird migrants. In fall this species is less numerous, but migration is protracted over a nearly three-month period.

Our winter records of this species now involve at least 14 individuals in 8 different years, from Yuma to Needles. Closer examination of these records teaches us two important lessons. First, it is often assumed that most December records of normally nonwintering species (especially those found on CBCs) are late or lingering migrants. Our close monitoring of specific sites has allowed us to conclude otherwise in several instances. For example, one bird found at BWD on 8 December 1977 was relocated in the same spot on four occasions until 9 March 1978. Another, found 14 December 1979 at Cibola NWR, was seen again three times until 3 March 1980. A second assumption is that wintering individuals represent birds from normally migrating populations that, for some reason, have stayed farther north than usual. However, the only winter specimen (13 December 1950, from BWD) has

been determined as the race *E. d. cineritius* Brewster from Baja California (Phillips et al. 1964). This suggests the possibility that at least some
of the wintering birds are actually northward dispersers from Mexico.

HABITAT This species is fond of dense, shaded vegetation, where it typically sallies for insects well within the foliage. All winter records and
a majority of migrants are from willow groves, usually near water. However, other transients may be found in other riparian woods and around
human habitations.

COMMENT The recent determination that coastal and interior populations of the "Western" Flycatcher are two distinct species presents a
unique challenge to modern field ornithologists (Johnson and Marten
1988). At present there are no established criteria for separating the two
species in the field other than by vocalizations. The Pacific-slope Flycatcher (to include *E. d. difficilis* Baird, *E. d. insulicola* Oberholser, and
E. d. crineritius, mentioned earlier), far outnumbers the Cordilleran
Flycatcher (*E. occidentalis*) during migration throughout southern Arizona, although the latter represents the breeding population at higher
elevations in the Southwest (Phillips et al. 1964). We know of no valid
record of the Cordilleran Flycatcher from the LCRV, although migrants
may occur. The lack of records for the interior breeding birds parallels
the pattern for many locally breeding populations of other species.
Further complicating the picture, *E. d. insulicola* from the Channel
Islands in California has been documented in Arizona only from the
lower Colorado River (12 April 1948 at BWD) and is possibly a separate
species in its own right. Additional specimens will be necessary to establish the status of the two, three, or even four "Western" Flycatchers
in the LCRV, and to resolve the issue of wintering birds discussed above.

Black Phoebe (*Sayornis nigricans*)

STATUS A common winter resident and recently a fairly common and
local summer breeder throughout the valley. An influx of nonbreeding
birds begins in July and these birds leave in March. Not recorded as
breeding before the 1950s. Max: 144; 22 December 1982, Parker CBC.

Grinnell (1914) did not find this species after 5 April in 1910, and he
concluded that it was only a winter visitor in the valley. The early
establishment of a breeding population is largely undocumented, but
Monson (pers. comm.) considered it a rare breeder during the 1950s.
When we began our work in 1974, it was one of the characteristic breeding species in its proper habitat throughout the valley. Whether our
large wintering population (up to 144 birds on recent CBCs) represents
an increase from former years is also not known. Contrary to comments in Phillips et al. (1964), the relatively small breeding population

along the lower Colorado River may be resident, at least at BWD (see Breeding below).

HABITAT This species has a very strong affinity for water and is found along riverbanks, agricultural canals, and other water areas shaded by riparian trees, canyon walls, or high banks. Few native Western birds have benefited more from Southwestern settlement than the Black Phoebe. It can be seen commonly around farm houses or flycatching over city lawns, and there is little doubt that the spread of irrigated agriculture into our region is at least partly responsible for this species' recent breeding success.

BREEDING Nests are placed low above water on overhanging tree trunks, branches, rock faces, in mine shafts, or frequently under bridges. Nest sites must be within carrying distance of a source of mud with which the nest is constructed. Four to five eggs constitute the normal clutch, with up to three broods raised in a season. It is not uncommon for a pair or later pairs to use the same nest site for many years, even though a new nest is constructed each year.

At BWD, pairs established linear territories along the Bill Williams River, with much intraspecific aggression and singing beginning as early as January. Nest construction begins in early March and the first eggs are laid in late March. First broods fledge in mid- to late April and second broods are completed in late May or early June. Young birds then disperse in midsummer, possibly to the mainstream of the Colorado River, although adults seem to remain territorial year round.

FOOD HABITS Arthropods formed 100% of the contents in six stomachs. Beetles, dragonflies, flies, bees, and wasps dominated the seven taxa represented. Most insects are captured in sallies above water or actually to the water surface, as determined from 297 observations. At other times Black Phoebes fly to the ground to capture terrestrial prey. Occasionally they perch on the bare riverbank to pick insects from the water's edge.

Eastern Phoebe (*Sayornis phoebe*)

STATUS A rare but regular transient and winter visitor from November through late March. About 10 records, all of lone individuals, since 1976; first recorded in 1952 (spec).

A territorial individual was seen regularly between 15 November 1976 and 21 March 1977 along a small stretch of the Bill Williams River. It was highly aggressive toward resident Black Phoebes. Probably the same individual returned on 13 September 1978 and was seen again on 9 November, but not later. This species has been found rather regularly throughout the Southwest in recent years.

HABITAT This species has been found in vegetation along riverbanks and agricultural canals.

Say's Phoebe (*Sayornis saya*)

STATUS A common winter resident throughout the valley. Fairly common in spring, with breeding only along the periphery of the river valley. Numbers drop markedly in late April and May and the few individuals remaining through summer are primarily juveniles. Winter arrivals begin in late August and are common by late September. Max: 244; 22 December 1980, Parker CBC.

The Say's Phoebe is among the most abundant species in winter in the extensive agricultural portions of the LCRV. The importance of the region to this species is illustrated by the national high counts recorded on the Parker CBC from 1978 to 1982. The 244 seen in 1980 represent an all-time high count from anywhere in North America (see Appendix 1).

HABITAT This species is found mainly in agricultural and sparse riparian habitats in fall and winter. It resides primarily in rocky desert areas and around human residences for breeding. It has no special fondness for water as do its congeners. This species favors more open, sunny habitats where its sandy-colored plumage makes it less conspicuous.

BREEDING Before humans and their buildings, nests were placed on rock shelves and in crevices in rocky cliffs protected by overhanging rock ledges. Old buildings, mine shafts, and wells quickly became surrogates for rock walls and broken cliff faces. Even in 1910, Grinnell noticed nesting at adobe ruins near Ehrenberg and Cibola.

Pairs begin investigating nest sites in late February. Eggs are laid in April and the first fledged young appear in mid-May. Nests with eggs have been found in late June at BWD, suggesting that two broods may be raised by at least some pairs.

FOOD HABITS The most characteristic foraging mode of this species is to sally to the ground from a low, exposed perch to capture terrestrial insects, as determined from 79 observations. Without low perches, the birds will often hover above the ground until prey is spotted. Aerial flycatching is also common, especially in fall. During winter, some are seen capturing insects around blooming cottonwoods and willows at BWD.

These behaviors are reflected in the diet, which was dominated by beetles, grasshoppers, and crickets. Flies, bees, wasps, butterflies, caterpillars, true bugs, dragonflies, and earwigs also were found in the 13 stomachs examined. More short-horned grasshoppers were taken in agricultural areas, and bees, wasps, and bugs were more frequent in riparian samples.

Vermilion Flycatcher (*Pyrocephalus rubinus*)

STATUS A rare and local resident, slightly more widespread in winter. Max: 21; 27 December 1950, Havasu NWR.

The abundance and distribution of this species have been drastically reduced over the past 50 years due to recent water management practices resulting in the loss of much suitable habitat. Grinnell (1914) reported this species as numerous from Ehrenberg south to Yuma. He especially noted the propensity of the species to occur where large clearings had been opened in the cottonwood stands near Yuma. He speculated that this was one species that would fare well with the coming habitat changes. Unfortunately, Grinnell, who was rarely wrong, underestimated the extent of the habitat changes to come (cottonwood loss and removal) and the sensitivity of this species to these changes. Cessation of annual floods, along with the concomitant disappearance of native cottonwoods and willows, has virtually eliminated suitable breeding and foraging habitats. Today, a few pairs remain along the BWD. Along the entire mainstream of the lower Colorado River, breeding pairs number no more than 10 and these are widely scattered. Areas of consistent occurrence include the Blythe golf course, Clark Ranch, Parker Dam residences, and Willow Valley Estates. During winter they are found more frequently and at a variety of other sites.

The species is presently listed as a Species of Special Concern by the California Department of Fish and Game. As in several other riparian species that have declined in recent years, the Vermilion Flycatcher is near the western edge of its range in the LCRV. In contrast to other declining species, it begins breeding much earlier in spring. However, the peak egg-laying period still overlaps the hot midsummer months in which Yellow-billed Cuckoo, Willow Flycatcher, Bell's Vireo, Yellow Warbler, and Summer Tanager also nest. In addition, the midsummer-breeding species listed above are all long-distance migrants, whereas this flycatcher probably migrates only short distances and is the only listed species to remain in winter. The relationship between wintering and breeding individuals is unknown, however, and these may represent separate populations. Regardless of the population components, the continued existence of this flycatcher as a breeder can be measured in a few years unless public pressure and management efforts are intensified.

HABITAT The Vermilion Flycatcher may be found in riparian woodland, residential areas, and along the margins of agricultural fields near accessible water. Breeding sites, especially at BWD, usually consist of cottonwood–willow groves with nearby honey mesquite, open water, and pastures. This species has never been known to use the extensive but dry honey mesquite woodlands found from Parker to Ehrenberg, even though it is a common breeder in honey mesquite in eastern and south-central Arizona. In winter, additional individuals are found in agricultural areas, usually around well-vegetated human habitations, well away from riparian woodlands.

BREEDING Nests are located on horizontal branches of mesquites, willows, or cottonwoods, 2–11 m above ground. Sometimes exotic trees are used for nest sites in residential areas. Singing posts are usually 4.5 m or higher above ground. Males frequently engage in entertaining and spectacular nuptial flights from these posts as they may rise 20–30 m into the air.

Historical data on breeding include two nests found near Palo Verde in 1916 that contained two eggs each on 7 and 16 April. One was in a screwbean mesquite above water. Second broods were raised in both nests with three young present in one and three eggs in the other on 1 June (Wiley 1916).

The following observations made on nesting at BWD during 1977 and 1978 compare favorably with observations from the early 1900s. Each year, a single overwintering male began his courtship flights in mid-February and was the first to pair and start breeding. Females and other males appeared in late February. Earliest copulation was observed on 28 February, on a nearly completed nest. However, due to much territorial squabbling some nests were not begun until late March. By April in both years, four pairs occupied a 500-m stretch of willows bordered by a pasture. Most nests found were in willow or cottonwood trees, 3–10 m above ground, but one nest was 1 m high in a saltcedar.

First fledged young were observed in 1977 on 9 May. In 1978, increased aggression and frequent nest-site changes delayed the appearance of family groups for nearly a month. Most pairs raised two broods in each season; one female was feeding her second brood in the nest on 28 June.

We may make several tentative conclusions from these observations. First, the breeding season appears to be very protracted, lasting from late winter through midsummer. Second, the maintenance of an overwintering territory appears to be important for early pairing and nesting (the habitat seemed to support many more birds in spring than in winter). Finally, the close packing of nests and the high levels of territorial aggression suggest that habitat and nest sites may be severely limiting to the remaining individuals in this area.

FOOD HABITS This flycatcher partakes in aerial sallying much more than most of our other local species. They typically perch atop a willow or mesquite (or telephone wire, when available) and fly out, often in long, upward, zigzagging pursuits of passing insects. At other times, however, they perch low on shaded branches or on fences, and either make short sallies just above the ground or actually go to the ground to capture prey. We have no specific information on this species' diet in the LCRV. However, it almost certainly includes bees, wasps, flies, beetles, and grasshoppers.

Dusky-capped Flycatcher (*Myiarchus tuberculifer*)

STATUS A casual visitor. Two records: 9–14 November 1977 at Cibola NWR (spec) and 9 January 1988 at Topock. Two individuals may have been involved in the first report, as the specimen was collected 8 km from where this species was first seen.

Some confusion surrounds previously published accounts of the first record, owing mostly to the complicated state boundary at Cibola. Both locations where the bird(s) occurred were on the Arizona side, although the initial sightings on 9, 11, and 12 November were on the *west* bank of the Colorado mainstream, which is in Arizona at Cibola. Thus, to our knowledge, this species has never occurred in the California portion of the region.

The specimen was identified as *M. t. olivascens* Ridgway, the northernmost (and Arizona breeding) race, thus ruling out long-distance dispersal from the south. This species has been recently detected as a late fall–winter visitor in California. The January report provided the first winter record for this species in Arizona.

Ash-throated Flycatcher (*Myiarchus cinerascens*)

STATUS A partially migratory resident and common breeder in summer from April through September; fairly common by March and into October, and uncommon to rare through winter. Max: 41; 20 July 1954, BWD.

This species is unique among our common breeders because it is resident for most of its annual cycle in the LCRV. However, a majority of individuals retreat to the south between late November and early March. Overwintering birds are uncharacteristically quiet and reclusive and are, therefore, easily overlooked. As many as 25–30 birds were detected between Cibola and BWD during winter 1977–78. From our average winter density estimates from all study transects, we estimate the total wintering population for the entire valley at roughly 250 birds, or about 3% of the summer population. The large influx of spring breeders begins in late February.

HABITAT These flycatchers are widely distributed through most riparian vegetation in the LCRV. Breeding season densities are uniformly high (10–15 birds/40 ha) in honey mesquite, screwbean mesquite, desert wash, and sparse willow habitats. Numbers are somewhat reduced (5–10 birds/40 ha) in saltcedar and mature cottonwood–willow groves. In the latter habitat, this species is outnumbered by the Brown-crested Flycatcher. Suitable habitat for Ash-throated Flycatchers is determined in part by the presence of woodpecker holes and other naturally occurring nest sites. In winter, most individuals are found in desert washes along the periphery of the riparian floodplain or, to a lesser extent, in sparse honey or screwbean mesquite woodlands.

BREEDING Our knowledge of the specific behavior of this species in the Colorado Valley comes largely from the observations of Timothy Brush from 1978 to 1980. Pair formation and territory establishment take place throughout March in both honey mesquite and willow habitats. Copulation by a pair in flight was observed as early as 27 March. Nest sites in honey mesquite are typically old Ladder-backed Woodpecker cavities in dead snag portions of live mesquite trees. At BWD, all nests were found in dead willow snags along the edge of the groves, between 4.5 and 8 m above ground. Eggs are laid in early April, and both members of the pair engage in nest defense and care of the young. Most pairs raised two broods in quick succession, with young fledging at the end of May and again in the beginning of July.

Ash-throated Flycatchers readily accepted wooden nest boxes, provided during experimental studies of their breeding behavior. Brush (1983) concluded that numbers were only partly limited by the availability of nest sites. Although densities could be artificially inflated with nest boxes, many potential cavities always remained unused. The need to maintain relatively large (3–4 ha) feeding territories also seems to be very important.

FOOD HABITS This species typically feeds by sallying to the foliage of low trees or shrubs to capture large insects, as determined from 287 observations. The birds move deliberately from perch to perch, slowly cocking the head to scan nearby substrates. Insects are taken less frequently from branches or trunks (about 15% of all foraging), from the ground (about 10%), or around flowers (about 5%). Only rarely does this species "flycatch" for aerial prey. As with other wintering insectivores in honey mesquite, they concentrate their foraging at evergreen saltbush and wolfberry shrubs until the mesquites leaf out in mid-March.

The overall diet (from 124 stomachs) was diverse, with at least 32 taxa of insects and spiders represented. In spring (31 stomachs), caterpillars, butterflies, flying beetles, wasps, true bugs, and flies were all important prey. In summer and late summer (89 stomachs) the most important foods were cicadas, short-horned grasshoppers, and beetles. Our small sample (4 stomachs) from fall and winter included caterpillars, bugs, wasps, spiders, a few fruits, and a small lizard.

Brown-crested Flycatcher (*Myiarchus tyrannulus*)

STATUS A fairly common but local breeder from late April (earliest arrival, 25 April 1978) through mid-August. A few individuals linger into early September. Max: 23; 26 May 1978, BWD.

This species' exact status in the early part of the twentieth century is poorly understood. The very observant Grinnell (1914) and all previous workers did not detect it. The first record was of two specimens collected near Bard on 17 May 1921. The next individuals were found by Monson (1949) below Parker in 1946 and again in 1947 (and subsequently at Parker Dam and BWD). At that same time, a nest was discovered north of Needles on 23 May 1949 (McLean 1969). This species, therefore, apparently has increased substantially since the 1950s. It has spread as far west as Morongo Valley and as far north as the Kern Valley in southern California (Banks and McCaskie 1964; Garrett and Dunn 1981). Why this species suddenly found its way to the lower Colorado River and reasons for its dramatic increase in the face of massive habitat reduction remain unknown.

Population size was estimated at 800 birds in 1976, with the largest number at BWD (Hunter 1984). Presently, with recent losses in habitat, the species is estimated to have dropped 46% to 435 individuals (Ohmart et al. 1988). The California Department of Fish and Game lists it as a Species of Special Concern.

HABITAT This species is found primarily in mature cottonwood–willow woodlands, but it will nest in residential areas with tall trees. Occasionally it is found in other riparian habitats with scattered cottonwood trees or snags. It nested in city parks at Yuma, at least from 1958 to 1961.

This flycatcher has adapted well to human habitations and its population remains stable in the river valley because of this affinity. Approximately 200 birds are estimated to use residential areas (75 estimated for California; Hunter 1984), although the actual number of breeding individuals in these habitats may be much smaller. In addition, this species was the only canopy-using species not to decline immediately with the reduced quality of cottonwood–willow habitat after flooding in BWD (Hunter et al. 1987a).

BREEDING Nest placement and construction are similar to those of the Ash-throated Flycatcher in that woodpecker holes or natural cavities in trunks or large branches are used. They differ from Ash-throated Flycatchers in being more dependent on the larger woodpeckers (especially

Gila Woodpecker) for cavity formation. Egg laying begins in May and a normal clutch is four to six eggs. A single brood of three or four is fledged in early or mid-July.

The following observations on breeding were made at BWD in 1977 and 1978. Of 12 nests found, 9 were 5–15 m above ground in dead or dying willows, mostly in the interior of the large cottonwood grove. The remaining three nests were in saguaros in the nearby desert scrub. Birds from these latter nests held territories in the adjacent cottonwoods. Nesting began immediately after the first birds arrived in late April. Most pairs were observed feeding young beginning in mid-June. Most broods were fledged by 9 July, and family groups remained together for about a month.

In the cottonwoods at BWD, breeding by this species is typified by very high levels of both intra- and interspecific aggression at all stages of the nesting cycle. Brown-crested Flycatchers arrive after all other cavity-nesting species have commenced breeding and they must frequently compete with these species for nest sites. We observed occasional aggressive encounters with Gila Woodpeckers, but the flycatchers were never successful in evicting an actively nesting pair of woodpeckers. However, in 1978, three pairs of flycatchers used cavities from which Gila Woodpeckers had recently fledged their first brood (Brush et al. 1983). Individual flycatchers frequently fought with and chased each other in territorial disputes. For example, two pairs were observed in physical contact, with two, then three, then all four birds tumbling together to the ground in a fluttering, bill-snapping mass. On another occasion, a pair dominated a smaller pair of Ash-throated Flycatchers in a territorial dispute.

European Starlings are thought to interfere with the nesting activities of Gila Woodpeckers and Brown-crested Flycatchers (Remsen 1978). Our observations indicate that these flycatchers often defend their cavities successfully from starlings in the LCRV, whereas Gila Woodpeckers do not (Hunter 1984). They are noticeably more aggressive than Gila Woodpeckers in defense attempts.

Although competition for individual nest sites may be frequent, the number of unused cavities in each territory remains high. We must, therefore, conclude that it is not the availability of nest sites that solely limits the population of this species, but rather the total area of suitable habitat available for both nesting and foraging (see Brush 1983; Brush et al. 1983).

FOOD HABITS This species' foraging behavior was extremely similar to that of the Ash-throated Flycatcher. Most prey items were captured by sallying to live foliage in the canopy of willows or cottonwoods, as determined from 109 observations. The primary difference between the two species is in feeding height. The Brown-crested Flycatcher rarely descended to feed in mesquite, saltcedar, or shrubs which were used heavily by Ash-throated Flycatchers.

Diet, based on 13 stomach samples, is also very similar to the mid-summer diet of Ash-throated Flycatchers. Cicadas, grasshoppers, and beetles were the most important foods. The two species overlapped heavily in prey size and type, but only the Brown-crested took large mantids or dragonflies. One individual was observed catching a small lizard (probably *Urosaurus*) from the trunk of a cottonwood tree.

At BWD, this is one of eight species that feast on the annually emerging cicadas in midsummer. The benefits of a reliable food supply apparently outweigh the possible ill effects of high temperatures and increased competition for nest sites (see Rosenberg et al. 1982).

Tropical Kingbird (*Tyrannus melancholicus*)

STATUS A casual summer and fall visitor. Six records during these seasons: 1 October 1947 at Topock; 16 August 1954 at BWD (spec); 10 June 1973 at Tacna, east of Yuma; 5 June 1978 at BWD; late May 1980 at Davis Dam; and 6–9 June 1980 in Dome Valley. Also, one early spring sighting, 23 March 1957, at Palo Verde.

This species regularly wanders up the Pacific Coast from Mexico in fall. Our early summer records may represent spring migrants wandering beyond southern Arizona breeding centers.

Cassin's Kingbird (*Tyrannus vociferans*)

STATUS A rare fall transient in very small numbers and a sporadic summer visitor. Nested in 1983, 1984, and 1985 north of Blythe. Two to five individuals were noted each fall from 1977 to 1981, with records extending from 31 August (1978) to 18 October (1980), both at Cibola NWR. Also, a casual spring transient with records on 17 May 1950 at Topock, 31 March 1952 at Parker, and 19 May 1977 near Poston (spec).

There are five summer records between 1950 and 1955 from BWD, Parker, and Imperial NWR. More recent summer records are of one at Bard on 8 June 1976 and one at Dome Valley on 4 June 1980. These seemed enigmatic, but it now appears that isolated individuals or breeding pairs may occur in any year.

HABITAT These kingbirds have been found at tall trees in riparian or agricultural areas where Western Kingbirds are common.

BREEDING A nest was built on the Clark Ranch by presumably the same pair in 1983, 1984, and 1985. Young were fledged in 1984 and 1985 (photos). This is the lowest elevational breeding record known for this species away from the Pacific Coast. The nest was constructed on the perimeter of a large cottonwood, with Western Kingbirds simultaneously nesting on the opposite side of the same tree.

Thick-billed Kingbird (*Tyrannus crassirostris*)

STATUS A casual fall and winter visitor. Seven records within 11 years: 17 December 1972–4 January 1973 at Laguna Dam (photo); 2 December 1977 at Parker (spec); 5 August–16 September 1978, north of Blythe (photo); late November–23 December 1979, southwest of Parker (Fig. 31); 6 September 1981 at BWD; 9 September 1982 at Cibola NWR; and 22 December 1983 at BWD.

The regularity with which this species has appeared in the LCRV is somewhat surprising, considering there are no other winter records elsewhere in Arizona. The species also remains unrecorded elsewhere in southeastern California. These birds represent possibly both wandering individuals from the south and migrants from local Arizona breeding populations.

Western Kingbird (*Tyrannus verticalis*)

STATUS A common summer breeder and transient between April and mid-September. The first arrivals appear in mid-March, and numbers decline steadily in late September and early October; casual after mid-October. Max: 50; 1 September 1978, Parker Valley.

The earliest arrival date is 14 March, but only a few are seen before April. This is one of the few common breeding species in the largely treeless agricultural valleys in summer. Fall migration begins in August, when birds begin to congregate on telephone wires and loose groups can be seen moving south through the valley. This movement continues until the last week of September, after which only individuals are usually seen. The latest acceptable record is 29 October 1953 at Topock (spec). There are a few later sightings of "yellow-bellied" kingbirds, such as 6 November 1977 south of Parker, 11 November 1979 north of Ehrenberg, and 29 November 1979 near Poston. However, assignment of these sightings to species was not possible. It may be equally possible, if not more likely, for Cassin's or Tropical Kingbirds rather than Western Kingbirds to occur on those dates.

HABITAT This species is most common at tall trees in agricultural areas and also around human residences. Riparian habitat use for nesting is mostly restricted to tall cottonwood and athel tamarisk trees. Also, open riparian areas, including recently burned sites, are used for foraging. Migrants require only a temporary perch, fulfilled by telephone wires or fences along open fields.

FIGURE 31. Thick-billed Kingbird at Lost Lake Resort, California, on 23 December 1979. (Photo by K. V. Rosenberg.)

BREEDING The bulky nest is built in the perimeter of tall cottonwood trees. Four to six eggs may be laid, and we noted recently fledged young in June and early July. The ability of this kingbird to breed in open areas during the middle of the summer remains an enigma and should be investigated.

FOOD HABITS Western Kingbirds are the most aerially foraging flycatchers. About 95% of their foraging (44 observations) consisted of long sallies, often upward, from the tops of tall trees or telephone wires. The remainder of their prey was taken from the ground by pouncing from an elevated perch. Sixty-one stomach samples contained a wide variety of flying insects. Beetles and short-horned grasshoppers were eaten in large numbers throughout the season. Cicadas and dragonflies were more important in mid- to late summer. Terrestrial prey included earwigs, ants, and spiders.

Eastern Kingbird (*Tyrannus tyrannus*)

STATUS A casual fall transient. Six records: 4–11 September 1949 at Parker; late August 1967 at Topock; 16 August 1976 at Davis Dam; 4–5

September 1976 north of Blythe; 7–8 September 1977 near Poston (spec); and 4 September 1979 at the same location near Poston.

The consistent timing of these records corresponds well with the rare but regular early fall passage of this species in eastern Arizona and the California deserts, where it was also noted in late spring.

Scissor-tailed Flycatcher (*Tyrannus forficatus*)

STATUS Recently a very rare but consistent transient or straggler from late May to mid-November. About 12 records (3 specs, 3 photos) between 1974 and 1983, including a female that paired with a Western Kingbird; built a nest at Needles, 26 May–16 July 1979; laid five eggs; but did not raise young. Probably the same individual was observed on a nest in the same area at Needles from 19 April to 31 May 1983.

Aside from the bird at Needles, half of our records are between 22 May and 14 July and half are from October. The latest sighting is 18 November 1982, south of Parker, which is also the latest record for Arizona. This species is a rare but regular transient through much of the Southwest, and a few have wintered in southern California.

HABITAT Most individuals were found on telephone wires adjacent to agricultural fields. The nesting attempt at Needles was in a scattered group of cottonwood trees on a golf course.

Horned Lark (*Eremophila alpestris*)

STATUS A fairly common but local spring breeder and a common to abundant winter resident from August to March. Nonbreeding birds often form large flocks. Max: 3,950; 18 December 1978, Parker CBC.

This is one of few species present during the heat of summer in the vast, treeless agricultural lands that now dominate the LCRV. Their abundance and affinity for cultivated fields (at least for nesting) are apparently recent and little-documented phenomena. Grinnell (1914) did not record this species from the immediate floodplain in 1910. It has been considered a sparse resident of the region, nesting primarily in open desert habitats (e.g., Phillips et al. 1964; Garrett and Dunn 1981).

We have estimated the present abundance of Horned Larks in late spring to be between 4 and 20 birds/40 ha in the Parker Valley, with the first influx of nonbreeders detected in August. Winter numbers may vary greatly from year to year, with densities reaching 925 birds/40 ha in optimal habitats, probably reflecting weather conditions (e.g., snow cover) in regions north of the LCRV.

The resident and breeding birds (*E. a. leucansiptila* (Oberholser)) in the LCRV are the palest of all races. Darker and more colorful individuals are seen in the large winter flocks. Presently, no series of specimens exists from which racial origins of these birds can be determined (see Monson and Phillips 1981; Rea 1983 for other Southwest regions).

HABITAT Horned Larks are found year round in agricultural lands. In order of importance, plowed fields, recently planted wheat fields, and recently planted truck crops (primarily lettuce) are the most heavily used agricultural habitats. A few also are present in the open creosote bush desert along the periphery of the valley.

BREEDING The nest is constructed in a hollow scrape in the ground, usually in cultivated fields or open desert. Pairs sing and display throughout March and April, with copulations observed in late April. Juveniles are noted usually by late May or early June.

FOOD HABITS Horned Larks feed entirely on the ground. They take seeds or insects off live plants or pluck and eat entire young seedlings. In spring–summer, insects accounted for 38% of the diet (from 48 stomachs) with dominant items being aphids, caterpillars, weevils, and wasp pupae. The remainder was plant material, with goosefoot, purslane, panic grass, mustard (Cruciferae), and smartweed dominating. By fall (45 stomachs) the birds virtually had shifted completely to plant material (99%) and were taking few insects. In winter (86 stomachs), 84% of the diet was vegetation and 16% arthropods. In total, the annual diet consisted of 13 insect taxa and 24 plant taxa (most of which were "weed" grasses and annuals).

This species has been persecuted by farmers because of depredation of newly planted lettuce, wheat, milo, and oats. Horned Larks feed on the new shoots by tugging them out of the ground or clipping the shoot at ground level. Armed farm workers patrol the fields to discourage large flocks from damaging crops. In lettuce fields the damage period is six to eight days after the fleshy shoots appear. Farmers claim that the production of an entire field is eliminated if depredation is not prevented. Conversely, the spring–summer insectivorous habits of this flocking species must be beneficial to the farmer as the most frequent prey are herbivorous pest species.

Purple Martin (*Progne subis*)

STATUS A rare but annual migrant, usually occurring singly from mid-April (exceptionally, 2 April 1977) to mid-May, and from late July to early October. Records outside these periods include a female at Martinez Lake on 22 June 1954 and a pair seen on the unusually early date of 25 February 1983 at Dome Valley.

Purple Martins are rarely seen anywhere in the Southwest away from their montane or saguaro desert breeding grounds. Although much earlier than any other regional records, the February sighting concurs with the period of peak arrival in the southeastern United States. Individuals are usually found migrating or foraging with other swallows over open river, marsh vegetation, or other habitats adjacent to the river.

Tree Swallow (*Tachycineta bicolor*)

STATUS A common to abundant transient and winter resident between late July and early April, occasionally in huge flocks. Numbers decrease rapidly through April with stragglers seen into early May. First fall arrivals begin appearing in the first week of July; a few individuals have been noted in June. Max: 750,000; 31 March 1976, Fort Mohave to Davis Dam.

The Tree Swallow is the LCRV's most numerous migratory species. Several million may pass through the valley each spring and fall, with many birds taking up temporary residence in roosts at Imperial Dam and Topock. Despite this species' overall abundance, their local, seasonal, and even annual status may be quite complex and variable. Away from large roosts, numbers may be deceptively small, especially in early winter when the fewest transients are passing through. Variation in winter is reflected in numbers on the CBCs. The largest wintering concentration noted was 6,900 at Imperial Dam, 27 December 1955.

Peak migration appears to be in August and early September, and again in March and early April. In some years there is a decline after early September, with a later influx of wintering birds before the appearance of spring migrants, typically in late January.

HABITAT Tree Swallows commonly forage over the open river, marshes, and agricultural fields. The species primarily roosts in extensive marsh vegetation.

Violet-green Swallow (*Tachycineta thalassina*)

STATUS A very local breeder in small colonies, at least at BWD and Parker Dam, where present from late January through summer. Also an uncommon to fairly common transient from February to April, and rare to uncommon from August through October. Very few substantiated records in November or December. Max: 400; 13 October 1954, Upper Imperial NWR.

This swallow, along with the Cinnamon Teal, is among the earliest spring migrants. First arrivals typically appear at local breeding sites, although the earliest record is of several migrants at Imperial Dam on 12–14 January 1977. As noted by other authors, this early arrival is responsible for mistaken impressions of wintering in the region. In spite of numerous "sightings," especially on early CBCs at Yuma, the only carefully documented sightings outside the normal migration periods are: 24 November 1947 at BWD, 5 January 1951 at BWD, 23 November 1979 (two) at West Pond, and 20 December 1979 (two) at Parker. Some additional records may indeed be correct, but any sighting during this period requires documentation. There are also a few breeding season records away from known nest sites, such as 18 May 1978 (seven birds) at Lake Mohave and 25 May 1980 at Willow Valley. These records may represent either very late migrants or perhaps wanderers from as yet undiscovered colonies in the northern part of the valley.

The Colorado River population is isolated from other breeding populations in montane forests in both Arizona and California. Phillips et al. (1964) have suggested that these birds are actually intermediate toward the smaller *T. t. brachyptera* Brewster from the deserts of northwestern Mexico.

HABITAT This species breeds along cliffs and forages over open water and marsh vegetation. They frequently travel and forage along the river with other kinds of swallows during migration.

BREEDING Nests are located in small crevices or holes along cliff faces. Evidence for local breeding rests largely on the consistent presence of birds in the vicinity of cliffs throughout late spring and summer, and the appearance of juveniles in these flocks in June. However, no study of their nesting habits has been undertaken.

Northern Rough-winged Swallow (*Stelgidopteryx serripennis*)

STATUS A common transient and resident from late January through early September, breeding in spring and early summer. Uncommon through fall and early winter, with numbers increasing in late December. This species has increased as a wintering bird since the 1950s. Max: 800; 1 May 1954, Topock.

This is the only swallow that may be found on any day of the year on virtually any stretch of the lower Colorado River. However, it is almost never seen in large concentrations, and there is no evidence that any individuals are truly resident. The largest numbers are consistently present where tall earthen bluffs line the riverbank, such as just downriver of Parker and again below Ehrenberg.

The relatively few specimens available from this region indicate that the breeding race is *S. s. psammochroa* Griscom, present at least by late January. The postbreeding status of this population is largely unknown. At least one overwintering bird, taken 10 January 1954, proved to be this race. The more northen, widespread race, *S. s. serripennis* (Audubon), is a confirmed migrant through the valley. It appears by early March and as late as 10 November 1953 (2 specs, Monson). Thus, either form may be represented in this small fall–winter population.

HABITAT This is the only Arizona swallow that nests in dirt banks. Breeding pairs may be present along river bluffs, dirt-lined canals, or even steep-walled desert washes some distance from the river. Foraging birds are principally found along the river or its backwaters, and over agricultural fields and canals.

BREEDING The nest is usually placed in a crevice or burrow in a vertical bank or wall, often along dirt-lined canals. Grinnell (1914) noticed that all nest sites along the river in 1910 were above the highest flood marks on the banks, even though plenty of otherwise suitable sites were present below these lines. It would be interesting to know if this habit persists today, long after the cessation of annual flooding. We have seen birds investigating nest sites by early February; eggs have been found in April and May, and recently fledged young were noted in early June.

Bank Swallow (*Riparia riparia*)

STATUS An uncommon to fairly common transient in April and early May, and from mid-July to early October; rare in late March, late May, and early July. Four careful winter sightings: 16 February 1974 (2) at Imperial Dam; 23 December 1977 at Parker, and possibly the same individual there on 1 March 1978; 24 January 1980 at Mittry Lake; and 19 December 1987 (2) on Yuma CBC. Max: 40; 3 May 1972, Imperial Dam.

Exceptionally early fall migrants were noted in 1977 on 29 June near Parker and 30 June at Ehrenberg; a very late migrant was at Cibola NWR on 5 November 1978. A report of 7 individuals on the Yuma CBC in 1972 was unsubstantiated. There are 6 winter records from the Salton Sea.

HABITAT This swallow forages over open water, marsh vegetation, and agricultural fields, usually in flocks with other migrating swallows.

Cliff Swallow (*Hirundo pyrrhonota*)

STATUS A common breeding resident and transient from March through August, with first birds arriving in mid-February. Breeding colonies are vacated in August, but migrants remain until early October. One bird seen 27 January 1980, east of Yuma, was probably an early spring migrant. Max: 6,000; 4 August 1942, east of Yuma (fields).

As with many other species, it is very difficult to delineate the status of resident and migrant populations. Elsewhere in Arizona, the presence of birds at nest colonies is not necessarily proof of residency (Phillips et al. 1964). We do know, however, that the first arrivals begin nest building in late February and are thus part of the breeding population. Continuing migrants must pass through later in spring. We have also noted large, nonbreeding congregations from mid-July to mid-August, indicating an early peak of migration when local breeders still are present at colonies. After the first week of September, only a few individuals may be noted daily, although 20 were seen north of Parker on 8 October 1981 (latest record).

Cliff Swallows have undoubtedly increased since historical times with the widespread construction of man-made structures that are ideal for nest sites. Grinnell (1914) noted the first use of such an artificial site at the headgates of the Imperial Canal at Andrade in 1910. Today this swallow is among the most numerous summering species in the valley, occupying nearly every dam, canal gate, and bridge. Concomitant with this numerical increase has been an apparent tendency toward earlier spring arrival since 1910, when Grinnell noted migrants on 17 March and 12 April and the first nesting activity on 5 May. Arrival in southeastern California is even earlier, however (in late January).

HABITAT This species nests along cliffs, at dams, canal gates, and under bridges spanning water. The birds forage widely over open water, agricultural areas, and even open desert.

BREEDING During nest building in March and April, they can be found gathering mud in densely packed groups adjacent to the river. This highly gregarious breeder plasters its mud nest to a rock or cement face. As such, a wide variety of natural and artificial structures are considered suitable.

Natural colonies include two cliff sites on Imperial NWR, the extensive rock walls high above BWD, and the stony bluffs above Lake Havasu and in Topock Gorge. Grinnell (1914) noted colonies on the more crumbly earthen bluffs, such as at Pilot Knob near Yuma. However, we have not observed this recently. Finally, we note that even the historic London Bridge, in its new home at Lake Havasu City, now hosts a large colony on the very rocks that were used by House Martins (*Delichon urbico*) when the bridge spanned the Thames River.

Barn Swallow (*Hirundo rustica*)

STATUS A common to abundant spring and fall transient. Periods of migration are extremely protracted, but large numbers occur only from mid-April to mid-May and from mid-September to mid-October. Small numbers are seen consistently throughout March, in early June, throughout August, and to mid-November. There are several December records, at least in 1973 and 1974, and annually from 1979 to 1988, mostly from near Yuma and Parker. Extreme dates in spring range from 27 February (1977) to 29 June (1961). There are no records for January or July. Max: 18,000; 30 September 1955, Martinez Lake.

Their passage along the Colorado River peaks a month or more later than the Tree Swallow's, in both spring and fall. Also, in contrast with the Tree Swallow, Barn Swallows are more abundant in fall than in spring. When present in numbers, they form temporary roosts in extensive marshes, such as at Imperial Dam, from which they range out to feed by day. These massive swarms, so characteristic of migration in the valley today, may not have occurred before the formation of large marshes behind major dams. Indeed, Grinnell (1914) noted only one bird on his entire journey down the river in spring 1910!

HABITAT These swallows forage mainly over agricultural fields, open water, and marshes, often in large flocks.

Steller's Jay (*Cyanocitta stelleri*)

STATUS A casual fall and winter visitor during flight years. Records include a flock of 50 on 23 February 1935 at Blythe (spec). Single birds were found at Palo Verde, 8 November 1950; in the winter of 1950–51 at Parker Dam (eight on 13 December), BWD (spec), and Topock; 21 November 1957 at Martinez Lake; and in the winter of 1960–61 at Lake Havasu, Parker Dam, and at Martinez Lake.

These occurrences correspond with times of major invasions of this and other corvids throughout the Southwestern deserts. The Blythe specimen was determined as *C. s. diademata* (Bonaparte), which breeds throughout Arizona. The BWD specimen was *C. s. macrolopha* Baird from farther north. Both races have been found in the same flocks elsewhere in Arizona (Phillips et al. 1964).

Scrub Jay (*Aphelocoma coerulescens*)

STATUS A rare and irregular transient and winter visitor from September through April; locally uncommon during flight years.

Unlike other jays that have reached the Colorado River, this species has been found almost annually. However, large numbers occur only during the same flights that bring other montane species to the lowlands. The most recent and well-documented flight was in 1978, with 1 near Parker on 3 September, at least 18 birds at BWD by early October, 15–20 individuals accounted for through March south to Yuma, and the last sighting at Cibola NWR on 19 April. In other years, singles usually appear in the northern parts of the valley in fall. At least one individual remained year round near Lost Lake Resort from about 1976 to 1978. The few specimens from the LCRV are from both Arizona (*A. c. woodhousei* (Baird)) and more northern breeding races (Phillips et al. 1964).

Pinyon Jay (*Gymnorhinus cyanocephalus*)

STATUS A casual fall and winter straggler during flight years. Recorded 25 September 1952 at Topock (3 birds), 8 October 1955 at Imperial Dam

(2), 28 October 1955 at Imperial NWR, 9 October 1978 at BWD, 23 November 1978 near Parker Dam (6), and 5 January 1990 at Blythe.

These dates of occurrence correspond with widespread irruptions of this species into the lowlands, but not necessarily with irruptions of other corvids.

Clark's Nutcracker (*Nucifraga columbiana*)

STATUS A casual fall straggler during flight years. As part of a major flight detected throughout the Southwest in 1972, individuals were sighted on 28 August at Needles, 14 September at Topock, 15 September at Bullhead City, 23 November near Yuma, and 17 December at Laguna Dam. The only other sighting was on 9 October 1955 near Imperial Dam.

As with other corvids, these years of occurrence were ones of major exoduses from its usual montane range.

American Crow (*Corvus brachyrhynchos*)

STATUS Until recently a rare or casual straggler in fall and winter. Since about 1976, a large winter roost containing 500–1,000 individuals has formed annually at Cibola NWR. Another winter roost formed in Mohave Valley near Topock, perhaps as early as 1973, with up to 700 birds in 1981.

The sudden and unprecedented establishment of two large aggregations of crows in a lowland desert region, where the species is otherwise virtually nonexistent, has been termed "astonishing" by Monson and Phillips (1981), and mistakenly interpreted as "major flights . . . in mid-winter" by Garrett and Dunn (1981). We therefore chronicle these events in some detail. The first crows seen at Cibola were 60 on 3 December 1975. About 600 wintered there in 1976–77, with 1,000 + birds nearly every year since then. These birds have been seen from 20 October to 9 March. Near Topock and Needles, where coverage has been much less complete, 70 crows were first noticed from January to March 1973. Since 1979, groups of between 180 and 700 individuals have been seen from 28 November to 23 February, documenting the presence of a large wintering flock of unknown total number.

Concomitant with the formation of these two flocks has been a dramatic increase in the number of migrant or wandering individuals seen away from Cibola or Topock. Notable among these records are 20 at BWD, 8 November 1977; 32 flying north at Parker, 1 March 1978; 25 west of Poston, 7–17 November 1981; and a specimen found dead north of Blythe, 27 November 1981. There are about 10 other records of 1–5 individuals since 1977, mostly from Parker or Poston.

Records before 1975 away from presently known roosts were 1 taken below Yuma, 14 March 1894; a flock of 8 at Parker, 14 December 1947, decreasing to 6 on 7 February 1948 and then to 3 on 4 March 1948; 2 at Imperial Dam, 17 November 1960; a flock of 300 north of Blythe, 28 November 1964; and 1 at Poston, 7 October 1974 (spec). The 1964 flock may indicate that large winter numbers were present before the discovery of birds at Cibola. At least one old specimen was determined as *C. b. hesperis* (Rea *in* Phillips 1986), probably originating in California. However, the origin of recently wintering crows is not known.

HABITAT Roosts form in riparian woodland, adjacent to agricultural land where flocks forage widely.

Common Raven (*Corvus corax*)

STATUS A fairly common resident and local breeder in northern parts of the valley but increasingly rare south of Blythe. More widely dispersed in winter; almost never in flocks. Max: 50; 21 February 1981, Parker Dam to Davis Dam.

Although ubiquitous year round in many parts of the valley, ravens have received little specific study here. We know of very few recent records below Blythe (e.g., only found twice on Yuma CBC), although Grinnell (1914) noted no change in abundance, "seen at practically every point of observation . . . from Needles to Pilot Knob."

HABITAT Ravens require desert cliffs for breeding, but range widely at other times over desert, riparian, and agricultural areas.

Mountain Chickadee (*Parus gambeli*)

STATUS A casual fall and winter visitor. Two seen 14 November 1976 at BWD (one remained until 19 March 1977, spec); one at Cibola NWR, 26 September–18 October 1977, with two there on 8 and 13 December; and one at Cibola NWR, 24 October 1983–21 February 1984.

These records are among the very few from the Lower Sonoran Desert. There are four records from the Imperial Valley in California and one from Regina, 26 February–4 March 1978, only 16 km west of the Colorado River. The specimen from BWD has been identified as *P. g. inyoensis* (Grinnell), the closest breeding population in northwestern Arizona and southeastern California. Our records are from tall cottonwood–willow riparian woodland. The BWD bird traveled frequently with a large flock of Yellow-rumped Warblers.

Bridled Titmouse (*Parus wollweberi*)

STATUS A casual winter visitor. One first seen on 17 February 1977 at BWD was collected on 20 March 1977.

Although this species wanders regularly in winter to river valleys near Phoenix, this represents the lowest elevation and westernmost record. At present, there are no records for California. This bird frequented tall cottonwoods and willows adjacent to the Bill Williams River.

Verdin (*Auriparus flaviceps*)

STATUS A common resident and breeder throughout the valley. Population numbers peak in August and are much reduced after severe winters. Max: 333; 18 December 1978, Parker CBC.

Verdin population fluctuations can be very dramatic from season to season and year to year. The fluctuation is probably related to the relative severity of the winters. For example, average breeding densities throughout honey mesquite habitats dropped from 40 birds/40 ha in 1978 (after a very mild winter; see Table 5, p. 44) to 10 birds/40 ha in 1979 after a severe freeze. In most years populations drop as much as 84% between the late summer peak and the following winter low. However, fecundity is high and recovery of populations is usually rapid. These population fluctuations are in contrast to stable population levels in the ecologically similar Black-tailed Gnatcatcher. Differences between these two species are further illustrated by their breeding and foraging biology, discussed below.

Verdins in the LCRV, and westward into California, are distinctly paler than those occurring farther east and have been assigned subspecific status, *A. f. acaciarum* Grinnell (Grinnell 1931; Phillips et al. 1964). Birds intermediate between *A. f. acaciarum* and *A. f. ornatus* (Lawrence) (the race occurring throughout the rest of Arizona) have been found from Davis Dam north into Nevada, in the Kofa Mountains, and along the Big Sandy River. The paler plumage is a characteristic shared with many other Colorado River populations of widespread, geographically variable species.

HABITAT Verdins are widespread in most riparian woodland, desert washes, and around human habitations. Honey mesquite supports the highest densities in the riparian zone (15–40 birds/40 ha), although the species is common in all other habitats. Verdins also maintain higher densities year round in pure saltcedar than any other permanent resident except Abert's Towhee.

The range of habitats used by Verdins is influenced by annual fluctuations in weather, especially in winter. For example, as populations shrink during or following exceptionally cold winters, habitat use becomes restricted to the most favorable sites. In the cottonwood groves at BWD, Verdins largely are seasonal residents only, withdrawing to the adjacent desert scrub habitats during the coldest periods. In contrast, this species is more frequently found foraging and nesting within the shaded understory of these same groves in midsummer. Coldest winter temperatures typically occur in dense riparian habitats adjacent to rivers, whereas nearby upland habitats may be considerably warmer (Hastings and Turner 1964).

This apparent response to local temperature differences among habitats in the LCRV mirrors a broader pattern of habitat use seen throughout the Verdin's range in Arizona (Hunter 1988). Although remaining a common desert upland species at higher elevations east of the Colorado River, Verdins become habitat specialists in the riparian zone. They use honey mesquite woodlands almost exclusively and avoid habitats such as cottonwood–willow and saltcedar, even in summer. Thus, on both a local and regional scale, the distribution of the Verdin, along with several other desert-adapted, permanent resident species, appears limited by extreme winter temperatures. This is in marked contrast to the pattern seen in migratory, midsummer-breeding species that are cottonwood–willow specialists in the LCRV, yet become habitat generalists across the same elevational gradient (Hunter et al. 1987b; Hunter 1988; see also Chapter 3).

BREEDING The covered, dome nests with restrictive openings are placed most frequently in mesquite, palo verde, or other spiny trees. Saltcedar also is used frequently for nest placement. Breeding extends primarily from late February through June, but in some years may persist into August. Two to three broods may be raised in a season, making the Verdin a highly fecund species in the LCRV.

The structure of the nest has been hypothesized to defend against cowbird parasitism, depredation, and to provide protection against the thermal environment. We have never observed cowbirds being raised by Verdins. Thus the Verdin may produce many more of its own young each season than do other similar-sized, but open-nesting species, such as the Black-tailed Gnatcatcher. Verdin nests used in summer are oriented toward prevailing winds for cooling, while those used for winter roosting are randomly oriented and spring breeding are oriented away from these winds (Austin 1976; Buttemer et al. 1987). Winter nests are similar to summer nests but are more extensively insulated. Several nests are often located close together, with one summer nest and one winter nest available for each member of a family group.

FOOD HABITS Twenty arthropod taxa and 3 plant taxa were found in 209 stomachs. Dominant foods throughout the year were scale insects, caterpillars, jumping spiders, and aphids. In spring, beetle and wasp larvae also were present in large numbers. In summer, many leafhoppers were taken. Verdins eat berries but it is not known how important fruit is in the overall diet. Other plant matter was probably taken incidentally.

About 1,600 foraging observations allow us to make a detailed assessment of resource use by the Verdin in the LCRV. Gleaning from leaf surfaces was the most important behavior during all seasons, ranging from 60 to 78% of all observations, including winter (60%) when leaves of most riparian plants have dropped. Foraging at flowers was frequently observed in cottonwood–willow groves during winter (55%) and spring (37%). Verdins around houses will often visit sugar-water feeders placed to attract hummingbirds. Other foraging substrates (air, ground, bark, etc.) were used to a lesser extent during all seasons.

This dependency on live foliage throughout the year is in contrast to the Black-tailed Gnatcatcher, which switches partially to bark foraging during winter. Most leaf foraging by Verdins in winter was on perennial shrubs such as quail bush or saltbush. These shrubs most frequently occur in honey mesquite woodlands, thus possibly explaining why such habitats are optimal for Verdins in winter. The interplay of climate, topography, habitat, diet, and foraging behavior suggests that winter is the season of most intense selection for the Verdin, and may influence later use of habitats during the breeding season (see Chapter 3).

Bushtit (*Psaltriparus minimus*)

STATUS A rare but regular winter visitor in small numbers, at least at BWD, from late October to early March. Recorded south to near Ehrenberg with a flock of 10, 25 February–2 March 1977 (two specs), and a flock of 15, all winter 1982–83. Max: 87; 19 December 1978, Bill Williams Delta CBC.

Bushtits wander regularly to the higher deserts of Arizona and California, and have been found at Phoenix and the Salton Sea. Closest breeding localities to the lower Colorado River are the Mohave and Hualapai Mountains in Arizona. All of our records pertain to the interior race, *P. m. lloydi* Sennett. The coastal race breeds as close as Joshua Tree National Monument, California, and has been seen at the Salton Sea.

HABITAT This species has been seen in willows and mesquite, as well as in adjacent desert scrub.

Red-breasted Nuthatch (*Sitta canadensis*)

STATUS A rare but possibly regular transient in September and October. A few individuals winter irregularly or appear as spring transients in late February to April. Max: 5; 8 October 1979, Parker to BWD.

This species has been considered to be an irruptive visitor, reaching the low deserts only in exceptional flight years. However, our recent records suggest a regular migration route through the LCRV in early fall. In the six-year period of 1976–81, two to six individuals were detected each year in September and October, with the earliest arrival on 29 August 1977, east of Topock. Individuals seen at BWD on 27 February and 13 April 1978, at Topock on 16 and 28 April 1978, and at Yuma 6 April 1988 were most likely spring migrants. The only winter records during this period were of one that frequented a single bottle-brush tree in a Yuma yard, 9 December 1979–28 February 1980 (not a flight year), and one seen at BWD, 28 January 1982.

White-breasted Nuthatch (*Sitta carolinensis*)

STATUS A casual fall and winter visitor during flight years; a few present at BWD during 1950–51 (23 October–2 February; spec, 10 November) and 1975–76 (spec, 31 January) until 18 March. Also one at Imperial NWR, 4–26 November 1961 (spec), and one seen 20 December 1979, southwest of Parker.

As in several corvids, these records correspond with major irruptions of montane species into the Southwestern deserts. Although White-breasted Nuthatches regularly move locally into lowland valleys elsewhere in Arizona, the few specimens from our area are *S. c. tenuissima* Grinnell from the Great Basin or eastern California.

Pygmy Nuthatch (*Sitta pygmaea*)

STATUS A casual visitor, one old specimen (now lost) collected by Herbert Brown from Yuma on 30 September 1902.

This particular record is virtually the only one from a lowland desert locality, and we must express some doubt about the certainty of the original collecting locality or identification. Two others were seen north of our region at Boulder City, Nevada, 8 August 1950. This species does stray regularly to the higher deserts of southeastern California.

Brown Creeper (*Certhia americana*)

STATUS An uncommon winter resident at BWD between mid-October and early April. Rare and local elsewhere, with records south to Imperial NWR. The most reliable areas of occurrence are Parker Oasis, Topock, and BWD. Max: 10; 10 December 1981, BWD.

Brown Creepers apparently migrate annually to tall trees in the Southwestern lowlands, but they appear in larger numbers during irruptive flight years. Specimens from our region represent birds from central Arizona highland, Rocky Mountain, and Alaskan populations (*C. a. montana* Ridgway).

HABITAT Most records are from areas with willow, cottonwood, or athel tamarisk trees, including residential areas.

Cactus Wren (*Campylorhynchus brunneicapillus*)

STATUS A fairly common resident and breeder throughout the valley, becoming local in the northern parts. Winter numbers are variable and may represent periodic local movements into and out of the valley. Max: 78; 18 December 1978, Parker CBC.

HABITAT Cactus Wrens are habitat generalists in the LCRV, with use of some habitats apparently influenced by variation in winter temperatures. Largest numbers are always found in honey mesquite woodland and adjacent desert washes (5–6 breeding pairs/40 ha in summer). Smaller numbers nest in denser willows, screwbean mesquite, saltcedar, or around human habitations.

Populations plummet in all habitats in fall and winter, suggesting dispersal out of riparian areas. This species may be restricted to honey mesquite habitats in the coldest winters. This may be countered in certain years (e.g., 1978), however, by an influx from the adjacent desert uplands. Interestingly, Grinnell (1914) did not consider the Cactus Wren to be a riparian species, finding it primarily on the peripheral desert and only locally in the mesquite belt. At higher elevations farther east in Arizona where winter temperatures are much lower, Cactus Wrens remain common in desert uplands, but become extreme habitat specialists in the riparian zone by completely avoiding riparian habitats other than honey mesquite year round (Hunter 1988).

BREEDING Nesting activities begin by late February and two broods are raised from late March through June. The well-known nest is a covered dome of sticks placed in trees, shrubs, or cacti. This species, like the Verdin, uses its domed nest during winter nights for heat conservation.

FOOD HABITS Fourteen arthropod taxa and 1 plant taxon were found in 23 stomach samples. Diet varied little seasonally, with beetles, ants, and stinkbugs making up the core. Short-horned grasshoppers were eaten in summer and late summer (28%), whereas spiders were important in fall and winter (27%). A few small lizards were also taken.

This wren forages primarily by probing bark crevices (especially mesquite and willow) or litter on the ground beneath trees. Bark foraging predominates year round (45–80% of 194 observations), but ground foraging becomes important in fall and winter (37–55% of 73 observations). Only in spring do they forage to any extent on live foliage (up to 33% of 34 observations).

Rock Wren (*Salpinctes obsoletus*)

STATUS A fairly common and local resident and breeder only along the drier edges of the valley. More widespread in winter. Max: 56; 19 December 1978, Bill Williams Delta CBC.

This is strictly a desert species, virtually never entering riparian habitats. However, we have detected a regular dispersal or migration in fall that brings individuals to areas where they do not breed. Such records from nonbreeding habitats extend from 19 September (1977) to 27 February (1980).

HABITAT This species is found primarily on desert mesas or in washes. Occasionally, they wander into agricultural areas during fall and winter. These latter individuals may take up temporary residence around stacked hay bales or unused farm equipment. A few Rock Wrens have even been found on artificial rock levees along the river.

Canyon Wren (*Catherpes mexicanus*)

STATUS A locally common resident and breeder at BWD, Parker Dam, and Topock Gorge. Very rare and irregular elsewhere. Max: 68; 24 December 1977, Bill Williams Delta CBC.

Our only definite records south of Parker Dam are 29 October 1942 at Imperial Dam, 3 December 1942 at Laguna Dam, and 5 December 1978 at Blythe. Additional reports on the Yuma CBC in 1979 and 1983 may be correct but were not documented. This species is much more sedentary than the Rock Wren, although local movements are observed occasionally elsewhere.

HABITAT Canyon Wrens require rocky cliffs for breeding. They seem especially fond of the dark reddish rock walls of the Buckskin Mountains which reach the Colorado River in the vicinity of Parker Dam. Unlike the Rock Wren, this species moves into dense riparian vegetation at BWD to forage from August through March.

Bewick's Wren (*Thryomanes bewickii*)

STATUS Fairly common in winter throughout the valley. Also a fairly common and local summer breeder from BWD northward, at least since 1975. Virtually absent farther south between April and September until very recently when a breeding pair was found 24 km north of Blythe in June–July 1981, and several pairs were at Cibola NWR in 1982 and 1983. Max: 108; 18 December 1978, Parker CBC.

This species was formerly only a winter visitor to the LCRV, with the only previous summer record from Parker, 4 August 1946. At present, migrants are detected south of known breeding sites from 19 August to 5 April (both in 1979). This species was evidently very rare in summer during the 1950s.

Apparently Bewick's Wren is slowly moving to lower elevation riparian habitats as a breeding species in the LCRV. However, this species is common to abundant in all riparian habitats at higher elevation sites on the upper Gila River in eastern Arizona but remains completely absent during summer in these same habitats along the lower Gila River (Hunter 1988).

The LCRV resident and breeding population belongs to the pale, long-tailed race, *T. b. eremophilus* Oberholser. At least a few winter specimens share characteristics with populations from the northwest Baja California–San Diego area, *T. b. charienturus* Oberholser, and from the southern San Joaquin Valley, *T. b. drymoecus* Oberholser (Monson and Phillips 1981).

HABITAT Breeding birds near Topock and Needles are in low brushy willows, sparse screwbean mesquite and saltcedar stands, and in athel tamarisks. At BWD, this species nests in the brushy understory (primarily saltcedar) of mature cottonwood groves. Recent breeders near Blythe and Cibola were in screwbean mesquite and willows. In fall and winter, Bewick's Wrens are found in most sparse riparian habitats as well as in desert washes. They are most numerous in mesquite but are outnumbered by House Wrens in dense willows and saltcedar.

Grinnell (1914) observed them in catclaw acacia, creosote bush, and saltbush, but not in the willow association where House Wrens were found. He inferred that the two species have inherent preferences for different habitats. Today, mesquite habitats are a zone of overlap for these two species, with House Wrens almost exclusively found in small patches of dense brush and saltcedar among the mesquites.

BREEDING This species nests in cavities in stumps, snags, or live trees. Some birds begin singing in February (even wintering birds), but breeding is somewhat later. The young fledge by May and June.

FOOD HABITS Three stomach samples contained only arthropods, including spiders, ants, and beetles. This species feeds primarily by probing into or under loose bark, or gleaning from bare branches or twigs. This behavior accounted for 79% of 96 observations in screwbean mesquite and willow habitats. In more open honey mesquite and desert wash habitat, 60% of 85 observed foraging attempts were on the ground.

House Wren (*Troglodytes aedon*)

STATUS A fairly common transient and winter resident between mid-August and the beginning of May. Max: 121; 23 December 1977, Parker CBC.

This species consistently arrives at BWD between 13 and 16 August. The earliest record is of one at Cibola NWR on 8 August 1979. House Wrens are locally abundant in September, but a drop in numbers later in fall suggests that many of these birds are transients. No equivalent influx is detectable in spring. Only a familiarity with its raspy scolds and chatters will reveal its true abundance, especially compared with the more conspicuous, but actually less numerous, Bewick's Wren.

HABITAT This wren is common in dense riparian woodland, especially willows and saltcedar, as well as in other brush and the margins of agricultural fields. They are outnumbered by Bewick's Wrens only in honey mesquite and desert wash habitats.

FOOD HABITS Three stomachs contained primarily spiders and caterpillars. Ten foraging observations showed that this species foraged on leaves (40%), bark (30%), and ground (30%).

Winter Wren (*Troglodytes troglodytes*)

STATUS A rare but regular winter resident in very small numbers between late October (exceptionally, 9 October 1978) and early April. Most frequent at BWD, but records extend from Topock to Yuma.

One of the many surprises of our recent fieldwork was the discovery of a small, but stable, wintering population of this wren. They have been found every winter since 1976, with up to nine individuals noted in a season (1978–79). These birds are only detected by those willing to crawl through the understory of the densest riparian thickets, and by those familiar with this species' subtle yet distinctive calls. Occasionally in early April, the cottonwood forest at BWD is transformed by the ethereal, trilling songs of Winter Wrens as they boldly assert their presence before departing for the north. The latest record is 7 April 1977. All birds seen well and heard in the LCRV appeared typical of the Pacific race, *T. t. pacificus* Baird, although the Eastern race, *T. t. hiemalis* Viellot, has been found as close as southeastern Arizona and Death Valley, California.

HABITAT Winter Wrens have been found in dense cottonwood–willow and saltcedar habitats, such as at BWD, and in athel tamarisk groves, such as at Topock and Dome Valley.

FOOD HABITS These wrens skulk and forage like shrews among fallen, rotting logs or mats of accumulated saltcedar litter. One bird, however, uncharacteristically climbed the trunk of a willow like a creeper, to rummage in a cluster of dead leaves 9 m above ground. It then fluttered to the bases of several other trees to climb and feed on the trunks and small branches in the canopy.

Marsh Wren (*Cistothorus palustris*)

STATUS A locally common summer breeder in extensive marshes, and common to abundant winter resident throughout the valley. Max: 430; 24 December 1977, Bill Williams Delta CBC.

As with several other marsh-nesting species, the historic breeding status of the Marsh Wren along the Colorado River is uncertain. Grinnell (1914) found no evidence of nesting in the few scattered marshes in 1910, but Grinnell and Miller (1944) considered it a breeding resident. Certainly the large populations present today at Topock, BWD, and Imperial Dam have colonized since the construction of dams.

These populations, along with those at the Salton Sea and east to near Phoenix, have been recently named as a distinct subspecies, *C. p. deserticola* Rea (Rea *in* Phillips 1986). Since most of the marshes within its range are of recent origin, we wonder whether this form has expanded from the Colorado River Delta, like the Yuma Clapper Rail, or whether differentiation has actually taken place only recently. In any event, it is interesting that in contrast with most other species with differentiated Southwestern races, our local Marsh Wrens are darker and more richly colored than northern migrants (Rea *in* Phillips 1986). Migrants to the LCRV are all *C. p. plesius* (Oberholser) from the southern Rocky Mountains and Great Basin, and occur from late August (based on an influx of birds into nonbreeding areas) until 7 May (1888, spec from Yuma).

HABITAT This species breeds only in extensive marshes, with the largest populations at Topock Marsh, BWD, and above Imperial Dam. Breeding densities are highest (238 birds/40 ha) in the densest stands of cattail and bulrush, and are lower in more open marshes or marshes dominated by reed. In fall and winter this species may be found in any habitat with wet soils and dense, low vegetation cover. Filling these requirements are the downed branches, logs, and debris inside the partially flooded cottonwood groves at BWD, weed-choked agricultural canals, and even tall, wet alfalfa fields. During September, Marsh Wrens are found occasionally in dry habitats, such as patches of dense quail bush.

BREEDING This wren breeds in marshes from March to July. The distinctive nest is a compact ball of vegetation intertwined with stems or cattail blades.

FOOD HABITS From 22 observations of foraging behavior, wintering birds at BWD gleaned insects from a variety of substrates, including emergent vegetation, dead branches, and the surface of shallow water.

American Dipper (*Cinclus mexicanus*)

STATUS A casual visitor; one seen south of Parker, 22 October 1980.

This represents virtually the only record of a dipper from a lowland desert locality, although they are present upstream on the Colorado River at the bottom of the Grand Canyon. This bird was on the rocky, riprapped riverbank.

Golden-crowned Kinglet (*Regulus satrapa*)

STATUS A rare but regular visitor from late October through December and irregular in January and February, very rarely later. Usually occurs in very small flocks (2–6), especially at BWD. Max: 13; 10 December 1975, southern tip of Nevada.

This species occurred annually from 1975 to 1985, with up to 11 sightings (of 28 birds) in a single season (1976–77). Only 5 of 47 sightings during this period were later than December, including a specimen, 28 January 1982, at BWD. The range of dates is from 18 October (1978) to 5 April (2, 1982). Prior to 1975, this species was found only once, 21 November 1957 at Martinez Lake (spec). *R. s. satrapa* (Lichtenstein), breeding in the northeastern United States, is the only race known from the desert lowlands, although the identity of the 2 Colorado River specimens has not been determined.

HABITAT This species has been found in dense riparian woodlands, especially willows, saltcedar, and athel tamarisk.

FOOD HABITS Of 30 observations in cottonwood–willow habitats at BWD, half involved gleaning from bare twigs, often while these kinglets were hanging from a perch, and the remainder were in live foliage or on the ground.

Ruby-crowned Kinglet (*Regulus calendula*)

STATUS A common to abundant winter resident from early October to mid-April. First arrivals are in late September and a few linger to early May. Occurs both singly and in small flocks. Max: 1,347; 24 December 1977, Bill Williams Delta CBC.

This is one of the LCRV's most abundant and well-studied riparian species in winter. Details of its population dynamics, habitat affinities, resource use, and how these relate to the winter environment along the Colorado River may be found in Laurenzi et al. (1982). In general, kinglets flood into the valley throughout October, usually reaching their peak densities in November (up to 120 birds/40 ha in favored habitats). Numbers then decline steadily through winter and early spring, with most birds departing by early April. The magnitude of the winter peak and subsequent decline is determined largely by the severity of the season. For example, in the very mild winter of 1977–78 (see Table 5, p. 44), kinglet numbers were much higher throughout the winter than in any other year of study.

HABITAT These kinglets may be found in virtually any stand of riparian vegetation, as well as in other trees and shrubs in towns and along agricultural margins. Highest densities, however, are always in mature cottonwood–willow habitat, such as at BWD.

They always seek the tallest or densest vegetation in any habitat. The extent of various habitats occupied also appears to be related to winter temperatures. For example, in the very cold, dry winter of 1975–76, Ruby-crowned Kinglets were largely restricted to cottonwood–willow stands after midwinter. In 1977–78 they were relatively common in all habitats including arrowweed. These patterns result, in part, from changes within the particular habitats, such as timing of leaf drop or insect productivity. However, the total population size in the valley also influences habitat use. As higher numbers in cottonwood–willow stands render those areas less suitable to other individuals, there is a "spill-over" into more marginal habitats.

FOOD HABITS Kinglets are very active foragers. They perpetually flick their wings as they hop rapidly from twig to twig, gleaning or hovering at leaves or branches. Foraging methods do not vary much through the winter or among habitats for this species, with gleaning the most frequent activity (60–80% of 2,704 observations), followed by stationary hovering (20–30%), and a small amount of aerial sallying. Use of various substrates, however, does vary markedly from site to site. These kinglets appear to seek live foliage when available. In addition, they feed heavily at new, opening cottonwood and willow flowers at BWD in late winter. Thus the feeding habits of these birds closely follow the phenology of the individual tree species in their habitats. When foliage is rare, as in screwbean mesquite and some willow groves in winter, kinglets feed predominantly at bare twigs and small branches.

The species is almost entirely insectivorous. Only 3 of 104 stomach samples contained seeds. Leafhoppers, beetles, and flies comprised 50–100% of the diet in any one month, and 95% of all prey were <5 mm in length. The proportions of insect types eaten each month corresponded well with sweep-net samples of available insects, indicating that kinglets opportunistically prey on the most abundant foods.

Blue-gray Gnatcatcher (*Polioptila caerulea*)

STATUS A fairly common winter resident from September through March, uncommon in late August and early April. Max: 102; 21 December 1983, Parker CBC.

They breed in scrub oak habitats as close as the Cerbat, Hualapai, Castle Dome, and Kofa Mountain ranges east of the Colorado River.

HABITAT This gnatcatcher is most common in screwbean mesquite woodland close to the river but is also widely dispersed in other riparian and desert habitats. They favor taller and denser vegetation than the Black-tailed Gnatcatcher does, although they are outnumbered by the latter in most parts of the valley.

FOOD HABITS Seven stomachs contained mainly stinkbugs, midges, beetles, and leafhoppers, plus a few spiders, ants, and caterpillars. This species is primarily a live-foliage gleaner (42–65%) and bark forager (22–25%), as determined from 233 observations. It differs from the more numerous Black-tailed Gnatcatcher primarily in its propensity for aerial sallying after flying insects (15–24% in winter and early spring). The feeding behavior of this species was the subject of a classic ecological study (Root 1967), part of which was conducted in a 10-ha screwbean mesquite stand along the lower Colorado River near Yuma.

Black-tailed Gnatcatcher (*Polioptila melanura*)

STATUS A common resident and breeder throughout the valley, becoming somewhat less common north of Lake Havasu. Found in pairs or family groups year round. Max: 142; 22 December 1980, Parker CBC.

Five years of continuous surveys in honey mesquite woodlands in the LCRV indicated that this species maintains very stable population sizes from season to season and year to year. This is in contrast to the wide population fluctuations found in the ecologically similar Verdin. This gnatcatcher is very territorial year-round, with breeding territories remaining intact through winter.

Birds in the LCRV are of the interior race, *P. m. lucida* van Rossem. California birders now have two "dark-tailed" gnatcatchers within their borders with the recent recognition of coastal populations as a separate species, the California Gnatcatcher (*P. californica*) (Atwood 1988).

HABITAT This gnatcatcher is widely distributed in all but the densest riparian habitats. It is also widespread in desert washes and adjacent uplands.

Honey mesquite constitutes the optimal habitat, supporting densities of 10–16 birds/40 ha, with screwbean mesquite ranked second in importance. The range of habitats occupied depends partly on the season and the severity of the winter. However, this range varies much less than in other "desert" species whose peak populations are also in honey mesquite.

This gnatcatcher becomes strictly associated with honey mesquite habitats in riparian zones at higher elevations (550–600 m) east and north of the LCRV. It is usually the first of the permanent residents to disappear in honey mesquite and subsequently in upland desert habitats as elevation, latitude, and longitude increase (Hunter 1988). Perhaps harsher winter temperatures affect this species more severely than others, as vegetation structure at these higher elevation sites does not differ radically from areas where all the desert birds are common. In fact, the Black-tailed Gnatcatcher was the only Southwestern desert permanent resident to have its northern winter distribution limits associated with a specific average minimum January temperature (− 1.1°C; Root 1988).

BREEDING The nest is an elaborate cup in a branch fork or mistletoe clump. Breeding activity begins in March and lasts into July. This species is a frequent host for the Brown-headed Cowbird, especially during later nesting attempts.

Conine (1982) studied the nesting biology of this and other species within a continuous honey mesquite stand and along a honey mesquite-agricultural edge. She found that there was virtually no chance for a nest to produce young after Brown-headed Cowbirds arrived in riparian habitats by late March. Only the earliest nesting attempts were successful. Therefore, estimates of overall nesting success were surprisingly low for this species, but were higher along habitat edges (with or without cowbirds present) than in continuous honey mesquite. For this and other spring-nesting species, increased food resources along agricultural edges appear to aid in nesting success before the arrival of cowbirds.

Heavy parasitism by cowbirds in the LCRV was documented as early as Grinnell's (1914) study. The persistence of this species in the face of such very low nesting success appears to be dependent on (1) a successful first brood, and (2) a continual dispersal into the valley by birds from the expansive Sonoran Desert habitat where cowbird parasitism is infrequent.

FOOD HABITS Thirty-three arthropod and 3 plant taxa were identified in 156 stomach samples. Diet composition varied little among seasons with caterpillars, beetles, and leafhoppers predominating. Plant material may be only incidental in the diet (3% overall).

This species feeds by gleaning from live foliage and small branches throughout the year, as determined from 918 observations. Compared with the similar Blue-gray Gnatcatcher, this species rarely sallies for aerial prey. The Black-tailed Gnatcatcher exhibits a marked shift from primarily foliage gleaning in summer and fall (60–80% of observations) to heavier use of bare twigs and branches in winter and early spring (50–60%). This shift is related to the period of leaf drop by the dominant tree species in mesquite habitats and is in contrast to foraging by the Verdin, which relies more heavily on perennial shrubs in these same habitats in winter.

Western Bluebird (*Sialia mexicana*)

STATUS An uncommon to fairly common winter visitor in the northern half of the valley, from November to mid-March, usually occurring in small nomadic flocks. Rare south of Blythe, at least in recent years. Max: 56; 22 December 1980, Parker CBC.

This is one of several frugivores that invades the LCRV in winter to feast on mistletoe berries in riparian woodlands. They are not territorial at the mistletoe clumps, as are resident Phainopeplas and Northern Mockingbirds, but wander from patch to patch in small groups of 5–10 birds. Although present every year, timing of arrival and population sizes vary from winter to winter. Variation does not seem related to local weather or to food supply. It more likely reflects conditions outside our region, from where the birds are dispersing. Their apparent rarity in the Yuma area is probably due to the nearly complete clearing during the past two decades of mesquite woodlands with berry-producing mistletoe. The largest populations remain between Poston and Ehrenberg.

HABITAT Western Bluebirds are found primarily in honey mesquite woodland and desert washes, and other riparian woods that support at least some mistletoe clumps.

FOOD HABITS Five stomach samples contained nearly all mistletoe berries (98%) and a few unidentified insects. Besides foraging for berries at mistletoe clumps (68% of 108 observations), they were also seen foraging on the ground (26%), presumably for insects, and occasionally flycatching (4%) from exposed snags.

Mountain Bluebird (*Sialia currucoides*)

STATUS An uncommon to fairly common winter visitor from late October through early March, usually in small nomadic flocks. One unseasonal sighting of a male near Poston, 6 May 1978. Max: 155; 22 December 1980, Parker CBC.

This species differs from other visiting thrushes in that it normally avoids wooded riparian habitats. It appears more often in open agricultural valleys or even in open desert. It is much less dependent on mistletoe berries than the Western Bluebird, and is also the more regular of the two species around Yuma where most mesquite has been cleared. Numbers are highly variable from year to year, largely independent of local food supplies. In February 1978 and again in January–February 1982, when Western Bluebirds were relatively scarce, Mountain Bluebirds invaded the mistletoe-infested mesquite bosques near Poston and Ehrenberg. At these times, single flocks of over 60 birds were noted and a total of 200 was near Poston on 19 January 1982. In other years when both bluebird species invaded simultaneously, Western Bluebirds were common in mesquite, while large flocks of Mountain Bluebirds were noted only in agricultural areas (e.g., 1980–81, 1984–85).

HABITAT This bluebird is seen most frequently in grassy or plowed agricultural fields or pastures. They also may occur in mesquite woods and desert washes with mistletoe-infested trees during some invasion years, but rarely during coinciding invasions of Western Bluebirds.

FOOD HABITS Three stomachs of birds from agricultural areas contained grasshoppers, beetles, and bugs. These insects are captured much in the manner of a Say's Phoebe, by pouncing to the ground from a low perch or from a stationary hover. The major food of this species in riparian areas appears to be mistletoe berries (81% of 31 observations).

Townsend's Solitaire (*Myadestes townsendi*)

STATUS A rare to uncommon and irregular transient and winter visitor from September to late March. One very late spring record at Blythe on 15 May 1988. Usually observed singly.

This is the least common of the berry-eating thrushes that regularly reach the LCRV. Although at least one or two were noted almost every fall from 1975 to 1982, they only wintered twice during this period; about 10 individuals in 1977–78 and about 8 in 1981–82. This species is more numerous in the slightly higher valleys and deserts to the east in Arizona and to the northwest in California, where it also appears as a spring transient.

HABITAT This species occurs primarily in honey mesquite, desert washes, and even mixed saltcedar–honey mesquite woodland when mistletoe is present.

FOOD HABITS We have observed them feeding only on mistletoe berries, although they may also eat some insects.

Swainson's Thrush (*Catharus ustulatus*)

STATUS A rare to uncommon transient in May and very early June. Casual in fall with one seen 20 September 1953 at Parker. Max: 13; 9 May 1979, Yuma.

Our recent records of this migrant are curiously concentrated from the Yuma area, despite our reduced field efforts in the southern part of the valley. For example, in eight years of fieldwork at BWD between 1975 and 1983, there were only two sightings. All specimens to date are from Pacific Coastal populations (the Russet-backed Thrushes), *C. u. ustulatus* (Nuttall) breeding on the northern Pacific Coast and *C. u. oedicus* (Oberholser) breeding on the southern Pacific Coast (Phillips et al. 1964).

HABITAT Nearly all records are from dense riparian woodland, especially cottonwood–willow groves in the Yuma area.

Hermit Thrush (*Catharus guttatus*)

STATUS An uncommon to fairly common transient and winter resident from early October until mid-April. A few individuals appear in mid- or late September and rarely stay until mid-May. Number of wintering birds fluctuates from year to year. Max: 188; 24 December 1977, Bill Williams Delta CBC.

This species was unusually abundant in the mild winter of 1977–78 (see Table 5, p. 44) when, in addition to the high count above, 77 were on the Parker CBC and a few were regularly encountered in nearly every patch of riparian vegetation. They may be quite scarce in other years; e.g., only 2 birds were detected in 1979–80, other than those on CBCs.

The Hermit Thrush is our best example of a species with a cryptically complex status, which is revealed only by close examination of collected specimens. The situation in Arizona was beautifully explained by Phillips et al. (1964; also see Monson and Phillips 1981). In the LCRV, no fewer than five distinct races have been identified, and a sixth probably also occurs. Most wintering birds are from Alaska (*C. g. guttatus* (Pallas)) and coastal British Columbia (*C. g. verecundus* (Osgood)) populations. Other Western populations represented by specimens in the LCRV are *C. g. vaccinius* (Cumming) and *C. g. slevini* Grinnell (including *C. g. jewetti* Phillips); *C. g. oromelus* (Oberholser) is suspected to winter but is not represented by typical specimens. One individual at Parker, 29 November 1953, proved to be from the northeastern United States (*C. g. nanus* (Audubon)). The closest breeding races, in the Rocky Mountains and Great Basin mountains (*C. g. auduboni* (Baird), including *C. g. polionotus* (Grinnell)) and Sierra Nevada (*C. g. sequoiensis* (Belding)), have not been found to migrate or winter in our area. This follows the pattern observed in several other species that have distinctive interior Southwestern mountain breeding populations that are rarely found in the Sonoran Desert during migration or winter.

HABITAT This species favors dense riparian woodland, especially willows, and may be seen occasionally in other dense vegetation, such as surrounding residential areas or the borders of fields. When large numbers wintered in 1977–78, this species was present in every riparian habitat type.

FOOD HABITS Beetles, ants, caterpillars, and bugs formed the major portion of eight stomach samples. Most foraging was in leaf litter on the ground (84% of 51 observations), but insects were occasionally taken from branches and fallen tree trunks (16%). Although known frugivores in many parts of their range, they apparently do not share in consumption of abundant mistletoe berries as do other visiting turdids.

Rufous-backed Robin (*Turdus rufopalliatus*)

STATUS A casual straggler from Mexico. One was present 17 December 1973–6 April 1974 at Imperial Dam (photo) and two at Parker on 15–17 November, with one remaining at least to 22 December 1982 (photos).

This species appears almost annually in southeastern Arizona but is extremely unusual farther west. The Imperial Dam record was the first for California. The two 1982 individuals were among a high of 11 birds found in Arizona, with 2 additional birds reaching the Pacific Coast. Our records were from residential parks where the birds frequented lawns and tall, fruiting trees.

American Robin (*Turdus migratorius*)

STATUS An uncommon to common winter visitor from October to March, rarely through April. In some years, large numbers arrive in the valley during midwinter. Also recently a rare but regular summer breeder at inhabited areas such as Blythe, Willow Valley Estates, Needles, and near Yuma. Max: 389; 23 December 1977, Parker CBC.

This is another fruit-eating thrush that irregularly invades the LCRV in winter to feast on mistletoe berries. However, unlike the bluebirds, robins may also swarm into towns and orchards to feed on other fruit. For example, 350 were in Yuma in late February and early March 1955, 200 were in Yuma in mid-December 1957, 300 were in an abandoned citrus orchard east of Vidal in December 1977, and 200 were in Blythe in late February 1979. In some winters, however, they are virtually absent from the valley. For instance, Rufous-backed Robins were almost as numerous as American Robins wintering in the Arizona lowlands as well as in the LCRV during 1982–83! The earliest fall record was 15 September 1977 at BWD, although 1 was seen at Parker, 16 August 1953.

HABITAT Robins are most often found in mesquite woodlands and desert washes, with trees supporting mistletoe, or in towns and orchards. However, they may be locally common in mature cottonwood–willow groves, such as at BWD, where fruit is scarce. Nesting takes place in tall trees near irrigated lawns in city parks, such as at the Blythe Municipal Golf Course and in Willow Valley Estates.

BREEDING Since 1977, robins were found to nest in tall trees in parks or near houses. Young have been noted in late May and June. The first confirmed breeding was at Todd Park in Blythe, where a pair was feeding recently fledged young on 30 May 1977.

FOOD HABITS Mistletoe berries (81%), beetles, caterpillars, and spiders were represented in eight stomach samples. When in cottonwood–willow groves, robins forage primarily on the moist ground in leaf litter or among downed branches and debris (87% of 91 observations). In mesquite woodlands, however, they are almost exclusively frugivorous, with 99% of the 119 observations at mistletoe clumps. In residential areas they feed heavily on pyracantha fruits.

Varied Thrush (*Ixoreus naevius*)

STATUS A rare and irregular winter visitor from the Northwest, primarily in flight years. At least 10 individuals were found between mid-November 1977 and early April 1978, including 6 together on 20 November 1977 at BWD (1 spec, 29 November). Other records include 12 April 1973 north of Blythe; late December 1973 near Yuma; 19 December 1978 south of Ehrenberg; 3 November 1979 north of Blythe; and 3 April 1988 near Needles.

Varied Thrushes are at the southern extreme of their distribution in Arizona and southeastern California, and they are absent from this region in most winters. All identifiable specimens this far south are of the northernmost breeding race, *I. n. meruloides* (Swainson).

HABITAT Most records are from cottonwood–willow groves, with a few seen in dense vegetation near houses or in athel tamarisks.

FOOD HABITS One stomach sample contained termites and beetles.

Gray Catbird (*Dumetella carolinensis*)

STATUS A casual transient, with one seen 26 September 1978 at BWD.

Although a local breeding resident in much of the interior western United States, with even a small population in eastern Arizona, this species is very rarely detected in migration in the Southwest. Most records away from the Pacific Coast are in late spring.

Northern Mockingbird (*Mimus polyglottos*)

STATUS A fairly common summer breeder and common winter resident, with an influx of birds into the valley between October and April. Max: 165; 23 December 1977, Parker CBC.

This is the only permanent resident frugivore. Local changes in distribution and abundance may be due to both dispersal of birds within the valley and migration from other regions. They occur solitarily or paired, and are highly territorial and aggressive at fruiting shrubs or trees. As such, a single mockingbird may prevent a flock of robins or Cedar Waxwings from feeding at any one particular site.

Grinnell (1914) did not find any evidence of the mockingbird breeding in this region. Monson did not note widespread breeding until the 1950s, after a series of exceptionally wet winters (Phillips et al. 1964). Therefore, it is likely that this species has changed from being only a winter resident to becoming a breeding permanent resident since 1910.

HABITAT The success of this bird at adapting to human-altered habitats is rivaled by few other native species in the Southwest. It is most common all year in towns, orchards, and around farm houses where fruiting shrubs are planted. In native habitats, this species exhibits marked seasonal changes in distribution unlike those of any other resident bird. Breeding takes place in low densities (1–2 pairs/40 ha) in honey mesquite woodland and open cottonwood–willow groves in spring and summer. In late summer, numbers decline in these habitats and birds are consistently detected in saltcedar and screwbean mesquite where they were absent in other seasons. Then, in fall and winter, the species occurs almost exclusively in honey mesquite and desert washes with trees supporting mistletoe, in densities 5–10 times those attained for breeding.

BREEDING Nests are in bushes, small trees, cacti, or tangles of vines. Breeding is from April to August, with most midsummer nesting restricted to suburban areas. Recently fledged young have been noted in mesquite habitats in early June.

FOOD HABITS This species' diet in our region consists primarily of mistletoe berries and beetles. However, ants, grass seeds, earwigs, pseudoscorpions, bugs, and caterpillars were represented in 13 stomach samples. In spring and summer, arthropods and berries are eaten in about equal frequency. By late summer, berries drop to 25%, increase to 60% in fall, and reach a high of 80% during winter.

Foraging for insects is much in the manner of other thrashers, running on the ground with tail cocked, pausing to pick prey from the bare ground or litter. In riparian habitats, nearly all foraging is on mistletoe berries during fall and winter (95% of 63 observations). At other seasons they may forage for insects among foliage, bare branches, or on the ground. When feeding on mistletoe they are silent and secretive, in contrast to Phainopeplas and other frugivores that are noisy and conspicuous at fruit clusters.

Sage Thrasher (*Oreoscoptes montanus*)

STATUS An uncommon to fairly common late winter and early spring transient from mid-January through early April. Also a rare fall transient and early winter resident from early September through December. Max: 12; 2 April 1910, opposite Cibola.

This species is unusual in that it arrives in midwinter, occasionally in numbers, sets up temporary territories in mistletoe-producing woodlands, and remains for one to two months before departing. This claim is based largely on individuals seen repeatedly at the same berry clumps, although the extent of turnover by transients is not known. Locally, at these times, they may be the second most abundant frugivore. During 1977, 96 sightings were made in February and March, and 87 were recorded during the same months in 1978. In other years very few are seen, even though the species is consistently present. The latest spring sighting is 12 April (1948 at BWD and 1982 near Poston).

In fall, usually only one or two are noted annually, most often in October and November, with a very early sighting on 1 September 1954 at BWD. This species is much less regular in early winter with most records from washes along the periphery of the valley.

HABITAT This thrasher is most numerous in honey mesquite woodland and in desert washes with mistletoe berries. They are occasionally found in other open habitats, including margins of agricultural fields.

FOOD HABITS Six stomachs contained mistletoe and other berries, beetles, ants, and earwigs.

Brown Thrasher (*Toxostoma rufum*)

STATUS A casual or very rare transient and winter visitor. Four sight records: 15–17 November 1978 at Ehrenberg; 3 October 1980, south of Parker; 22 December 1980, near Poston; and 6 March 1987 at Cibola. Also, distinctive call heard 29 September 1979 at BWD.

These records fit a pattern of rare but regular fall and winter occurrence throughout the Southwest. This species has been found in dense riparian woodland and other shrubby vegetation.

Bendire's Thrasher (*Toxostoma bendirei*)

STATUS A rare and irregular visitor from adjacent desert uplands, primarily in winter and spring; possibly breeds along the periphery of the valley. About 48 records since 1946.

Records of wintering individuals include 17 December 1973–1 February 1974 at Bard; late fall 1978 to 21 January 1979, northeast of Yuma; and winter 1981–82 (until March) at Cibola NWR. Additional transients were 13 April 1947 (2) at BWD; 19 April 1957 at Ferguson Lake; 26 March 1953 at Havasu Lake; 8 April 1961 at Imperial Dam; 9 April 1971 at Parker Dam; 21 March 1977 at BWD (spec); and 20 August 1978 near Poston. Other records, suggestive of local breeding, were 22 June 1948 below Parker Dam; 3 June 1955 at Imperial NWR (Adobe Lake); 30 July 1959 at Martinez Lake; a singing bird on 20 May 1977 north of Needles; a pair with a fledged juvenile on 12 June 1978 in a desert wash north of Ehrenberg; and an immature on 16 July 1979 at BWD (photo). This species breeds as close as the Black and Kofa Mountain ranges in western Arizona and in the eastern Mohave Desert of California.

HABITAT This thrasher has been seen in desert washes, sparse riparian woodland, and agricultural land.

Curve-billed Thrasher (*Toxostoma curvirostre*)

STATUS A rare and irregular visitor from desert areas to the east, primarily in fall and winter.

This species has been recorded about nine times between 8 August (1978) and 9 March (1974) from near Yuma and Bard. Three reports were from BWD, and one was from the California shore of Lake Havasu, 26 December 1952. It has strayed westward to the Salton Sea at least five times.

Most records have come from desert scrub along the periphery of the valley and in sparse riparian woodland. This species is a common breeding bird in the Sonoran Desert, 30 km east of the Colorado River and north of the Gila River.

Crissal Thrasher (*Toxostoma crissale*)

STATUS A common but secretive resident and breeder throughout the valley. Max: 106; 18 December 1978, Parker CBC.

The Crissal Thrasher is near the western edge of its range in the LCRV. Although it is currently listed as a Species of Special Concern by the California Department of Fish and Game, we have not seen evidence of a serious population decline. As in many other permanent resident species, population size peaks in late summer and is lowest in winter in all habitats. The initiation of singing in midwinter is one of the first indications of breeding activity by our resident avifauna.

Phillips et al. (1964) describe the paler birds of the LCRV as *T. c. coloradense* van Rossem. This subspecies also occurs along the Bill Williams River, Big Sandy River, and possibly the lower Gila River, the Little Colorado River, and in the Grand Canyon. *T. c. coloradense* is among several local populations with paler plumage than conspecifics elsewhere.

HABITAT This species is present in most riparian woodlands, especially those on sandy soils. Honey mesquite habitats support the largest populations throughout the year. Other habitats used include desert washes, citrus orchards, and agricultural edges. Unlike other "desert" thrashers, the Crissal is rarely found far away from extensive cover.

Subtle differences in relative population size among habitats suggest that some individuals may disperse from their natal territories in mid- and late summer. For example, population densities tend to increase slightly in saltcedar and screwbean mesquite areas in late summer and fall, when declines are first noted in honey mesquite. Whether this indicates later breeding in suboptimal habitats or actual habitat shifts is unknown, however.

This is the most common thrasher in riparian habitats throughout its range, and some populations also occur in upland chaparral. Along higher elevation floodplains in eastern Arizona, this species is largely restricted to honey mesquite habitats year round (Hunter 1988). However, the Crissal Thrasher was found to use saltcedar habitats to a significant degree on the Pecos River in southeastern New Mexico, in numbers similar to those found in LCRV habitats (Hunter et al. 1988).

BREEDING The nest of twigs, placed 0.5–2.0 m high, is usually in mesquite trees and well concealed. Nests are found occasionally in saltcedar and quail bush. Breeding activity begins in late January, when thrashers begin to sing throughout the valley. Recently fledged young have been seen as early as the end of April.

Hatching success has been estimated to be 51% and 45% for this species at honey mesquite-agricultural edges and in interior honey mesquite habitats, respectively (Conine 1982; Finch 1982). The average number of young produced per nest was estimated at 1.5 for edge habitat (Conine 1982). As with Black-tailed Gnatcatcher and Abert's Towhee, Crissal Thrasher appears to do well along these interfaced habitats that have an abundance of food resources.

In an interesting experiment, Finch (1982) artificially parasitized 9 of 15 thrasher nests with Brown-headed Cowbird eggs and reported that in all cases the cowbird eggs were ejected. Egg ejection occurred almost immediately, regardless of nesting stage. This may explain the very low incidence of documented parasitism and the relatively high fecundity of this thrasher compared with other open-nesting species in honey mesquite.

FOOD HABITS Twenty-one arthropod and 2 plant taxa were found in 32 stomach samples. Beetles were the most important food throughout the year (30–50% of diet). Caterpillars were preyed on heavily in fall, winter, and spring (20–25%), and many maggots were eaten in summer (25%). Ants were most important in winter and spring (18%). Short-horned grasshoppers were eaten primarily in late summer (37%). This diet reflects the exclusively ground-foraging habits of this species. The long, curved bill is used to probe soft ground and to "thrash" through leaf litter and debris on the surface.

Le Conte's Thrasher (*Toxostoma lecontei*)

STATUS Possibly a resident or transient along the periphery of the valley; very few records in the immediate floodplain.

Recent records include one collected 22 January 1947 west of Havasu Lake; one, 8 December 1973 at Imperial Dam; one, 9 February 1976 near Blythe; one collected 13 November 1979 in a harvested cotton field near Poston; and one or two seen in the same desert wash southwest of Parker in December 1980, 1982, and 1983, and on 18 July 1982.

American Pipit (*Anthus rubescens*)

STATUS A common to abundant winter resident from October through mid-April. A few arrive by mid-September (exceptionally, 7 September 1976) and linger rarely until early May (exceptionally, 27 May 1980, two). Often occurs in large flocks. Max: 6,723; 20 December 1978, Parker CBC.

This is among the most numerous species wintering in the valley today. Undoubtedly their abundance is because of the expansion of agricultural land in recent years. This species is highly gregarious, with flocks commonly of 100 or more individuals.

Although American Pipits breed south to the mountains of northern Arizona and in the Sierra Nevada (*A. r. alticola* Todd), the LCRV wintering birds come from much farther north, primarily Alaska (*A. r. geophilus* Oberholser). One specimen from Yuma (22 December 1902, collected by Herbert Brown; Phillips et al. 1964) is probably referable to populations from northeastern North America (*A. r. rubescens* Turnstall).

HABITAT This species is most numerous in irrigated alfalfa fields (35–66 birds/40 ha) and recently plowed fields (24–39 birds/40 ha). Smaller numbers are present in grass (20 birds/40 ha) and other cultivated crops (but not cotton). Flocks are also seen on sandbars and riverbanks. Individuals may be scattered along agricultural canals and shores of pools and ponds.

FOOD HABITS American Pipits forage on the ground, often gleaning from low, herbaceous crop plants. Overall, they take weed seeds and insects in about equal frequency. The most common dietary items in 45 stomach samples were bugs, beetles, flies, caterpillars, and the seeds of Amaranthaceae, Portulacaceae, and *Panicum* spp. A few wheat and other cultivated crop seeds were also present.

Sprague's Pipit (*Anthus spragueii*)

STATUS A rare but probably regular winter visitor in small numbers from late November to March.

Since the discovery of this species near Parker in 1977, it has occurred in the valley almost every winter, with up to 16 sightings in December 1982. A specimen was collected west of Poston on 18 December 1978, and another north of Ehrenberg on 2 December 1982. Concentration of records between Parker and Ehrenberg reflects increased coverage in this area. Other records are from near Bullhead City, Needles, Blythe, Cibola NWR, and east of Yuma. A sighting at Topock, 27 September 1949 (the only older record), was unusually early for our area, but was within the period of peak migration for eastern Arizona and New Mexico. The LCRV represents the westernmost fringe of this species' winter range; it is considered a casual transient farther west in southern California.

The key to locating this pipit is knowing its preferred habitat (discussed below) as well as its distinctive behavior when flushed. This species occurs singly and remains concealed until approached closely. It does not flock or walk in the open with American Pipits or sparrows. When disturbed, an individual rises abruptly in a series of stepped undulations, uttering its sharp single or double call note at each step. Then it may fly a considerable distance, often circling high above the fields. The bird usually returns near to where it was flushed but away from the original disturbance. The flight is always terminated by a steep dive, with wings folded, as the bird plummets like a stone into cover.

HABITAT They are very particular about the fields they inhabit in the LCRV and, with practice, their presence can be predicted with some degree of accuracy. The most important ingredient is dry Bermuda grass. Pure grass fields are rare but these will invariably support this species. Alfalfa fields mixed with patches of dry grass are also suitable. Rarely, one may be flushed from tall alfalfa without much grass.

As with several other agricultural-adapted species, Sprague's Pipit may be a recent addition to the local avifauna. It is unlikely that any native habitat supported this species before the development of agriculture in the valley.

Bohemian Waxwing (*Bombycilla garrulus*)

STATUS A casual visitor in late winter and spring, with records only from the northernmost parts of the valley. Three records: 10 January 1861 at Fort Mohave (spec), 6 March 1959 at Davis Dam (2 spec), and 12–30 January 1977 at Katherine Landing (flock of 10–14 birds). There is an additional record north of our region from Willow Beach below Hoover Dam, 30 April 1938.

These records delimit the southernmost extension of this species' invasive winter range. At least during the January 1977 sightings, they were abundant in many other parts of western North America, south to New Mexico. Cedar Waxwings, however, were very scarce in our region that winter. The Bohemian Waxwing should be looked for at berry-producing shrubs and trees, especially pyracantha bushes.

Cedar Waxwing (*Bombycilla cedrorum*)

STATUS An erratic transient and winter visitor; rare to uncommon from the end of August through December, uncommon to common from January through May. A few individuals linger into early June, and at least one stayed at Parker until 12 July 1953. Max: 460; 30 October 1961, Yuma.

When present in the LCRV, they are often found in large nomadic flocks, but numbers and timing of occurrence are unpredictable from year to year. Like the robin, but unlike other frugivores reaching the region, waxwings are as prone to invade cities and residential parks as they are berry-producing riparian woodlands. For example, single flocks of 316 birds at Yuma, 19 February 1960, and 200 at Blythe on 17 April 1978 were in addition to the Max count above. The largest concentration in riparian habitat was 256 near Poston, 6 February 1978.

Small numbers are noted each year in September and October, even in years when wintering birds are few or nonexistent. This suggests a regular fall migration route through our area that is independent of winter status, perhaps involving individuals that normally winter south into Central America.

HABITAT Waxwings occur in riparian woodland and desert washes with mistletoe, and also around human dwellings with pyracantha and other berry-producing shrubs.

FOOD HABITS In our region, the diet consists entirely of mistletoe and other berries, depending on habitat, and figs.

Phainopepla (*Phainopepla nitens*)

STATUS A common to abundant winter resident and early spring breeder from late September through early May. Numbers increase and decrease rapidly in October and May, respectively. A few can be found all summer, but most of the population is absent between early June and early September. Max: 798; 23 December 1977, Parker CBC.

Phainopeplas are unique among Colorado River birds in that they arrive in fall, overwinter, and breed in spring, thus completely avoiding the hot summer period. In other parts of their range, such as southeastern Arizona and coastal California, they primarily arrive during summer and breed. Because the overwintering population in the LCRV is greater than the breeding population, it is likely that some of the wintering birds migrate westward in early spring rather than breed locally. Some individuals may breed sequentially in both areas, but this, as yet, is unproven.

The general biology of the species in the LCRV has been summarized by Anderson and Ohmart (1978) and in California by Walsberg (1977). The most important aspect of this species' biology is its close association with mistletoe. This involves adaptations for specialized feeding

FIGURE 32. Immature Northern Shrike at Topock, 15 January 1982. (Photo by G. H. Rosenberg.)

and dispersal of the mistletoe seeds (Walsberg 1975). In fact, there is probably no better example in North America of such a strong interdependency between a bird and plant species. This is reflected in their habitat selection and diet. However, this specialization also makes the species vulnerable to habitat loss and climatic fluctuations.

The largest remaining breeding population in the LCRV exists in the large mesquite bosques near Poston and Ehrenberg. In 1977 it was estimated at close to 1,500 pairs. Historically, this population was much larger, and even this last stronghold is steadily declining as mesquites are cleared for agricultural development. In winter 1978–79 this species suffered a drastic decline, when a severe freeze in late December destroyed most of the mistletoe berries in the northern two-thirds of the valley. A large influx was noted that winter in the Yuma area where, however, little nesting habitat remains. Breeding populations north of Ehrenberg remained reduced for the next five winters but recovered completely by 1983–84. The mistletoe crop had recovered somewhat by 1982–83 but was still well below pre-1978 levels (see Chapter 3).

HABITAT Highest densities are in honey mesquite woodland and desert washes with mistletoe (up to 70 birds/40 ha). They occur elsewhere mainly as transients, although summering individuals are often found adjacent to the river in cottonwoods and other tall trees.

BREEDING The nest is usually placed in a mistletoe clump. One brood is raised in March and April. Recently fledged young are seen in late April or May, and occasionally into early June and very rarely into early July. Breeding activity is well timed with the emergence of flying insects during early to mid-March (Walsberg 1977).

FOOD HABITS As determined from 143 stomach samples, mistletoe berries formed the bulk of this species' diet, even in fall when unripe berries were consumed. In spring, as mistletoe berries become scarcer, wolfberry becomes an important food. In late spring and early summer, insects form as much as 37% of the diet, with beetles, flies, bugs, and caterpillars taken most frequently.

Northern Shrike (*Lanius excubitor*)

STATUS A casual visitor in winter. Four records: one immature seen 21 November 1977, one collected 9 January 1978, and one seen 23 December 1981, all near Poston; and an immature at Topock, 15 January 1982 (Fig. 32).

These are among the southernmost occurrences of this erratic visitor from the far north. All records are from agricultural land, along riparian or other brushy margins.

Loggerhead Shrike (*Lanius ludovicianus*)

STATUS A common winter resident throughout the valley; also a fairly common breeder during spring in drier habitats. Max: 194; 20 December 1978, Parker CBC.

As a breeding bird in our region, it is more numerous in the open Sonoran Desert surrounding the LCRV than in the riparian floodplain. The agricultural lands that now fill much of the valley are entirely suitable for nonbreeding shrikes, and an influx is noted as soon as the young birds disperse in July. Our local population (*L. l. excubitordes* Swainson) is augmented by migrants from farther north (*L. l. gambeli* Ridgway), as determined by racially identifiable specimens.

Although this shrike has declined in many parts of its range, populations in our area have been stable in recent years. It is possible that use of agriculture habitats in the nonbreeding season has served to increase winter survivorship, perhaps offsetting some loss in breeding productivity. Direct comparisons with historic numbers in the Southwest are not possible, however.

HABITAT This species breeds mainly in sparse riparian woodland and desert washes. It is more widespread in winter in other open habitats, especially agricultural land.

FOOD HABITS Shrikes have a very diverse diet, as evidenced by seven stomach samples containing rodents, lizards, small birds, beetles, grasshoppers, and wasps. These prey are captured either in aerial pursuit or by pouncing to the ground like a kestrel or Say's Phoebe.

European Starling (*Sturnus vulgaris*)

STATUS A common breeder and abundant winter resident throughout the valley, often found in large flocks. Max: 6,830; 21 December 1983, Parker CBC.

Perhaps the most unwelcome, first Arizona specimen from the LCRV was of this exotic species, at Parker on 16 November 1946 (Monson 1948b). It was only a winter visitor until 1958; it soon became a widespread breeder and is still probably increasing in numbers.

HABITAT Starlings breed in towns and riparian woodland with tall trees and in adjacent saguaros. They are widespread in agricultural and urban areas at other times.

BREEDING Nests are in cavities in trees, saguaros, or buildings. Two to three broods are raised in a season, making this recent invader more productive than most native cavity-nesting species.

Starlings are extremely aggressive at nest holes, often driving away other species to claim the nest cavity. Only the equally aggressive Brown-crested Flycatcher may successfully defend its nest sites against starlings.

FOOD HABITS Eleven arthropod and two plant taxa have been identified in seven stomach samples. Arthropods formed 63% and plants 37% of the bird's diet. Weevils, caterpillars, and ground beetles were the most common insects eaten. Starlings are ground foragers, probing their sharp bills into the soil to capture prey.

Bell's Vireo (*Vireo bellii*)

STATUS At present a rare to locally uncommon summer resident and breeder between late March and late September. The earliest arrival date is 8 March ("many" in 1910; Grinnell 1914) and the latest departure dates are from late November. Four winter records: 7 February–7 March 1951 at Topock (spec), 18 December 1978 and 24 December 1981 at Parker, and 15 December 1984 near Laguna Dam.

FIGURE 33. Singing Bell's Vireo, one of the most severely threatened breeding birds in the lower Colorado River Valley. (Photo by K. V. Rosenberg.)

The Bell's Vireo (Fig. 33), until the 1950s, was quite abundant and much more widespread. In fact, Grinnell (1914) called this vireo "one of the most characteristic avifaunal elements in the riparian strip." Its near elimination as a common breeding resident has been attributed to a combination of both loss of preferred willow habitats and increased pressure from parasitism by Brown-headed Cowbirds, concomitant with agricultural development (discussed below). Recent population estimates, based on our inventories, indicate that there was a 57% population decline between 1976 and 1986 (203 to 88 individuals). Most of the decline occurred after heavy floods in 1983 (see Chapter 2). This is the most dramatic drop seen in any riparian breeding species besides Yellow-billed Cuckoo during this 10-year period.

Breeding during 1974–84 occurred only in the vicinity of Needles, Topock Marsh, BWD, and Cibola NWR, with a few scattered pairs near Parker and Poston. Most of the habitat, along with the breeding birds, at Cibola NWR disappeared after the 1983–86 flooding. A small population nested at the southern tip of Nevada in 1975 and 1976, but no later. The last definite evidence of successful breeding south of Cibola was of a pair at Laguna Dam in June 1974. A few transients still appear in other areas, as at Laguna Dam and Yuma.

Because of the dramatic decline of this subspecies (*V. b. arizonae* Ridgway), it was proposed for federal listing in 1981 as Endangered. The listing process was complicated, however, by the presence of stable populations of this subspecies in central and southeastern Arizona and in northern Mexico. In these areas, they are found to breed in high numbers in a variety of riparian habitats, including desert washes. Presently, the California Department of Fish and Game is the only agency to list this subspecies as Endangered. At present, there is little hope of preventing its complete extirpation from the LCRV.

HABITAT   Early accounts (e.g., Grinnell 1914) indicate that this species was most common in willow habitats, where it occupied the understory shrubs such as "guatemote" (seepwillow). Most remnant LCRV populations breed only in tall screwbean or honey mesquite woodlands near water. These may be mixed with scattered willows or saltcedars and must have a well-developed shrub layer. Along the Bill Williams River (where the largest remnant population persists) and near Needles, large stands of recently regenerated willows (mixed with screwbean mesquite) are used (Serena 1986).

Although it is very specialized in habitat use in the LCRV, this species becomes a broad habitat generalist east of the valley. In fact, the Bell's Vireo occurs in higher densities in honey mesquite and saltcedar than in cottonwood–willow habitats at higher elevations (Hunter

1988). The recent spread of this species into the Grand Canyon is apparently concomitant with the invasion of saltcedar, a habitat not used to any extent in the LCRV (Brown et al. 1983; Brown and Trosset 1989).

BREEDING The nest is a hanging cup in a willow, mesquite, or understory shrub. Typical clutch size is four eggs, with possibly two broods raised from May to August. This species is a frequent host of the Brown-headed Cowbird, which may have expedited its demise.

Serena (1986) found that Goodding willow was the most important plant contributing to cover around nest sites in 18 of 35 breeding territories. Mesquite and saltcedar usually ranked second or third in importance. In all territories willows occurred in small patches and were interspersed with other plants. This supports the view that willows are still widely selected over other types of vegetation, and that the absence of willows contributes largely to this vireo's rarity in the LCRV.

Repeated claims that Brown-headed Cowbirds "caused" the declines of many open-nesting passerines, including the Bell's Vireo, often do not consider other factors that may affect these species' populations. Brown-headed Cowbirds have been considered an abundant bird along the lower Colorado River since at least Cooper's time (1869). Heavy parasitism of Bell's Vireo has been documented since at least 1900 (Brown 1903). Serious declines in Bell's Vireo (and Yellow Warbler) populations were not noted, however, until the 1950s. This was a period of massive habitat destruction and cessation of flood-induced regeneration. Apparently these species coexisted with cowbirds until severe habitat fragmentation exacerbated the effects of sustained nest parasitism. In theory, the local extirpation of isolated populations is compounded by a lack of colonists from other similarly affected areas, eventually leading to an overall population crash. This process is continuing today in species such as Yellow-billed Cuckoo and Summer Tanager, which are not heavily parasitized by cowbirds. Therefore, it is most likely that cowbird parasitism only served to hasten the demise of those species most susceptible, but may not have caused such a response without concomitant habitat loss. In any case, cowbirds remain a serious threat to the few Bell's Vireos that struggle to persist in the LCRV today.

Gray Vireo (*Vireo vicinior*)

STATUS A casual transient with four sight records: 29 April 1952 south of Needles; 6 September 1976 near Yuma; 1 October 1977 north of Ehrenberg; and 9 May 1983 at Cibola NWR.

Although a local breeder in the mountains of Arizona and southwestern California, this vireo is almost unknown as a migrant away from these areas. This species also winters in small numbers within 30 km of the Colorado River in southwestern Arizona.

Solitary Vireo (*Vireo solitarius*)

STATUS A fairly common transient from early April to late May and from late August through October, uncommon until mid-November. Also a rare but regular winter resident north to BWD.

   Two distinct races occur in our region, and these have been considered to be separate species by some authors. The breeding form in Arizona, *V. s. plumbeus* Coues, has expanded its range into the mountains of southeastern California in recent years, after first appearing in that state, near Needles, on 26 November 1960. It and *V. s. cassini* Xantus, of the Pacific Coast, presently occur in roughly equal numbers in the LCRV, during both migrations and winter. For example, of about 31 records of wintering birds between 1975 and 1987, 13 were *V. s. plumbeus*, 8 were *V. s. cassini*, and the remainder were unidentified to race. In contrast, Monson and Phillips (1981) comment that *V. s. cassini* is the most common of the two in winter in southern Arizona. Our records as yet do not indicate any differences in the timing of migration of the two races. *V. s. cassini* has occurred from 28 August (1976) to 17 May (1978), and *V. s. plumbeus* has occurred from 25 August (1982) to 26 May (1978). It is not uncommon to find both together in the same area.

HABITAT Wintering individuals are usually found in cottonwoods at BWD or around human residences. Migrants may appear in any tall vegetation.

Yellow-throated Vireo (*Vireo flavifrons*)

STATUS A casual transient. One specimen, 10 October 1953 at BWD.

   This species has been found casually throughout Arizona and southern California, although most records have been in spring.

Hutton's Vireo (*Vireo huttoni*)

STATUS A casual or very rare fall transient and winter visitor in November and December, with four records from BWD (spec, 24 November 1953), one from Parker (22 December 1980), and two from opposite Picacho (spec, 13 November 1960). Also, one seen 5 May 1980 in Dome Valley, northeast of Yuma.

   The LCRV specimens belong to the race *V. h. stephensi* Brewster, which breeds in the mountains of southeastern Arizona, west to near Prescott, and is very rarely seen in the desert lowlands. Whether these records represent a small but regular migration or erratic local dispersal is unclear.

Warbling Vireo (*Vireo gilvus*)

STATUS A fairly common to common spring and fall transient with very protracted migration periods. Occurs between mid-March and mid-June, and from mid-July to mid-November, but seen regularly only from April to mid-May, and from early August to mid-October. Unseasonal sightings include: 28 June 1977 near Poston, 17 June 1978 at Cibola NWR, and 19 December 1978 at BWD. Max: 40; 5 September 1981, Ehrenberg to BWD.

This is one of the most numerous of our landbird migrants (see Fig. 12, p. 75). As with many other species that breed in the nearby Arizona mountains, the local breeding race (*V. g. brewsteri* (Ridgway)) is not known to migrate through the LCRV. Instead our migrants come from the Pacific Coast (*V. g. swainsoni* Baird) and Great Basin (*V. g. leucopolius* Oberholser) regions, with individuals from the former region perhaps passing through earlier in spring (according to Grinnell and Miller 1944).

HABITAT This vireo is most often encountered in tall riparian woodlands, especially willows, and in other tall trees.

Red-eyed Vireo (*Vireo olivaceus*)

STATUS A casual or very rare transient and summer visitor. Five records: 28 May 1956 at Imperial NWR; 5 June 1964 at Imperial Dam; a singing male at Blythe, 23 June–6 July 1976; 7–10 September 1977 at Blythe; and 28 June 1978 at BWD.

The singing bird at Blythe was apparently territorial in a tall citrus orchard; it was seen regularly during the period indicated above and may well have been present longer. In general, this species is a rare but regular migrant throughout the lowland deserts of the Southwest, with most records in early fall.

Blue-winged Warbler (*Vermivora pinus*)

STATUS A casual transient. One specimen taken 5 September 1953 at BWD.

This species is among the rarest Eastern migrants to be found in the Southwest, with only 1 other record for Arizona, and only about 10 in southern California through spring 1987.

Golden-winged Warbler (*Vermivora chrysoptera*)

STATUS A casual transient. One male seen 8 October 1978 at BWD.

This species has been found almost annually in recent years in Arizona and California, in both spring and fall, with possible breeding occurring in central Arizona.

Tennessee Warbler (*Vermivora peregrina*)

STATUS A casual transient and winter visitor; one singing male seen 23 April 1983 and one seen 20 December 1986, both near Laguna Dam.

This species is a rare but regular transient in southern California and Arizona and may have been overlooked in the LCRV. There are a few early winter records from southeastern California and near Phoenix.

Orange-crowned Warbler (*Vermivora celata*)

STATUS A fairly common to common transient and winter resident from late August to early May. Uncommon in early August and late May; one seen 1 July 1957 at Imperial NWR. Max: 352; 18 December 1978, Parker CBC.

Although common throughout its period of stay in the LCRV, the various geographical populations of this species exhibit a more complex pattern of occurrence. This is revealed, as in the Hermit Thrush, only by the subtle characters apparent on collected specimens. The majority of the LCRV wintering birds are *V. c. orestera* (Oberholser) from the Rocky Mountain region, including Arizona. This is one of the few cases in which the Arizona breeding race of a migratory species also winters in our region. Birds from the Pacific Coast, *V. c. lutescens* (Ridgway) migrate through the LCRV and are especially numerous in early spring when they overlap with the lingering winter residents. Very few *V. c. lutescens* remain here in winter. *V. c. celata* (Say), from the Northeast, occurs rarely in migration.

Numbers build through the fall, peaking in November or December, then decline steadily through winter. A second, smaller peak is noticeable in March and April when migrants pass through the valley. The overwintering population also varies in size from year to year depending on winter temperatures, with numbers severely reduced during coldest months. The size of the winter population, in turn, influences its distribution within the valley and the variety of habitats occupied.

HABITAT This species is always most abundant in riparian willows and cottonwoods, and in very cold winters may be restricted to these habitats. Orange-crowned Warblers also frequent low shrubby patches, such as saltbushes, close to the river or along agricultural margins. In mild winters and during migration the species may be more widespread, occupying saltcedar and mesquite habitats, especially those with saltbush or other shrubs.

FOOD HABITS This warbler is a characteristic member of the small foliage-gleaning insectivore guild in the LCRV. It is often associated with groups of Yellow-rumped Warblers and Ruby-crowned Kinglets.

Most foraging (854 observations) was in dense green foliage in all habitats, although they will frequently inspect bark as well as clusters of dead leaves suspended on branches. Flowering trees and shrubs are exploited heavily when available and account for 40% (of 336 observations) of this species' winter foraging in cottonwood–willow habitats. In mesquite and other brushy habitats, this warbler feeds heavily at perennial shrubs such as saltbush and quail bush.

In BWD they feed sequentially at the blooming cottonwoods and then willows in late winter, eating nectar as well as insects. Nectar feeding is also evidenced by individuals that visit sugar-water feeders provided for hummingbirds.

The largest portion of the diet was leafhoppers and other small bugs (85% in 52 stomach samples). A variety of other arthropods was eaten, including beetles, ants, spiders, and caterpillars. A few seeds found in the stomachs were probably ingested accidentally.

Nashville Warbler (*Vermivora ruficapilla*)

STATUS An uncommon to fairly common transient from late March to early May, and from early August to early October. Also one winter record, 27–28 December 1985 at Parker Dam (photo). Max: 21; 11 April 1977, Parker and BWD.

Spring migration is earlier than in most other landbirds (see Fig. 12, p. 75), reflecting its early arrival on the breeding grounds in northern California (by late April) and the lack of a breeding population in the Rocky Mountains or farther north. The only later records are 15 May 1910 below Yuma (Grinnell) and 17 May 1978 at Cibola NWR. It has been found a few times in early winter in southeastern California and southern Arizona.

HABITAT Migrants may be found in any riparian woodland or other tall vegetation where they feed heavily at flowering trees and shrubs in spring. The wintering individual frequented blooming palo verde trees in a trailer resort.

Virginia's Warbler (*Vermivora virginiae*)

STATUS A rare but regular transient from mid-April through mid-May, and from early August through mid-September. One specimen taken 23 July 1946 at Topock Marsh and one seen 2 November 1977 at Cibola NWR were unseasonal.

Although this is a common breeding species throughout the Southwestern mountains, the bulk of its migration is east of our region. It is actually seen less frequently in the LCRV than some typically Eastern warblers, such as American Redstart and Northern Parula.

Lucy's Warbler (*Vermivora luciae*)

STATUS A locally common spring breeder from mid-March to mid-July; numbers then decrease rapidly, but a few always linger into September. Max: 50; 19 March 1978, Poston.

This is one of several species first described to science from the LCRV, by J. G. Cooper in 1861. It is common to the east across southern Arizona but barely penetrates westward into California. This species' arrival, en masse, into the valley each spring is one of the spectacular avian events of the year, coinciding roughly with the leafing out of the mesquite trees. Although the timing may vary by a week or more from year to year (earliest 7 March), the woods are filled with singing and territorial birds within four or five days. The early departure of this species in midsummer, after producing only one brood, is also unusual among Colorado River birds.

This species has become scarce in some areas due to habitat destruction, such as near Yuma. Grinnell (1914) noted that Lucy's Warblers were completely absent from Picacho to Pilot Knob and ascribed this to the massive removal of honey mesquite for fuelwood in the Yuma Valley. Even the large mesquite bosques remaining today near Ehrenberg and Poston, which were estimated to support as many as 750 pairs (1976–80), are gradually being cleared for agricultural development. This species decreased rapidly throughout the valley in the 1950s during a period of widespread habitat loss. It was virtually absent between 1954 and 1960 (Monson, pers. comm.). Why the Lucy's Warbler declined and then recovered in remaining habitat, while the Yellow Warbler and Bell's Vireo had similar declines and did not recover, is unclear (see Bell's Vireo species account). The nesting habits of the Lucy's Warbler and its ability to use habitats exposed to extreme summer temperatures (like honey mesquite and saltcedar) probably were involved in this species' recovery.

HABITAT Grinnell (1914) found this warbler to be "very closely confined to the narrow belt of mesquite," and even today the highest densities (25–30 birds/40 ha) occur in remaining honey mesquite woodlands. However, this species is now somewhat of a generalist in using the new assortment of habitats available. Moderate breeding densities (15–20 birds/40 ha) are found in saltcedar (especially athel tamarisk), screwbean mesquite, and cottonwood–willow (but not the tallest stands at BWD). Late lingering birds in August and early September are most frequently found in dense screwbean mesquite close to the river and may represent dispersing juveniles or transients. Although a common bird in the LCRV, Lucy's Warbler numbers pale compared with those in riparian habitats in central and eastern Arizona (80–150 birds/ 40 ha; Higgins and Ohmart 1981; Hunter 1988).

BREEDING This species is nearly unique among wood warblers in being a cavity nester; the only other such species is the Prothonotary Warbler. Along the Colorado River, Lucy's Warblers use unfinished or abandoned Ladder-backed Woodpecker holes and natural cavities such as knotholes and openings behind strips of loose bark. In saltcedar they use matted clumps of dead leaves for nest sites. The cavity-nesting habit largely frees this species from parasitism by Brown-headed Cowbirds, thought to be partly responsible for the demise of some other small open-nesting species. Use of cavities adds additional protection from insolation in relatively open habitats such as honey mesquite. Why this successful warbler raises only one brood in May and early June and then departs while insect abundance is still high remains a mystery.

FOOD HABITS Lucy's Warbler forages primarily by gleaning insects from live foliage (60–90% of 553 observations) or from twigs and small branches (up to 30%). They feed at flowers when available, especially in cottonwood–willow habitats in early spring (30%). The diet, based on 68 stomach samples, consists mainly of caterpillars, beetles, and leaf-hoppers, with a smaller number of bugs, spiders, ants, bees, and wasps.

Northern Parula (*Parula americana*)

STATUS A rare but probably regular transient and winter visitor with 18 records between mid-October and early June: 4 in fall, 7 in spring, and 7 of probably wintering birds. Also one breeding record. A female discovered north of Blythe on 9 June 1981 was later found feeding two fledglings.

This primarily Southeastern-breeding species appears regularly throughout the Southwest, most often in late spring. At the Clark Ranch, where the 1981 female (above) was found, a singing male accompanied by a female was seen from 22 to 31 May 1980. Also, a pair was present together on 27 May 1977 and, oddly, on 11 October 1978. These records suggest that previous nesting attempts may have gone unnoticed at this locality. There is one additional breeding record near the Arizona border in northern Sonora, Mexico.

Yellow Warbler (*Dendroica petechia*)

STATUS A fairly common transient from mid-April (rarely earlier) to early June and from early August to early October. Also a rare but regular winter resident in very small numbers. Casual in summer; formerly bred (only one recent record, 1986). Max: 29; 14 May 1977, Parker and BWD.

This species was formerly an abundant breeder throughout the valley, with local birds (*D. p. sonorana* Brewster) arriving by early April. Grinnell (1914) estimated that from one to four singing males occurred in every 0.4 ha of willow–cottonwood habitat. Given the expansiveness of the willow–cottonwood forests at the turn of the century, the total population size in the LCRV was enormous. Their disappearance was sudden, as breeding was noted commonly at BWD and Topock in 1952, but the species was absent in summer after 1955 (Monson, pers. comm.). Although loss of willow habitats and parasitism by Brown-headed Cowbirds are cited as causes of the Yellow Warbler's local extinction, its demise coincided exactly with that of the cavity-nesting, mesquite-inhabiting Lucy's Warbler, which subsequently recovered (also see Bell's Vireo species account).

In recent years, isolated singing males have occurred in June and July and there may have been sporadic attempts at local breeding. However, no populations have become reestablished, even where suitable habitat remains, such as at BWD. Most summer records come from the northernmost sections of the valley, where several singing males were present in 1977 and 1978 at Willow Valley Estates, one was at the Davis Dam residences in July 1978, and one was near Needles from 9 June to 21 July 1983. Most recently, in June and July 1986, up to five singing males were in the Needles–Topock area, three males and one female feeding a juvenile (28 May) were near Blythe, and three males were at BWD. All of these combined constituted the largest number of breeding season sightings since our study began. The sighting of a dependent juvenile provided the first evidence of successful nesting since 1955. These and several very early records of migrants, such as 19 March 1978 at Topock, suggest that individuals of the local breeding race may consistently stray westward from central Arizona where populations are still healthy. Specimen evidence of this is lacking, however.

As a migrant, this warbler is equally numerous in spring and fall, with peaks in mid-May and early September (see Fig. 12, p. 75). Most of the LCRV migrants are *D. p. morcomi* Coale (including *D. p. brewsteri* Grinnell), which originate in the Rocky Mountain region and the Sierra Nevada. Two specimens of wintering birds were also *D. p. morcomi*. Alaskan and other northern birds (*D. p. rubiginosa* (Pallas) and *D. p. amnicola* (Batchelder)) are also represented in specimens from our region. Northern birds likely make up most of the late spring migrants in late May and June.

This species has been found in every winter since 1976, from Lake Havasu City to Imperial Dam. Most records are from near Parker, however, where found on 9 of 10 CBCs (up to 4 birds in 1984). At least some of these pertain to birds found repeatedly through winter, at least to early March.

HABITAT The Yellow Warbler was formerly a characteristic breeder in the willows and cottonwoods that lined the Colorado River. Recent summering has also been in areas with many regenerating willows. Transients are found in any dense riparian vegetation, including salt-cedar, as well as other exotic tall trees. Most wintering birds have used planted trees in trailer resorts, such as at Parker, Earp, and Lost Lake.

BREEDING This species was documented to nest abundantly from mid-May to July by Cooper (1861) and Brown (1903). Brown (1903) also noted heavy parasitism by Brown-headed Cowbirds. Many active nests consisted of up to five layers, with the lower layers containing cowbird eggs.

The continuing abundance of Yellow Warblers in the early 1900s, despite a high frequency of cowbird parasitism, makes it difficult to accept that they suddenly declined to extirpation 50 years later from cowbird parasitism alone, as stated elsewhere (Monson and Phillips 1981). A more likely scenario would indicate massive loss of suitable habitat (willow–cottonwood) first, then breeding failure in replacement habitats (screwbean mesquite and saltcedar), and finally cowbird pressure in remaining stands of suitable habitat (as at BWD).

Chestnut-sided Warbler (*Dendroica pensylvanica*)

STATUS A casual transient and early winter visitor. Five records: 11 October 1978 north of Blythe; 29 November 1981 at Earp; 18 December 1978 near Poston; 22 December 1983 below Parker Dam; and 11–20 November 1988 at Parker.

This species is found annually in Arizona and California as a spring and fall transient and also somewhat regularly in December, primarily near Phoenix and the Salton Sea. Very few of these latter birds have been seen later in winter.

Magnolia Warbler (*Dendroica magnolia*)

STATUS A casual transient and winter visitor. Five records: 5 October 1949 at BWD; 11 November 1951 at Topock (spec); 12 October 1968 at Imperial Dam; 26–27 March 1975 at Earp; and 24 December 1977–24 January 1978 at BWD (photo).

As with many other typically Eastern migrants, our few records fit a consistent pattern of recent annual occurrence in California and Arizona.

Cape May Warbler (*Dendroica tigrina*)

STATUS A casual transient. One specimen taken 23 September 1924 at Laguna Dam and an immature female observed 6 October 1988 at Parker.

This species is less frequently encountered in the interior Southwest than other Eastern warblers, with only about six Arizona records. Interestingly, they feed on fruit and nectar during migration, and many records are from planted exotic trees, such as date palms where the bird at Laguna Dam was feeding.

Black-throated Blue Warbler (*Dendroica caerulescens*)

STATUS A casual transient and winter visitor. Four records: a male near Davis Dam, 15 October 1975; a female seen 10 October 1978 at BWD; a male for at least two weeks until 23 December 1978 at Yuma; and a female, 12–23 December 1987, at Lost Lake.

Unlike most other Eastern warblers that migrate through our region, this species is regular only in fall. However, there are several other winter records from Arizona.

Yellow-rumped Warbler (*Dendroica coronata*)

STATUS A common to abundant transient and winter resident between late September and mid-April. A few individuals arrive as early as late August and linger to late May. Sometimes occurs in large flocks. Max: 3,981; 23 December 1977 Parker CBC.

This is one of the LCRV's most abundant nonbreeding resident landbirds. It is second perhaps only to the White-crowned Sparrow in total numbers. In all but the coldest winters it may occupy virtually every patch of vegetation available. After infrequent severe cold spells in midwinter, this species may make a mass exodus from many parts of the valley. Recent evidence indicates that these movements represent true southbound migration rather than random dispersal (Terrill and Ohmart 1984; Terrill and Crawford 1988). This is one of few highly gregarious warbler species; flocks may associate with other insectivorous birds, including American Pipits and Western Bluebirds.

As in many other Arizona birds, the local breeding race, *D. c. memorabilis* (Oberholser), is rare or absent in winter. The LCRV birds are *D. c. auduboni* (Townsend), originating in the northern Rocky Mountains. The few Myrtle Warblers that winter in our area are *D. c. hooveri* McGregor from Alaska, rather than the Eastern race.

The Myrtle Warbler is a rare but regular winter resident, seen primarily from mid-October to March. Most records are of single birds in flocks of other Yellow-rumped Warblers, and it is likely that nearly every large flock contains at least one Myrtle. As many as 10 individuals have been carefully identified on a CBC (1980 at Parker).

HABITAT Overall, this species is an extreme habitat generalist. It occurs in virtually all riparian woodlands, in towns, marshes, brushy agricultural land, and even on open desert mesas.

Densities are always highest in tall cottonwood–willow groves, such as at BWD, and lowest in honey mesquite woodland. This species does range into agricultural areas far from riparian habitats. Agricultural habitats typically used include isolated inhabited areas, canal banks, field margins, and occasionally even cotton fields. Some flocks may roost at night in extensive cattail marshes, such as at BWD, and range out by day to forage in other habitats.

FOOD HABITS The diversity of habitats occupied by this warbler is paralleled by its diverse foraging habits, as evidenced by 2,042 observations. It is highly opportunistic, taking advantage of the seasonal availability of live foliage, flowers, or emerging insects. In general, gleaning from green leaves was the dominant feeding mode in all riparian habitats in fall (48–83% of 591 observations). Later in the season, foraging behavior diverged among the various habitats. In the cottonwood–willow forests at BWD, the birds first invaded the blooming cottonwoods in late January. They then went into the blooming willows in February, when >50% (of 730 observations) of foraging was at flowers (presumably they eat both nectar and insects). In contrast, birds in honey mesquite gleaned the leaves of perennial shrubs all winter. Also, up to 70% (of 132 observations) of foraging in spring was on the ground. Flycatching for aerial insects was also consistently employed in all habitats but was more common in spring than in fall, especially in saltcedar and screwbean mesquite (57–80% of 100 observations).

Diet selection is also highly varied, with 11 arthropod orders represented in 195 stomach samples. Small leafhoppers and aphids were the most important food in both riparian (134 stomachs) and agricultural (61 stomachs) habitats across seasons (23–47% of diet). Caterpillars were most frequently eaten in fall (25% of stomach samples), reflecting heavy foliage gleaning at that time. Similarly, the abundance of flies (20%), small bees and wasps (15%), beetles (21%), and ants (8%) in the spring diets (57 stomachs) corresponded with the increased aerial, flower, and ground foraging observed. Spiders, moths, bugs, and caddis flies also were consumed and a few weed seeds were found in the 103 winter samples.

Black-throated Gray Warbler (*Dendroica nigrescens*)

STATUS An uncommon transient from late March to early May, and in August; fairly common in September to mid-October. Also a rare but regular winter resident, primarily at BWD. Max: 25; 1 October 1982, Blythe.

This species is much more numerous in fall than in spring, when fewer than 10 are usually encountered in a given season (see Fig. 12, p. 75). Most spring records are in mid-April, with the latest being 13 May (1979). In fall, the earliest sighting was 8 August (1978) and it is most numerous after mid-September. The extremes of late fall and early spring migration are difficult to determine because of the presence of wintering individuals. There have been more than 30 such individuals found since 1972, including 11 birds counted at BWD and Parker Dam on 24 December 1977.

Most migrant and wintering individuals are referable to *D. n. nigrescens* (Townsend). The locally breeding *D. n. halsei* (Giraud) is known to nest as close as the Hualapai Mountains but is barely known as a lowland migrant throughout Arizona (Monson and Phillips 1981).

HABITAT Most records are from riparian woodland, especially cottonwoods, and other tall trees such as athel tamarisk and pecans. A few are also found regularly in desert washes, even in winter.

Townsend's Warbler (*Dendroica townsendi*)

STATUS An uncommon to fairly common transient from mid-April to late May and from August through mid-October. Four winter records: 22 December 1975 north of Yuma; 13 January 1977 at Imperial NWR; 2 on 23 December 1977 near Vidal; and 26 January 1978 at Parker. Max: 9; 9 May 1977, Parker and BWD.

Unlike the preceding species, the Townsend's Warbler is usually more numerous in spring than in fall, although normally, only 10–20 birds are noted each season (see Fig. 12, p. 75). Extreme dates of occurrence are from 8 April (1979) to 3 June (1980) and 1 August (1977) to 30 October (1953, BWD).

There are a few additional winter records of this species in southeastern California and Arizona. Our region, however, lies about midway between the two disjunct winter ranges, along the central Pacific Coast and in the mountains of western Mexico.

HABITAT This warbler occurs in tall riparian woodlands and other tall trees, especially athel tamarisk and pecan orchards.

Hermit Warbler (*Dendroica occidentalis*)

STATUS A rare to uncommon transient from late April to late May and mid-August to mid-October.

Nearly all spring records fall within the short period of 20 April–14 May, with two later records on 23 May 1951 (2) at Topock and 23 May

1985 near Yuma. Extreme fall dates are more widely spaced from 31 July (1977, an immature at BWD) to 25 October (1958 at Martinez Lake) and, exceptionally, 28 November (1981, an adult male north of Blythe). This species is about equally frequent in spring and fall, with four to eight individuals normally noted each season. The bulk of the Hermit Warbler's fall migrations through the Southwest occurs in the high mountains of central and southeastern Arizona.

HABITAT Most records are from cottonwood–willow groves, large mesquites, and pecan orchards.

Black-throated Green Warbler (*Dendroica virens*)

STATUS A casual or very rare fall transient. Six records: 21 October 1952 (spec) and 18 October 1953 at Parker; 30 October 1952 (spec) and 11 November 1978 at BWD; 23 October 1978 at Cibola NWR; and 14 October 1979 at Blythe.

These records fit a pattern of recent, consistent occurrence in late fall throughout the Southwest.

Yellow-throated Warbler (*Dendroica dominica*)

STATUS Casual transient and winter visitor. One photographed at Ehrenberg, 5 September 1982, and one specimen collected at Needles, 28 February 1984.

Both of these birds were the more "western" race *D. d. albilora* Ridgway, as are virtually all records of this Southeastern species in the far West. A majority of these records have been in spring. There are only two other winter records for the southwestern United States.

Grace's Warbler (*Dendroica graciae*)

STATUS A casual winter visitor. One seen 27 February 1977 at the southern tip of Nevada.

Although very rarely seen away from its montane breeding range in Arizona, this species has been found to winter occasionally in coastal southern California in recent years. Interestingly, the LCRV bird was found in an isolated cottonwood oasis alongside a wintering Painted Redstart, a species with which it commonly occurs in summer.

Prairie Warbler (*Dendroica discolor*)

STATUS A casual winter visitor or transient; one seen 25 December 1981, south of Parker Dam.

This species is a very rare but annual migrant along the southern Pacific Coast in fall. However, the few sightings in Arizona are for winter.

Palm Warbler (*Dendroica palmarum*)

STATUS A casual transient. Four records: 22 September 1942 at Ferguson Lake, north of Imperial Dam; 9 October 1979, south of Davis Dam (Nevada side); 22 December 1980, south of Parker; and 17 December 1988, Yuma CBC.

This species is a regular migrant along the Pacific Coast, but it is much rarer in the interior Southwest. There are several other winter records from Arizona and southeastern California, plus an exceptional sighting of 10 together in the Colorado River Delta at El Gulfo, Mexico, on 4 December 1983.

HABITAT Unlike most other migrant warblers, it is fond of open brushy habitats and has been found at agricultural margins and in open desert.

Bay-breasted Warbler (*Dendroica castanea*)

STATUS A casual transient. One seen 9 October 1978 at BWD and a female seen 11 June 1986 near Bullhead City.

This is another long-distance Eastern migrant that has been found to stray regularly to the Southwest, a pattern to which our local records contribute.

Blackpoll Warbler (*Dendroica striata*)

STATUS A casual transient. Five records: a male at West Pond, 15 May 1955; a female, 23 June 1977, at Imperial NWR (California side); a singing male, 10 May 1979, at Blythe; one 11 October 1980 at Parker (photo); and a male seen, 29 April 1983, at Cibola NWR.

Because most records of this species along the West Coast are in fall, it is somewhat surprising that most of our records are from spring. The fall of 1980 was exceptional for this warbler in the Southwest, with at least seven other sightings in Arizona. The 1955 record was the first for California.

Black-and-white Warbler (*Mniotilta varia*)

STATUS A rare but regular transient and winter visitor between early September and late May.

Of roughly 30 records between 1976 and 1983, half were of wintering birds, 11 were fall migrants, and 4 were in May. As with several other rare wintering species, individuals are rarely detected after early January, suggesting either low winter survivorship or midwinter dispersal to warmer regions. At least one bird remained at Laguna Dam until 9 April 1966.

HABITAT This species has been found in tall riparian woodland, especially willows, and also occasionally at planted trees around houses.

American Redstart (*Setophaga ruticilla*)

STATUS A rare but regular transient and winter resident, with records for every month except July and perhaps March. Max: 3; 16 September 1978, BWD.

This primarily Eastern species migrates in small numbers throughout the western United States and has nested as close as Prescott in west-central Arizona. There are at present over 60 records for our region, all but 22 since 1976. About 23 of these are of wintering birds between late November and February, 30 were fall transients from 17 August to 16 November, and 11 were spring migrants in May. Two additional summer records, 24 June 1973 below Parker Dam and 25 June 1977 north of Blythe, were possibly of late spring migrants. Most sightings, in all seasons, are of females or immatures. However, one adult male (presumably the same) returned to a small patch of willows near Imperial Dam for three winters in a row (1980–82).

HABITAT This species is found primarily in tall riparian woodland and other tall trees, especially cottonwoods planted around human residences.

Prothonotary Warbler (*Protonotaria citrea*)

STATUS A casual transient. Two sight records: 10 May 1977 at BWD and 8–16 October 1979 north of Blythe (Clark Ranch).

The first of these was part of a large wave of migrant warblers at BWD that also included a Worm-eating Warbler, Ovenbird, American Redstart, and Virginia's Warbler.

Worm-eating Warbler (*Helmitheros vermivorus*)

STATUS A casual transient. Two records: a singing male seen 10 May 1977 at BWD and one photographed near Parker Dam (Arizona side), 5 September 1981.

This species has occurred more frequently in Arizona in recent years, with records annually since about 1977.

Ovenbird (*Seiurus aurocapillus*)

STATUS A casual transient and winter visitor. Six records: 17 April 1973 at Imperial Dam; 10 May 1977, 5 September 1977, 14 October 1977, and 13 June 1979 at BWD; and 23 December 1977 near Parker.

Ovenbirds were found with unusual frequency throughout the Southwest in 1977, including 10 additional records in Arizona and about 15 in southern California. It occurs annually, however, in these areas during migration, and a few have wintered.

HABITAT Most records are from dense riparian willows and saltcedar. The winter individual was in an isolated clump of athel tamarisks adjacent to agricultural fields.

Northern Waterthrush (*Seiurus noveboracensis*)

STATUS A rare and irregular transient and winter visitor from mid-August through March. Max: 4; 20 August 1977 at BWD.

Winter records include two near Parker, 20 December 1976; one at the Clark Ranch north of Blythe, 21 February–12 March 1980, possibly the same individual returning 19–23 November 1980; and one at Blythe on 1 April 1978 that likely wintered locally. About six additional records are of early fall transients, as early as 16 August 1977 at Clark Ranch. This species occurs regularly in both fall and spring throughout the interior Southwest, and will probably prove to be more numerous, especially in early fall, in the LCRV.

HABITAT This species favors wet riparian woodland and marshes, where it forages on the ground near standing water.

Louisiana Waterthrush (*Seiurus motacilla*)

STATUS A casual transient. One careful sight record 31 July–15 August 1977 at BWD.

This individual was studied repeatedly at close range as it frequented a beaver dam and adjacent wooded pond edge. The unstreaked throat, broad white supercilliary, and different calls were noted in direct comparison with a Northern Waterthrush that arrived at the same pond during the latter part of this bird's stay. It was believed heard at the same location on 20 August. There are only two records of this species in California. However, there are up to nine other early fall sightings (late July) in Arizona through 1987. This species occurs as a rare but regular winter visitor near Nogales on the Arizona–Mexico border.

Kentucky Warbler (*Oporornis formosus*)

STATUS A casual transient. One caught, 20 June 1976, north of Yuma.

This individual was examined in hand and then released after it collided with a window at night. Kentucky Warblers have been found annually in southern Arizona in recent years.

MacGillivray's Warbler (*Oporornis tolmiei*)

STATUS A fairly common transient from late March through May and late August through early October. Uncommon outside these periods with extreme dates 18 March to 3 June, and 7 August (exceptionally, 29 July in 1978) to 29 October (1978). Later records suggesting possible wintering are 27 November 1971 and 22 December 1977 at Laguna Dam and 20 December 1976 at Parker. Max: 25; 26 September 1978, BWD.

This warbler is secretive and is easily overlooked on its migrations through our area. Only an observer familiar with its loud, dry call notes will appreciate its true abundance. Spring and fall migrations are extremely protracted, but most occur in fall, usually peaking in mid-September (see Fig. 12, p. 75).

All spring and most fall specimens are from Pacific and Northwestern breeding populations (*O. t. tolmiei* (Townsend)). At least a few fall migrants are (*O. t. austinsmithi* Phillips) from the Rocky Mountain–Great Basin region, and as in many other Arizona breeding species, the local breeding race (*O. t. monticola* Phillips) is unknown in our area during migration.

HABITAT This species may be found in any riparian woodland or other dense vegetation, including taller cotton and alfalfa fields and weedy agricultural margins. They apparently prefer foraging in dense vegetation close to the ground.

Common Yellowthroat (*Geothlypis trichas*)

STATUS A locally common summer breeder and fairly common transient and winter resident. May be found year-round in extensive marshes, and occurs at other breeding sites, such as BWD, from March through September. Max: 70; 5 July 1978, Imperial NWR.

Although the breeding race is the same that winters, there is much evidence of local movements, with the local breeding population possibly being partly or wholly migratory. Grinnell (1914) called these birds *G. t. scirpicola* (Grinnell), although Phillips et al. (1964) consider this to be part of the more widespread Western race *G. t. occidentalis* Brewster. Grinnell noted more Northern migrants ("*occidentalis*") between 23 March and 12 April 1910. Phillips et al. (1964) have determined three additional Northern races to occur, including a winter specimen of *G. t. campicola* Behle and Aldrich from Yuma, 28 December 1902; an early fall migrant *G. t. arizela* Oberholser (from the northern Pacific Coast), 28 August 1902; and a spring migrant *G. t. yukonicola* (Godfrey), 3 April 1930. The breeding race found in most of Arizona away from the LCRV, *G. t. chryseola* van Rossem, does not reach the LCRV during migration or winter.

Recently, transients often have been noted in upland and agricultural habitats in March, including a large influx near Parker on 22–23 March 1977. This period coincides with an influx of singing birds into suitable marshes, with additional breeders continuing to fill flooded riparian habitats, such as at BWD, later in April. According to Phillips et al. (1964), most wintering individuals are males. Thus most females and immatures may migrate out of the valley as transients. A more thorough study of specimens from throughout the year is necessary to disentangle the complicated status of this species along the Colorado River.

An interesting comparison may be made between this species and the ecologically similar Marsh Wren in the extensive marshes and riparian groves of BWD. Here the yellowthroat is resident (at least a few) in the open marshes along the periphery of BWD. They increase in these marshes in early spring, but the highest breeding densities occur in the partially flooded cottonwoods and willows in late spring and summer. In contrast, Marsh Wrens are common residents and breeders only in open marshes. Marsh Wrens occur in flooded cottonwood–willow groves as abundant winter visitors, but none remain there to breed.

HABITAT Common Yellowthroats nest in both marshes and flooded riparian woodland, the key requirement being emergent vegetation. In winter they are restricted to extensive marshes, although migrants may appear in any dense, low vegetation, including agricultural fields and patches of quail bush.

BREEDING Nests are located on or near the ground in emergent vegetation such as cattails and bulrushes. Two broods are raised, with fledged young noted at BWD in late June and again in late July.

FOOD HABITS Twelve arthropod taxa were identified from 13 stomach samples. These included flies, ants, spiders, assassin bugs, and Lepidoptera. Based on 54 observations, this species forages primarily on leaves (e.g., cattail blades) and branches close to the water and may pick insects directly off the wet ground.

Hooded Warbler (*Wilsonia citrina*)

STATUS A casual transient or summer visitor. One female seen 29 June 1979 at BWD.

Most records of this Southeastern species in Arizona are in midsummer and include singing birds and pairs seen together. In contrast, most southern California records are of spring migrants, with a few additional sightings in fall.

Wilson's Warbler (*Wilsonia pusilla*)

STATUS A common transient throughout April and May, and from late August through September; uncommon outside these periods from 8 March to 9 June, and from 5 August to 13 November. Rare but probably regular in winter, at least in southern parts of the valley. Max: 110; 10 May 1977, Parker and BWD.

This is perhaps the LCRV's most numerous landbird migrant, except the swallows. Unlike most other common transients, it is roughly twice as abundant in spring than in fall (see Fig. 12, p. 75). Peak passage is usually during the first half of May (up to 50–100/day), but large numbers may be seen earlier, i.e., 100 on 18 April 1979. The largest fall counts are from early September (20–30/day). This warbler was found in six of nine winters (late December) between 1976 and 1984 in the Yuma area or at Parker. In the mild winter of 1977–78, three individuals were found, with one seen as late as 19 February at Martinez Lake.

Two races of Wilson's Warbler illustrate well the differential timing of migration between populations breeding along the Pacific Coast and those from the western interior of North America (Phillips et al. 1964). In general, the Pacific race, *W. p. chryseola* Ridgway, migrates earlier than the interior birds, *W. p. pileolata* (Pallas). In spring, all specimens before 19 April are *W. p. chryseola* and all those after 2 May are *W. p. pileolata*. In fall, all migrants before the last week of August are probably *W. p. chryseola*, although both races may be equally common later in the season. We have no winter specimens from the LCRV, but the one Arizona winter bird was *W. p. pileolata*.

HABITAT This species may be found in any riparian woodland, as well as at other trees and shrubs in towns or agricultural areas.

FOOD HABITS Seven arthropod taxa were identified in nine stomach samples, dominated by bees, wasps, beetles, and caterpillars. This warbler forages actively, frequently flycatching or sallying to leaves and terminal branchlets.

Painted Redstart (*Myioborus pictus*)

STATUS A casual or very rare transient and winter visitor. Four records: 16 April 1975 at BWD, 25–27 February 1976 at the southern tip of Nevada (spec), 14–16 September 1978 at Ehrenberg, and 31 August 1979 at Imperial Dam (California side).

This montane species nests as close as the Hualapai Mountains and possibly in the mountains of southern California. It is rarely detected as a migrant in the adjacent lowlands, however. A few individuals winter occasionally in southeastern Arizona canyons and on the southern California coast.

Yellow-breasted Chat (*Icteria virens*)

STATUS A fairly common but local summer breeder from late April (earliest, 17 April 1979) to mid-September; a few lingering individuals or transients appear through mid-October. Max: 46; 13 July 1978, Imperial NWR.

Grinnell (1914) considered this to be one of the five most abundant species breeding in the cottonwood–willow forests of the LCRV, along with Bell's Vireo, Yellow Warbler, Summer Tanager, and Brown-headed Cowbird. Only the cowbird has increased in abundance since then. The Yellow-breasted Chat appears to be persisting, probably because it has spread into habitats other than cottonwood–willow. This species has declined 30% since 1976, however, from a population estimate of 600–1,000 individuals. Most of the decline can be attributed to habitat loss due to flooding from 1983 to 1986. It is listed as a Species of Special Concern by the California Department of Fish and Game, although there is little danger of its imminent extirpation from the LCRV.

HABITAT Chats are most numerous in tall riparian woodland, especially willows, with a well-developed shrub layer. They are one of the few migratory, midsummer-nesting species that can be found in mature saltcedar habitats in the LCRV.

Chats have been increasing in recent years in habitats other than cottonwood-willow. For example, since 1979 they have been detected with increasing frequency in local areas where they were absent from 1972 to 1978, especially in saltcedar and screwbean mesquite woodlands. In addition, extensive work at Cibola NWR from 1981 to 1983 revealed that chats were also more numerous in mixed saltcedar–honey mesquite habitats with an arrowweed understory.

Therefore, a cottonwood–willow overstory appears to be important in attracting large numbers, but apparently is not a necessity for chat occurrence along the lower Colorado River. Densities in saltcedar in the LCRV, however, still pale in comparison with densities in this habitat at higher elevations in eastern Arizona (5–10 birds/40 ha versus 60–90 birds/40 ha; Hunter 1988).

BREEDING Nests usually are placed low in thickets of willows or other shrubs, often overgrown with tangles. Breeding extends from May to August; however, very little is known about the breeding biology of this secretive bird in the LCRV.

FOOD HABITS Four stomach samples were dominated by cicadas, shorthorned grasshoppers, ants, and beetles.

Hepatic Tanager (*Piranga flava*)

STATUS A casual transient and winter visitor. Recorded: 18 November 1960 (spec) at Imperial NWR; 19 December 1973 and 28 December 1974 at Imperial Dam; 27 March–12 April 1975 at Parker Dam; and 2–5 April 1979 at Blythe.

The distribution of this montane species is similar to that of the Painted Redstart, but the tanager is even less frequently detected in the desert lowlands during migration. It is possible that most of our records represent locally wintering birds. This species has wintered near Phoenix and does so somewhat regularly in southeastern Arizona.

Summer Tanager (*Piranga rubra*)

STATUS A rare to uncommon summer breeder from late April to early October. Formerly much more abundant, especially at BWD, until the early 1980s. There are a few late fall and winter records such as 21 January 1944 near Needles, 19 December 1977–8 February 1978 and 11–13 November 1978 at BWD, and 9 April 1966 at Bard (possibly wintering).

The earliest arrival of a breeding bird was 13 April (at Bard in 1915). The latest migrant was 22 October at Imperial NWR in 1956. The breeding race is *P. r. cooperi* Ridgway, which is found throughout the Southwest. The only race known to winter in the southwestern U.S. is *P. r. rubra* (Linnaeus) from the East. Phillips also separates birds from the adjacent Big Sandy River and possibly the northern part of the LCRV on the basis of strongly ochraceous plumage in immatures (*P. r. ochracea* Phillips; *in* Monson and Phillips 1981). We have seen a few such individuals at BWD in fall.

Cooper (1861) considered this species to be common throughout the LCRV, north to Fort Mohave. Grinnell (1914) listed the Summer Tanager as among the most characteristic species of the willow–cottonwood association. By 1976, numbers had declined tremendously, with only 216 individuals estimated to occur in the LCRV. Nearly half of these were restricted to BWD. Small populations were also known in the mid-1970s from Havasu NWR (Topock Division), Cibola NWR, Imperial NWR, and between Imperial and Laguna Dams.

Only 5 birds were found in field surveys in 1983 on the California side. This was well below the 46 individuals estimated to occur, based on available habitat (Hunter 1984). Habitat losses associated with the flooding of 1979–81 at BWD and 1983–84 along the lower Colorado River mainstream resulted in a 36% decline, with 138 individuals estimated for the entire valley. Population densities in cottonwood–willow habitat at BWD dropped from 16 to 24 birds/40 ha in 1976–78 to 6 to

10 birds/40 ha in 1980–83. All of this points to the eventual disappearance of this species from riparian habitats along the Colorado mainstream. A field survey in June 1986 located only 3 males on the California side, 6 males on the Arizona side, and up to 13 males at BWD (Laymon and Halterman, pers. comm.), further indicating the severe decline of the species since the mid-1970s.

Hunter (1984) recommended that this species be considered expeditiously for Threatened or Endangered status in California. However, the Summer Tanager presently remains listed as a Species of Special Concern, providing no legal protection by the California Department of Fish and Game. The only other large population in California is found along the Kern River (Remsen 1978). Arizona does not include Summer Tanager on its Threatened Native Wildlife List as it remains common at higher-elevation riparian habitats in central and southern Arizona.

HABITAT Summer Tanagers prefer structurally well-developed cottonwood–willow stands in the LCRV. Here they attain densities of 20–30 birds/40 ha. It is very rare in all other LCRV habitats except a 69-ha of athel tamarisk near Topock (8–10 pairs estimated to occur). Athel tamarisks attain heights of up to 12 m (only cottonwoods average taller), indicating that height of canopy may be more important than tree species to these birds. A stand of athel tamarisk at Dome Valley, near Yuma, also attracted Summer Tanagers until this stand was cleared in 1985.

This species expands into honey mesquite and saltcedar habitats at higher elevations in eastern Arizona where midsummer temperatures are not as severe as in the LCRV (Hunter 1988). This pattern is essentially identical to that seen in other midsummer, open-nesting species, although only the tanager makes extensive use of exotic athel tamarisk in the LCRV.

BREEDING Observations of nesting were made at BWD during 1977 and 1978. Males arrived first and were extremely aggressive as they established their territories at the end of April and beginning of May. Nest building and courtship feeding took place in mid-May, and the first eggs hatched by 1 June. Even during this stage, males continued to be very aggressive around the nest site. For example, on 29 May 1978 a male tanager physically attacked a Northern Oriole and an Abert's Towhee near the nest. He then knocked an unsuspecting White-winged Dove out of the nest tree.

The first broods fledged in mid- or late June and the males immediately began to sing again to reestablish their territories. Most pairs then renested, with the second brood fledging in late July. Of seven nests found, four were in cottonwoods and three were in willow trees, 8–15 m above ground.

FOOD HABITS Summer Tanagers forage primarily for large insects in the canopy of tall riparian trees. There they most often sallied for aerial prey (41% of 131 observations), or snatched insects from the foliage (25%) or branches (25%) while in flight. They hunted very deliberately, slowly hopping from perch to perch, cocking the head, and scanning the area. Seven stomachs from BWD in midsummer contained many cicadas (43%), bees and wasps (26%), grasshoppers (11%), as well as a few spiders, beetles, flies, and bugs. We have seen this species hovering at flowering saltcedars to catch the abundant bees in late spring.

Scarlet Tanager (*Piranga olivacea*)

STATUS A casual transient. One female banded 18 October 1970 at Bard (photo).

This species occurs as a very rare migrant in both fall and spring in Arizona and southern California.

Western Tanager (*Piranga ludoviciana*)

STATUS A fairly common transient from late April through early June and in August and September. Early fall migrants arrive by mid-July and late individuals occur into mid-October. Max: 40; 5 September 1981, BWD and Parker Valley.

The species migrates later in spring than many other regular transients (see Fig. 12, p. 75), reflecting its high montane-breeding distribution and the lack of a population breeding along the Pacific Coast. It is quite rare before the last week of April, with only nine earlier records since 1975 (earliest, 5 April 1982). Numbers may still pass through at the end of May and stragglers have been noted as late as 14 June (1950 and 1976). There is barely a one-month gap after the last spring migrants pass and before first fall migrants appear. The earliest record is 14 July (1946 at BWD). At least one male remained through the summer at Parker (1953; Monson, pers. comm.). Our latest fall record is 18 October (1978 near Poston).

HABITAT This tanager is attracted to tall riparian woodland, orchards, and other tall trees. In spring, they frequently congregate at flowering trees or shrubs where bees and other insects are abundant.

Northern Cardinal (*Cardinalis cardinalis*)

STATUS A rare and local resident between Ehrenberg and BWD and recently recorded at Needles, Cibola NWR, Laguna Dam, and Yuma. Apparently spreading gradually. Max: 5; 23 December 1977, Parker CBC.

Cardinals were first discovered along the Colorado River in 1943 near

Earp and in the BWD. They were believed to have spread along the Bill Williams River from the Big Sandy Valley (Van Rossem 1946; Monson 1949). A resident population became established south to Parker, including at Earp on the California side of the river. Between 1976 and 1983, they were resident in small numbers in mesquite bosques between Poston and Ehrenberg (these were not "vagrants," as stated in Monson and Phillips 1981). Cardinals continued to be seen regularly at Parker, Earp, and BWD. In summer 1978, a major expansion was detected with a singing male at Laguna Dam, 10–23 May, and one at the north end of Imperial NWR on 10 June. This was followed by two singing males at Yuma in June and July, and one at Cibola NWR in early August. We do not know, however, if any of these pioneers established populations south of Ehrenberg.

The Northern Cardinal is listed as a Species of Special Concern in California. This species has never been common on the California side of the river and it is not likely to continue to exist in California with the present rate of habitat loss.

HABITAT On the western edge of the species' range, they are restricted to dense riparian woodland, especially honey mesquite and brush along the river edge. Clearing of this woodland now threatens this small population, especially on the California side south of Earp.

Pyrrhuloxia (*Cardinalis sinuatus*)

STATUS A casual summer visitor. A male was caught at Palo Verde on 14 July 1974, and a singing male was north of Ehrenberg on 4 July 1981. Also, a pair nested in Chemehuevi Wash (24 km west of Lake Havasu) in June 1977 and a female was resighted at that location on 14 May 1983.

Most extralimital records of this species have been in winter, when individuals disperse north and west from their southeastern Arizona breeding range. Several have reached the Imperial Valley in California. It is somewhat surprising, therefore, that our sightings have been in midsummer; these may potentially involve a very small resident population.

Rose-breasted Grosbeak (*Pheucticus ludovicianus*)

STATUS A rare but regular transient from mid-May through late June; casual in summer, late fall, and early winter.

This species was exceptionally numerous in June 1977 with about 10 reports, including 3 together at Parker on 7 June and up to 4 at the Clark Ranch from 19 to 25 June. Also, a pair is reported to have nested that summer at Lost Lake Resort near Vidal.

Other midsummer records include single males at Parker, 27 June 1953 (spec); BWD, 30 July 1977 and 9 July 1978; and a female at Clark Ranch, 27 June–3 July 1982. There are only four fall records, between 1 November (1978) and 8 December (1977). These dates of occurrence are consistent with this species' status elsewhere in Arizona and southern California. In addition, a hybrid male Rose-breasted × Black-headed Grosbeak was at BWD on 31 March 1982, much earlier than other records, but within the normal migration period of Black-headed Grosbeaks.

HABITAT This species has been found in tall riparian woodland and tall trees around houses.

Black-headed Grosbeak (*Pheucticus melanocephalus*)

STATUS A fairly common transient in April and May, and from late July through September. Spring migration extends from late March (earliest, 16 March 1978) to early June. Fall migration is from mid-July (earliest migrant, 11 July 1954 and 1979) to early October (exceptionally, 6 November 1981). Individuals noted in late June and early July may be nonbreeding summer residents. Also casual in winter; an immature male banded at Cibola NWR, 4 February 1980. Max: 20; 5 September 1981, BWD and Parker Valley.

This is yet another species for which it is difficult to precisely delimit the migration periods because of occasional birds remaining through midsummer (see Fig. 12, p. 75). For example, a male sang as if on territory in a citrus orchard at Blythe from 23 June through early July 1976. Also, a subadult male and a female were in a cottonwood grove at Yuma from 13 June until at least 15 July 1978. Other records may extend the start of fall migration to the beginning of July. This species breeds in Upper Sonoran Desert valleys in Arizona and along the Pacific Coast. Therefore, lingering individuals may occasionally be expected to attempt nesting in the LCRV.

HABITAT Migrants may be found in riparian woodland and other trees and shrubs.

Blue Grosbeak (*Guiraca caerulea*)

STATUS A common breeder from early May through September. Males first appear in late April and the latest migrants may occur through mid-October (latest, 19 October 1977). One winter specimen, 18 February 1951 at Parker. Max: 50; 13 July 1978, Imperial NWR.

For five consecutive springs (1976–80) the first singing birds were detected on either 21 or 22 April. Females are rarely seen before the first week in May, however. These arrival dates are earlier than those in central and southeastern Arizona, but are later than those farther west in California. A female seen at Imperial Dam on 30 March 1978 was much too early for a migrant anywhere in the Southwest and most likely wintered locally. Grinnell (1914) thought that his specimens from the LCRV were the large-billed race (*G. c. interfusa* Dwight and Griscom) that "differed markedly from the race summering in central and southern California west of the Sierran divide" (= *G. c. salicaria* Grinnell). However, Phillips et al. (1964) state that the Colorado River birds, including the winter specimen, are *G. c. salicaria* and not *G. c. interfusa*.

Because of its loud, vibrant song and tolerance of recent habitat alterations and disturbance, the Blue Grosbeak is a conspicuous riparian summer resident in the LCRV today. Its success is in contrast with many other summering species that have declined considerably in recent years. This success appears to be related to its broad habitat tolerances.

HABITAT   This species is a habitat generalist along the Colorado River, occurring in almost any riparian woodland, including saltcedar. Highest densities (4–6 pairs/40 ha) are in cottonwood–willow groves (especially recently burned sites) and moderately dense saltcedar. Smallest numbers breed in open honey mesquite woodlands. Key habitat ingredients for this species are a dense shrubby understory and at least a few taller trees for song perches. During migration they may appear in milo or other agricultural fields or on weedy roadsides with buntings and sparrows.

This grosbeak is unique among midsummer-breeding, migratory insectivores in being able to breed in stands of pure saltcedar. This ability allows it to remain common within the LCRV, while other midsummer-breeding, migratory insectivores have declined tremendously. Blue Grosbeaks are associated with disturbed and open second-growth habitats, habitat edges, and isolated hedgerows throughout its range.

BREEDING   Males establish territories shortly after arriving in late April and nest building takes place in May. The nest is an open cup concealed 1–3 m high in a dense shrub. Three nests found at BWD in 1977 were in dense saltcedar along the edge of a cottonwood grove. Fledged young first appear in late June and second broods may be raised in July and August.

FOOD HABITS   Despite its heavy, conical bill, this grosbeak is primarily insectivorous in the LCRV. Twenty-two stomachs collected from throughout the valley contained almost entirely short-horned grasshoppers and cicadas, with a few caterpillars, spiders, beetles, flies, and lacewings. Grass seeds were found in several stomachs. Foraging was typically around flowering annuals (46% of 39 observations), with insects also taken from live branches (21%) and the ground (18%).

Lazuli Bunting (*Passerina amoena*)

STATUS A fairly common transient from mid-April through May, and again from August through September. Extreme dates of migrants in spring are from 2 April (1978) to 5 June (1951, BWD), and in fall from 24 July (1976) to 2 October (exceptionally, 28 November 1982). One was seen on 5 and 12 July 1953 at Parker. The status of singing males noted in June and July at Topock and Blythe is unclear. One winter sighting, 18 December 1978 at Parker. Max: 200; 14 May 1977, Needles.

This showy migrant usually occurs in our area singly or in small groups with sparrows or other finches. Females especially may be easily overlooked by observers unfamiliar with the buzzy calls. The presence of singing birds in midsummer is problematical, primarily because these may represent late spring or early fall transients. However, fall migrants would not be expected to sing as if on territory, as was noted near Topock on 21 July 1976 and 24–26 July 1977. Also, a singing male was accompanied by a female at a citrus orchard near Blythe on 24–25 June 1978. At least one male sang in the same orchard from 2 June to 11 August 1979.

This species breeds at fairly low elevations in central and northern Arizona, including along the Colorado River in the Grand Canyon and, therefore, may be expected to attempt nesting occasionally in our region. It is also possible that the recent presence of Indigo Buntings in summer may lure a few Lazuli Buntings to linger farther south than usual. It is significant that in the locations where midsummer Lazuli Buntings were found, singing Indigo Buntings were also present. This situation deserves further study.

Lazuli Buntings winter somewhat regularly in southeastern Arizona and at Phoenix; however, there are no accepted winter records in southern California. The female seen 28 November 1982 at Katherine Landing above Davis Dam also may have been attempting to winter.

HABITAT This species occurs in open or burned riparian woodland, agricultural margins, citrus orchards, and other brush.

Indigo Bunting (*Passerina cyanea*)

STATUS An uncommon and local summer resident from May through August; rare transient in late April and September. Breeding confirmed only at Havasu NWR, Needles, and BWD, but is suspected elsewhere. Max: 12; 3 July 1977, BWD and Parker.

The Indigo Bunting was first recorded in the LCRV on 5 May 1974 at Yuma (spec). The species had been steadily expanding its range westward across Arizona, reaching southern Utah, Nevada, and California. By 1976, a small population was present (singing males) in recently burned sites at the Topock and BWD sections of Havasu NWR and at Cibola NWR. In June and July 1977, detailed surveys detected a minimum of 55 singing birds at these sites as well as near Needles, throughout the Parker Valley, and south to Imperial NWR. In later years, fewer birds have been found, although coverage of suitable sites also has been less complete. Nearly all records are of males, found between 28 April (1978) and 29 September (1977).

HABITAT The invasion of this species may be, in part, a response to large-scale disturbance of native riparian habitats. The largest number has been found where willows and saltcedar are partly regenerating after recent burning. Others occur along the edges of cottonwood groves, in open screwbean mesquite and arrowweed associations, other river-edge vegetation, and in citrus orchards.

BREEDING Evidence of breeding came from only three sites, all in 1977. However, this species undoubtedly breeds at most locations where territorial males are present through summer. No nests have been found, although two recently fledged young were being fed at Topock on 24–26 July, a male was feeding a fledgling north of Needles on 24 July, and family groups were noted at BWD. The presence of a few singing Lazuli Buntings at sites where Indigo Buntings are also present suggests the possibility of occasional hybridization between these two closely related species. Hybridization has been detected elsewhere in the Southwest.

Varied Bunting (*Passerina versicolor*)

STATUS A casual fall and winter visitor, with a flock of 15 seen at Blythe (two specs), 8–9 February 1914, and one seen 20 September 1952 at BWD.

Although reviewed by other authors and apparently valid, the Blythe record falls far outside any known pattern of occurrence for this species. It rarely is found in flocks, even within its normal range. However, there are several other extralimital records in fall, with the species reaching Death Valley and with one taken in the Mohave Mountains, just east of the Colorado River, 27 October 1949.

Painted Bunting (*Passerina ciris*)

STATUS A casual visitor. An immature female collected 6 November 1976, north of Ehrenberg, and another female-plumaged bird seen 13

November 1978 near Vidal. Also, a male was north of Blythe from 2 July to early October 1982.

This species wanders westward in fall, regularly to southern Arizona and irregularly to southern California. There are very few summer records and few records of males not suspected of being escapees from captivity. One was believed heard singing at Cibola NWR, 19 July 1982.

Dickcissel (*Spiza americana*)

STATUS A rare and irregular fall transient. About eight records: 26 September 1949 and 19 September 1954 (2) at Parker; 18 September 1952 at Topock; four individuals between 20 August and 25 September 1982 near Ehrenberg and Cibola NWR; and 21 August 1983 at Blythe.

This species is a regular fall transient in very small numbers throughout the Southwest. It is easily detected by its unique buzzing call note, given frequently in flight.

Green-tailed Towhee (*Pipilo chlorurus*)

STATUS A fairly common transient and uncommon winter resident. Largest influxes occur in September and early October, and again from early April to early May. Max: 24; 18 December 1978, Parker CBC.

This species has been recorded between 27 August (1950) and 23 May (1980). Numbers remaining through winter are variable from year to year and are usually highest in the southern half of the valley. It is a solitary and reclusive bird that is best detected by its distinctive cat-like calls. It is easily overlooked.

HABITAT This towhee favors the shrubby understory of open-canopied riparian woodland, especially arrowweed associations close to the river and saltbush along the edges of agricultural fields and canals.

Rufous-sided Towhee (*Pipilo erythrophthalmus*)

STATUS An uncommon winter resident from October through late March. Numbers may be highly variable from year to year. Max: 38; 19 December 1978, Bill Williams Delta CBC.

This species has arrived in the LCRV as early as 25 September (1977), but is rarely seen before the middle of October. Lingering individuals have been heard singing as late as 7 April. This species was unusually common during the winter 1978–79 (over 100 seen), but was extremely scarce during the following two winters (fewer than 10 reports each). This variation does not seem related to local weather conditions along the Colorado River and more likely represents influxes from populations normally wintering farther north or west.

The few specimens of this highly variable species from the LCRV have been identified as *P. e. curtatus* Grinnell from the Great Basin. However, unpredictable variation in winter numbers suggests that LCRV birds may originate from several regions. The breeding subspecies in Arizona, *P. e. montanus* Swarth, is not known to winter in the LCRV, although it is found wintering in central and southern Arizona and as close as the Kofa Mountains. *P. e. montanus* breeds as close to the LCRV as the Cerbat and Hualapai Mountains. A very brown female observed 23 February 1983 at BWD may have been from an Eastern race.

HABITAT This towhee is found primarily in dense understory of tall riparian woodland, especially willows, and other dense vegetation, including marshes.

Abert's Towhee (*Pipilo aberti*)

STATUS A common permanent resident and breeder throughout the valley. Often occurs in small family groups from late summer through early winter. Max: 594; 18 December 1978, Parker CBC.

Abert's Towhees are typical of permanent resident riparian species in the LCRV in exhibiting lowest overall population size and narrowest breadth of habitat use during the late winter and spring months. This species expands in both population size and breadth of habitats used from summer through fall, after breeding has been completed (Meents et al. 1981). Depending on the severity of the winter, populations decline to varying degrees but rebound again after the onset of the next breeding cycle.

Several studies have provided insights into the possible evolutionary history of this and other towhee species in the arid Southwest. The Abert's Towhee is postulated to have differentiated evolutionarily from other towhees in the vicinity of the LCRV (Davis 1951; Hubbard 1973). It shows its closest affinities in morphology, calls, and egg coloration to the California Towhees (*P. crissalis*) of the Pacific Slope (Davis 1951; Marshall 1960). The time of their divergence was estimated by Zink (1988) to have been about two million years before present, probably during a period of extreme aridity. Cottonwood–willow habitats were well represented then and probably served as important refugia for evolving towhees in an otherwise barren environment (Axelrod 1950; Davis 1951; Hubbard 1973; Brown 1982). Once differentiated, Abert's Towhees are thought to have then spread north to the Virgin River drainage in Nevada and east along the Gila River drainage where they are found today. Along the upper Gila River drainage, the Abert's Towhee comes into contact with the Canyon Towhee (*P. fuscus*).

These historical factors have resulted in the present-day distribution of towhees in the Southwest. Abert's Towhees are restricted to riparian habitats throughout their range, whereas Canyon Towhees are restricted to the surrounding desert uplands at low elevations. In the higher-elevation floodplains of eastern Arizona and southwestern New Mexico, Abert's and Canyon Towhees overlap in habitat use in mesquite woodlands and mesquite–grassland mixes (Marshall 1960).

California and Canyon Towhees have never been recorded in the LCRV or Imperial Valley, even though populations of the former occur as close as the Santa Rosa Mountains overlooking the Salton Sea, while populations of the latter occur in the Kofa, Mohave, Black, and Cerbat mountain ranges east of the Colorado River. The latter populations are thought to represent remnants from the last glacial age and are made up of extremely sedentary individuals (Marshall 1960; Phillips et al. 1964). As Abert's Towhee, which evolved in the hot floodplains, expanded its distribution and maintained gene flow along riparian corridors, upland desert Canyon Towhee populations became increasingly isolated in small montane pockets, especially in western Arizona.

The LCRV population, *P. a. aberti* Baird, is decidedly paler and more cinnamon than the only other recognized race, *P. a. vorhiesi* (Phillips), of the eastern part of the species' range. Paler plumage is a characteristic seen in most birds having distinguishable populations in the LCRV.

HABITAT Overall, Abert's Towhee is a habitat generalist within riparian areas of the LCRV. All riparian, urban, and even marsh habitats may harbor this species at any time. In addition, it also may be found a short distance from the LCRV proper along the better-vegetated desert washes.

Highest numbers are consistently found in mature, structurally well-developed cottonwood–willow groves (up to 55 birds/40 ha). Secondarily, well-developed saltcedar and honey mesquite support densities up to 30 birds/40 ha, followed in importance by saltcedar–honey mesquite and screwbean mesquite woodlands. During milder winters (1975–76 and 1977–78; see Table 5, p. 44), Abert's Towhees maintained stronger affinities to these habitat configurations than in harsher years (1976–77 and 1978–79; Meents et al. 1981). Relative breadth of habitats used during winter was always lower than during summer and did not differ among years. However, milder winters which supported higher food resources apparently resulted in less dispersal from preferred habitats.

Abert's Towhees are ground foragers but require extensive cover from a well-developed canopy layer. Within the cottonwood–willow habitat at BWD, this species occurred in areas with a dense, shaded understory (usually saltcedar). In mesquite woodlands they are fond of dense patches of saltbush or other shrubs. The cottonwood–willow habitats at BWD harbored very high numbers until floods in 1983 eliminated much of the canopy cover and inundated the ground. Densities immediately after the flood were very low (2–9/40 ha), with recovery occurring slowly after floodwaters receded three years later.

Low numbers may occur year round in all other riparian habitats, including arrowweed. Use of secondary habitats for breeding is probably determined by overwintering survival in those habitats. Fewer individuals may disperse from primary habitats in milder years, but individuals that do disperse may have a better chance of winter survival.

In addition, large concentrations may be found along riparian-agricultural edges. However, the species does not venture farther than 0.5 km into pure agricultural areas (Conine et al. 1978). An essential element along edge habitats is readily available cover, often dense stands of quail bush, as well as the abundant food provided by adjacent fields. Feeders in suburban areas may also attract this species, but, again, only where nearby escape cover is provided.

BREEDING The breeding biology of Abert's Towhee has been the subject of several studies and is, therefore, better understood than any other species in the LCRV (Finch 1981, 1983a, b, 1984; Conine 1982). Nesting towhees are exposed to extreme heat, predation, and cowbird parasitism. Finch (1983a) found that the first nests, in March, were placed on the southeast side of trees. There the nests received the highest incident radiation in the morning hours when ambient temperatures were low. Later in the season, by early June, nest orientation became more random but tended to be toward the north or northwest, which ameliorated against extremely high midday temperatures (>43°C). Shifts in nest orientation also were associated with cool winds early in the season, with late nests placed somewhat randomly with respect to wind. However, late-nesting birds almost invariably failed (approx. 95%), mostly due to predation and brood parasitism, thus making the process of nest placement late in the season almost academic (Finch 1983a, b). Therefore, weather and, consequently, nest placement early in the breeding season may have a profound effect on overall nesting success.

Finch (1981) noted at least three cases in which predators successfully disrupted a nesting attempt. She suspected that predation by coachwhips, Greater Roadrunners, and rodents may be quite frequent. A more serious problem, however, is the frequency of Brown-headed Cowbird brood parasitism.

Nests initiated from mid-March to late April were largely free of cowbird parasitism (2 of 34 nests), and early nests produced 88% of all nestlings (Finch 1983b). However, nests begun from early May to early June were heavily parasitized (58% of 36 nests) and produced only 12% of the total nestling population. Conine (1982) similarly observed a high incidence of nest parasitism, but also noted that nests in interior honey mesquite stands were parasitized more frequently than nests along an agricultural-riparian edge (50% vs. 20%). Apparently, first nesting attempts must be successful for towhees to produce any young at all, yet they continue to breed late into the season (June–July) when climate, predation, and parasitism cause almost certain failure. This pattern was also observed in the Black-tailed Gnatcatcher (Conine 1982).

Time–energy budgets during the breeding season showed that total energy expenditures are reduced in the hotter months (Finch 1984). Despite increases in day length and food resources, birds proportionally reduced their foraging activities to avoid thermal stress. They also increased perching time (males) and nest attendance (females) to reduce predation and parasitism (Finch 1981, 1983b, 1984).

FOOD HABITS This species is omnivorous, as opposed to granivorous, in the LCRV. Twenty insect and 9 plant taxa were identified from 108 stomach samples. Insects dominated all seasonal samples, with a low in winter of 73% and a high in late summer of 96%. Beetles consistently made up 20–30% of the diet year round, and 5–10% in every season consisted of ants. Seasonal foods included caterpillars in fall, winter, and spring (15–25%), and grasshoppers and cicadas in summer and late summer (up to 38%). Most seeds eaten were of Chenopodiaceae, with a few other weed and grass species represented. Although males and females select similar food items, females consume significantly larger items in all seasons. This species is primarily a ground forager in all seasons and habitat types (at least 70% of 747 observations).

Cassin's Sparrow (*Aimophila cassinii*)

STATUS A casual visitor. A male was singing and "skylarking" near Poston, 26 April 1981.

This species occurs sporadically westward into southern California after exceptionally wet winters. However, no such invasion occurred in 1981.

American Tree Sparrow (*Spizella arborea*)

STATUS A casual winter visitor. One seen 23 November 1968 at Bard; two seen south of Parker, 11 February 1977; and one seen 28 January 1981 at Hunter's Hole, south of Yuma.

These are among the southernmost records of this species, which winters regularly to northern Arizona. There are several additional lowland desert records from the Salton Sea area in California.

Chipping Sparrow (*Spizella passerina*)

STATUS An uncommon to fairly common transient and winter resident from late July to early May, often in small flocks with other sparrows. Max: 240; 21 December 1983, Parker CBC.

This species has been found in the LCRV as early as 19 July (1947) and as late as 19 May (1947). Winter numbers are somewhat variable from year to year and sometimes an influx of transients is detectable in September or October. Nearly all records are of the Western race, *S. p. arizonae* Coues, which also breeds in the Arizona mountains. However, one specimen from Yuma, 1 November 1902, was of the nominate Eastern race.

HABITAT These sparrows are generally more abundant in wooded or more mesic habitats than Brewer's Sparrows, although both species may appear in the same mixed flock. Habitats include open or grassy riparian woodland, brushy margins of agricultural fields, and citrus orchards.

FOOD HABITS This species eats both seeds and insects while in the LCRV. A sample of 26 stomachs contained primarily grass and weed seeds (about 75%) and a few caterpillars, bugs, and beetles.

Clay-colored Sparrow (*Spizella pallida*)

STATUS A casual transient. One seen with other sparrows, 4 September 1981, near Parker Dam and one at Cibola NWR, 28 August and 12 September 1982.

This species is regularly recorded elsewhere in the Southwest, especially in fall, and may be overlooked here.

Brewer's Sparrow (*Spizella breweri*)

STATUS A common winter resident from September to mid-April. First migrants may arrive in mid-August and some linger rarely to late May. Often in large mixed flocks with other sparrows. Max: 934; 20 December 1979, Parker CBC.

This is usually one of the most common wintering sparrows, often associating with large flocks of White-crowned Sparrows. It may be scarce, or at least very local, in some years, however. Small flocks may appear by the first week in September and single concentrations of >200 birds have been noted in October and November. Extreme dates of occurrence are from 12 August to 2 June (both 1977).

HABITAT These sparrows are most numerous at brushy margins of agricultural fields and desert washes. However, flocks may be found in shrubby openings of riparian woodlands, and even in open creosote bush desert along the edge of the valley.

FOOD HABITS In the LCRV, this species is primarily granivorous. In 106 stomach samples, grass and weed seeds (Portulacaceae, Amaranthaceae, and Chenopodiaceae) made up about 80% of the diet, with caterpillars, bugs, and a few beetles also eaten.

Black-chinned Sparrow (*Spizella atrogularis*)

STATUS A casual transient or winter visitor. Four records: 3 March 1910 at Needles (spec), 30 August 1955 at Imperial Dam, 7 February 1965 at Blythe, and 17 December 1973 at Mittry Lake (3).

This sparrow breeds in the mountains of southern California and central Arizona (as close as the Cerbat and Hualapai Mountains) and winters regularly in the Kofa Mountains, just east of the Colorado River. However, there are very few records in the adjacent desert lowlands.

Vesper Sparrow (*Pooecetes gramineus*)

STATUS A common winter resident from mid-September through mid-April (latest, 20 April 1986). Initially arrives in late August (earliest, 25 August). Max: 1,064; 18 December 1978, Parker CBC.

This is one of several sparrow species that has adapted very well to agricultural development in the LCRV. The maximum noted above represented a national high count among all CBCs conducted in North America that winter, indicating the potential importance of our area to this species. They are typically found in small loose flocks or scattered among larger flocks of Savannah or other sparrows. The LCRV wintering birds are the race *P. g. confinis* Baird from the Great Plains and Great Basin, distinct from the breeding birds in northern Arizona (*P. g. altus* Marshall).

HABITAT Vesper Sparrows are most common in weedy or grassy agricultural fields and margins. Some are found in sparse riparian woodland or desert flats where grass is available.

FOOD HABITS Based on 178 stomach samples, this species is primarily granivorous, eating a variety of weed seeds (Portulacaceae, Amaranthaceae, and Cruciferae) as well as grass seeds. A few beetles, bugs, and spiders were also eaten.

Lark Sparrow (*Chondestes grammacus*)

STATUS An uncommon transient and winter resident from late August to May. A few remain to breed in summer in citrus orchards at Blythe and Yuma and locally near Parker. Max: 121; 23 December 1977, Parker CBC.

Lark Sparrows were not suspected of breeding at such a lowland locality until 1976, when they were found summering near Blythe. Breeding was confirmed in 1977, and additional pairs were located at farmhouses south of Parker. In 1978, they nested at several sites near Blythe and Yuma, as well as at Tacna in the lower Gila Valley. This recent population expansion into previously unsuitable breeding areas is most likely in response to the presence of permanent irrigation water in agricultural habitats.

Locally nesting birds are present at least from April through August, obscuring the arrival and departure dates of nonbreeding migrants. Such migrants have been noted away from known nesting sites from 16 August (1954, BWD) to 12 May. Elsewhere in Arizona, definite migrants have been found much earlier in July. Wintering birds normally occur in small flocks, occasionally in association with other flocking sparrows.

HABITAT The nesting habitat appears to be exclusively citrus orchards and other cultivated vegetation at farmhouses. Wintering birds favor agricultural land with scattered trees or hedgerows. They are less likely to occur in open grassy or weedy fields far from taller cover.

BREEDING Probable breeding was first noted in 1976 at Blythe. Copulation has been observed in mid-May and recently fledged young were first noted 25 June 1977. No specific studies of the nesting of this sparrow in the LCRV have been undertaken, however.

Black-throated Sparrow (*Amphispiza bilineata*)

STATUS An uncommon and local spring and summer breeder from mid-February through August along the sparsely vegetated periphery of the LCRV. Rare or irregular in fall and early winter in the riparian floodplain. Max: 74; 21 December 1979, Bill Williams Delta CBC.

This typically common desert species is of somewhat marginal status in the LCRV proper, as well as around the Imperial Valley (Garrett and Dunn 1981). However, it is always present along the edge of the valley, sometimes in numbers, and is a fairly common to common permanent resident in the desert uplands bordering BWD. Flocks may appear in desert washes in winter, and a few may be interspersed with Sage Sparrows or flocks of White-crowned Sparrows in brushy riparian areas. This species also appears consistently as a postbreeding wanderer or migrant (mostly immatures), chiefly in September, and occasionally close to the river or in agricultural areas. This pattern is reminiscent of similar dispersal by the Rock Wren.

HABITAT In the LCRV, this species primarily nests in open, sandy, honey mesquite woodland, such as near Poston, and in desert washes extending out of the riparian zone. It is always more common in the open desert, well away from the floodplain. The low breeding season densities (3/40 ha) of this species in open honey mesquite habitats within the LCRV is in contrast with the much higher densities (35–45/ 40 ha) found in similar habitats along the upper Gila River drainage (Hunter 1988).

Sage Sparrow (*Amphispiza belli*)

STATUS An uncommon to fairly common and local winter resident from early September to late March. Often in small flocks. Max: 237; 22 December 1980, Parker CBC.

This is one of our more specialized wintering sparrows, with rather specific habitat requirements and, therefore, a patchy distribution. The loosely aggregated groups rarely mingle with other sparrows. Extreme dates of occurrence are from 5 September (1976) to 2 April (1978), both near Ehrenberg, with as many as 20 seen as early as 17 September (1976). The wintering population is a mixture of birds from the Great Basin (*A. b. nevadensis* (Ridgway)) and interior southern California (*A. b. canescens* Grinnell), although the extent to which the two races segregate or interact is unknown.

HABITAT Sage Sparrows are common only where dense patches of inkweed are interspersed with bare ground in moderately dense honey mesquite woodland or desert washes. Largest populations occur where these conditions are met north of Ehrenberg and east of Poston. This species has not been found in agricultural areas or other extensively disturbed habitats in the LCRV.

The singular importance of inkweed shrubs was demonstrated experimentally on revegetated plots. The sparrows strongly selected sites with the greatest density of these shrubs. Details of these experiments and other aspects of habitat use by this sparrow may be found in Meents et al. (1982).

FOOD HABITS A sample of 26 stomachs contained mainly seeds of Chenopodiaceae, Compositae, and Amaranthaceae, as well as a few beetles, bugs, ants, caterpillars, and grasshoppers. More insects were eaten in fall than in winter.

Lark Bunting (*Calamospiza melanocorys*)

STATUS A rare and irregular transient or winter visitor from October through February. Usually found singly in the company of sparrows, but flocks have occurred in some years. Max: 450; 19 December 1951, west of Lake Havasu.

Individuals probably reach the LCRV almost every year, primarily in late fall and early winter. The most recent flight was in 1977–78 when 74 were seen on the Parker CBC, a single flock of 60 was seen west of Poston on 25 December, and single birds were widespread from 28 October to 17 February. This flight reached well into southeastern California, where unusually wet and lush desert conditions enticed some birds to remain and breed, far from their normal range. Similar conditions existed in 1951, at the time of the high count noted above (Monson, pers. comm.). In this remarkable 1951–52 invasion, >90 were still present west of Havasu Landing on 26 April. Surprisingly, the only other spring records are those of Grinnell in 1910, including a flock of 12 near Needles on 8 March (4 male specs taken) and a female specimen at Cibola on 8 April.

HABITAT Lark Buntings occur primarily in agricultural fields and occasionally in sparse riparian woodland or desert flats.

Savannah Sparrow (*Passerculus sandwichensis*)

STATUS A common to locally abundant winter resident from mid-September to mid-April. First individuals arrive in late August and a few linger to mid-May. Max: 4,965; 22 December 1980, Parker CBC.

This sparrow has adapted remarkably well to agricultural development in the LCRV, even in areas where all trees and shrubs have been removed. It is probably the most abundant wintering species in the expansive alfalfa-producing farmland that now fills much of the valley near Poston, Blythe, and Yuma. Several hundred sparrows may occupy a single large field, although they flush singly and not in flocks. Extreme dates of occurrence are from 13 August (1976, 1977) to 22 May (1952), but the usual span of occurrence is from about 24 August to 2 May. As many as 17 have been seen as early as 29 August (1979). A very unseasonal individual was seen near Poston on 27 June 1978.

In this species, we see an excellent example of what can be learned about separate origins of individuals within a single wintering population by examining geographic variation of specimens. The bulk of the LCRV birds is a mixture of two races; *P. s. nevadensis* Grinnell from the Great Basin and Great Plains, and *P. s. anthinus* Bonaparte from Alaska and northwestern Canada. At least one late fall specimen from Yuma was *P. s. brooksii* Bishop from the northern Pacific Coast. In addition, the large-billed race from Baja California, *P. s. rostratus* (Cassin), was formerly a postbreeding wanderer north to the Salton Sea and southern California coast. This race may have regularly dispersed northward from the Colorado River Delta, where one was taken at Yuma, 15 August 1902. An individual seen at BWD in January 1977 was also believed to be of this distinctive race.

HABITAT This sparrow inhabits agricultural fields, especially tall dense alfalfa, Bermuda grass, and weedy margins of other crops. Some are found also in dense marshes. Grinnell (1914) found this species primarily at patches of naturally occurring grass along the drier edge of the valley.

FOOD HABITS Savannah Sparrows are omnivorous while in the LCRV, as evidenced by the contents of 259 stomach samples. About 66% of the diet was seeds from a variety of grasses and other weeds (mostly Portulacaceae, Cruciferae, and Amaranthaceae). Many aphids and other bugs were taken (15%), as well as caterpillars, beetles, ants, and small snails. About 5% of the contents consisted of milo, which is a grain crop.

Grasshopper Sparrow (*Ammodramus savannarum*)

STATUS A rare but probably regular transient and winter resident from late September (earliest, 26 September 1947, BWD, spec) to early May.

The Grasshopper Sparrow is represented in the LCRV by the same race (*A. s. perpallidus* (Coues)) found throughout California and most of Arizona. Records of this sparrow are evenly divided among the fall, winter, and spring migration periods. Four occurrences were of specimens from BWD, Cibola, Imperial NWR, and Parker. An individual found west of Poston on 23 December 1977 was relocated in the same weedy patch on 6 February 1978. The latest record is 7 May at the southern tip of Nevada (1934; spec). This species is extremely secretive and easily overlooked.

HABITAT This species was found in grassy or weedy agricultural fields and margins.

FOOD HABITS One stomach contained only Bermuda grass seeds.

Le Conte's Sparrow (*Ammodramus leconteii*)

STATUS A casual visitor. One seen at Topock, 30 November 1981.

This sparrow flushed repeatedly from a field of tall, dry Bermuda grass before perching in plain view on a low fence. This is the only record for Arizona, although there are several for southeastern California in fall and late spring.

Sharp-tailed Sparrow (*Ammodramus caudacutus*)

STATUS A casual visitor. One seen at West Pond, 29 March 1975, may have wintered locally.

This secretive sparrow winters in very small numbers in marshes along the California coast. There are several other inland records for southeastern California in late May, but none for Arizona.

Fox Sparrow (*Passerella iliaca*)

STATUS A rare but regular transient and winter resident from the end of September to early April.

The Fox Sparrow is almost always encountered singly in the LCRV, and rarely associates with other sparrows. Three were together at feeders at Clark Ranch, north of Blythe, on 31 January 1982. This species comes in three distinct color patterns. Most are of grayish-headed races from the Rocky Mountain and Sierra Nevada regions (primarily *P. i. schistacea* Baird). Within this group, *P. i. olivacea* Aldrich from the mountains of Washington has been taken once. The large-billed California race (*P. i. megarhynchus* Baird) was taken east of the LCRV in the Sierra Pinta Mountains, Arizona. The brighter, reddish-brown individuals that occasionally occur are from Alaska and northwestern Canada (*P. i. zaboria* Oberholser), or interior British Columbia and southwestern Alberta (*P. i. altivagans* Riley), rather than from Eastern populations. Darker brown, Northwest coastal races also should be looked for, especially in late fall.

HABITAT This sparrow favors dense or marshy riparian woodland, as well as dense vegetation in residential areas.

Song Sparrow (*Melospiza melodia*)

STATUS A common permanent resident; local as a spring and summer breeder, more widespread in winter and during migrations. Max: 234; 23 December 1977, Parker CBC.

This is one of several species in which local resident populations represent distinct geographic races, in this case, shared with birds from the nearby Salton Sea. These birds, *M. m. saltonis* Grinnell, are the palest of all races. This condition parallels that of several desert species, but not other marsh-nesting species, such as Marsh Wren and Clapper Rail, which are dark in the LCRV. The breeding population in the LCRV appears to be stable at present, inhabiting large marshes behind dams and impoundments and extensive flooded riparian habitat at BWD. This apparently represents a large shift in local distribution since 1910 when Grinnell encountered this species abundantly, but only away from the few marshes present at that time.

Although nonbreeding migrants from the north are easily recognized by their darker coloration, we have little specific information on extreme dates of occurrence or migrations of various races. Most are certainly *M. m. montana* Henshaw, from the Rocky Mountains and also northern Arizona. They are present from at least 24 September (1952) to 3 April (1954). One specimen of *M. m. fisherella* Oberholser was taken below Needles on 6 December 1946. At least three other races are known from specimens from southern Arizona in winter, and all may occur in the LCRV.

HABITAT Residents are common where emergent marsh vegetation occurs, whether in open, extensive marshes or in the understory of partially inundated riparian woodland. At BWD, breeding densities may reach 40–50 pairs/40 ha in the flooded cottonwood–willow groves. Here they also occupy the dense saltcedar understory that forms an impenetrable mass of stems and debris above wet ground. They are common in the similar understory of the tall athel tamarisk groves near Topock.

Grinnell (1914) described the Song Sparrow as a characteristic species of the arrowweed and young willow associations along the lowest terrace of the natural floodplain. During extensive surveys since 1975, we never found this species in arrowweed, which today is seldom inundated. Thus the important habitat feature seems to be dense cover above or close to water, regardless of the type of vegetation or the structure of the canopy above it. When many saltcedar stands at Cibola NWR were inundated in 1983, these were later colonized by breeding birds, thus supporting this theory. In roughly the past 60 years, a large segment of the population in the LCRV has shifted its primary habitat from arrowweed to recently created marshes and exotic saltcedar stands.

Migrants are less particular and may also occupy other dense, unflooded vegetation. These habitats include saltcedar, screwbean mesquite, and brushy borders of agricultural canals or fields.

BREEDING Recent nesting has been examined only at BWD, where territorial birds begin to sing in February or even late January. Nests are within 1 m of the ground in a shrub or marsh vegetation. Fledged young have been seen from as early as 17 April through July, and it is likely that two broods are raised in a season.

If the above observations are typical of the species in the LCRV, then there has also been a change in breeding behavior since 1910. Grinnell (1914) found no evidence of nesting activity until 15 May, when he left the valley. Old nests found were unusually high in shrubs, above the previous year's high-water line. This apparent delayed nesting, with elevated nest sites, appeared to be an adaptation to annual natural flooding. This behavior was not typical of the birds at the Salton Sea, and it is unnecessary under conditions that prevail in the LCRV today.

FOOD HABITS Song Sparrows are primarily insectivorous, at least during the breeding season. Fourteen stomach samples from July and August contained a wide variety of invertebrates (80% of the diet) including caterpillars, aquatic fly larvae, earwigs, beetles, bugs, spiders, ants, and small molluscs. One winter specimen contained equal amounts of weed seeds and beetle remains. Prey are captured primarily on the ground or from the surface of standing shallow water.

Lincoln's Sparrow (*Melospiza lincolnii*)

STATUS A common winter resident from September through March. A few linger through April into early May. Max: 360; 23 December 1977, Parker CBC.

This is a common sparrow in dense or wet brushy situations throughout the valley, although it is not gregarious and rarely occurs in association with other species. Its true abundance will usually be greatly underestimated by the observer unfamiliar with its thin buzzy calls. Extreme dates of occurrence are from 2 September (1976) to 11 May (1950). Most wintering birds are of the nominate race, probably from the Rocky Mountain or Sierra Nevada regions. At least two spring specimens (migrants) are believed to be *M. l. gracilis* (Kittlitz) from the Pacific Northwest coast.

HABITAT This sparrow is most common in the dense wet understory of riparian woodland, in marshes, and along margins of agricultural canals and fields.

FOOD HABITS Our small sample of 13 stomachs is sufficient to illustrate the diverse diet of this species in the LCRV. About 70% was grass and weed seeds (mainly linear-leaf cambess). The rest included spiders, caterpillars, beetles, bugs, flies, and ants.

Swamp Sparrow (*Melospiza georgiana*)

STATUS A rare to locally uncommon winter resident from November to mid-April. Max: 8; 23 December 1977, Parker CBC.

One must be familiar with the high, sharp call of this sparrow to appreciate its true abundance. At least 20 individuals were found during winter 1977–78, with most records from BWD and marshes below Parker. Not as many were seen in other winters, however. Dates of occurrence are from 14 November (1976) to 22 April (1977), with the last remaining birds at BWD heard to sing full songs. Although most southern Arizona specimens are *M. g. ericrypta* Oberholser from throughout Canada and the extreme northern United States, one from BWD (28 November 1952) is of the more southern, nominate race.

HABITAT This sparrow has been found in marshes and wet riparian groves.

White-throated Sparrow (*Zonotrichia albicollis*)

STATUS A rare but regular winter resident from late October to April. Usually occurs singly in flocks of White-crowned Sparrows or juncos.

One or two can usually be found each winter by diligently searching through the hordes of White-crowned Sparrows that are always present. However, this species is not as numerous here as it is farther east across southern Arizona. There are about 30 records from 10 different winters, spanning the dates 20 October (two seen at Imperial Dam in 1968) to 20 April (one lingering at a feeder north of Blythe in 1977).

HABITAT Although they may occur in any habitat where sparrow flocks are present, most records are from dense or marshy riparian woodlands and inhabited areas.

Golden-crowned Sparrow (*Zonotrichia atricapilla*)

STATUS A rare to uncommon winter resident from October to early May, occurring singly in large flocks of White-crowned Sparrows. Max: 8; 19 December 1978, Bill Williams Delta CBC.

This species is easily overlooked among the numerous flocks of White-crowned Sparrows, but the patient observer can usually find 1 in nearly every large flock. Most fall and winter records are of immatures, but by April many birds have acquired full alternate plumage. With careful searching, as many as 15–20 individuals have been located in recent winters, from 21 October (1976) to 11 May (1977). This species decreases sharply in abundance to the east of the LCRV across southern Arizona.

HABITAT This sparrow is most numerous in sparse riparian and desert habitats, but may be found anywhere White-crowned Sparrows occur.

White-crowned Sparrow (*Zonotrichia leucophrys*)

STATUS An abundant winter resident from late September to early April, often in large flocks. Numbers build slowly in early September and decrease through April and May. An unseasonal bird was at Parker, 8 July 1978. Max: 19,529; 22 December 1980, Parker CBC.

This is by far the most numerous wintering landbird in the LCRV. The Parker CBC recorded a national high count for the species in five consecutive years, from 1977 to 1982 (see Appendix 1), illustrating the relative importance of this area to the overall wintering population.

Wintering-site fidelity was studied on a 20-ha revegetated plot on the Cibola NWR from 1980 to 1983. From a banding study, we found that individuals returned to the same or adjacent 2-ha subplots year after year, even though the vegetation was changing in quality during development. These data suggest that unestablished birds were attracted to the vegetation characteristics of the entire 20-ha site. Once individuals established in the area they became behaviorally tied to specific 2-ha locations within the site, regardless of yearly habitat quality changes.

In our region, the two common races, distinguishable in part by the color of the lores, follow the same pattern of occurrence found throughout the Southwest. Nearly all wintering birds are the white-lored *Z. l. gambelii* (Nuttall) (synonymous with *Z. l. leucophrys* (Forster), according to Phillips et al. 1964) from Alaska and western Canada. These usually depart before the end of April, although the July individual was of this race. The dark-lored race, *Z. l. oriantha* Oberholser from the western United States, is primarily a transient in September, April, and May, when it is often the only race present. These migrants have occurred from 27 August to at least 30 October and from at least 11 March to 2 June. A few dark-lored birds may occasionally winter, and at least one specimen from Topock was identified as the Eastern race, *Z. l. nigrilora* (Todd), which also has dark lores.

HABITAT Before agricultural development, this species was abundant in mesquite and quail bush associations, ranging up brushy desert washes away from the floodplain (Grinnell 1914). Today they are still common in these same situations, but the largest flocks are found along the shrubby borders of agricultural fields. Smaller flocks may be present in nearly any habitat, including marshes, orchards, open saltcedar stands, and even creosote bush flats during wet years.

FOOD HABITS These sparrows are mainly granivorous, with seeds of linear-leaved cambess being the most abundant food in 318 stomach samples. Grass and a wide variety of other weed seeds (especially Portulacaceae, Chenopodiaceae, Polygonaceae, and Solanaceae) were also eaten. About 6% of all food taken was milo grain. Aphids were the most frequent insect food, especially in spring. Many ants were taken in fall, scale insects in winter, and caterpillars were an important spring food.

Harris' Sparrow (*Zonotrichia querula*)

STATUS A casual winter visitor. About nine records, all since 1968, including up to six individuals near Needles and Topock during 1972–73 (December through March).

Other records include an adult at Imperial Dam, 20 October 1968; individuals between Parker and Poston, 15 January and 18 December 1978 and 22 December 1984; and one at a feeder at Martinez Lake all winter until 22 April 1980. This is the rarest of the *Zonotrichia* sparrows in our region, but one to be looked for in any flock of White-crowned Sparrows.

Dark-eyed Junco (*Junco hyemalis*)

STATUS An uncommon winter resident from mid-October to early April, primarily in small flocks. This species exhibits marked geographic variation and the several distinct races will be treated separately below. Max: 134; 23 December 1977, Parker CBC.

Overall, numbers of juncos may vary considerably in the LCRV, undoubtedly reflecting conditions in the various regions of origin of the different races. These small flocks usually contain a mixture of several forms, some being distinct enough to have formerly been considered separate species.

A majority of the LCRV birds are black-hooded Oregon Juncos, including several subtly different races. Most of these (based on specimens) are from the northern Rocky Mountains (*J. h. montanus* Ridgway and *J. h. shufeldti* Coale), but birds from the Sierra Nevada and northern Pacific Coast (*J. h. thurberi* Anthony) are present in varying numbers, at least in some years. Oregon Juncos have been found from 28 September (1980) to 10 April (1950, 1979).

Three other distinct forms are about equal in abundance, being rare but regular visitors, usually mixed with other juncos. The Pink-sided Junco, *J. h. mearnsi* Ridgway, from the central Rocky Mountains, is perhaps the most numerous of these. One or several may be found in nearly any junco flock. The Slate-colored Junco, *J. h. hyemalis* (Linnaeus) and *J. h. henshawi* (Phillips), from the far north and east, has occurred from 27 September (1982) to 24 March (1977). Several are usually found each winter and as many as five were on the Parker CBC in 1977. Gray-headed Juncos (*J. h. caniceps* (Woodhouse)) from the southern Rocky Mountains have been recorded from 5 October (1952) to mid-April (1979), and exceptionally to 11 May (1957) at Cibola. Although most Gray-headed Juncos occur singly, it is not unusual to see two together in a flock.

One junco studied carefully at Ehrenberg on 7 March 1979 appeared typical of *J. h. aikeni* Ridgway, the White-winged Junco, from the Black Hills of Wyoming and South Dakota. This race is extremely rare in Arizona (and unrecorded in California), although there is an additional sight record from Roll, east of Yuma, on 13–14 November 1971.

HABITAT Juncos may be found in any riparian woodland, in orchards, and around human residences where food and brushy cover are available.

FOOD HABITS Four stomachs contained a mixture of weed seeds (mostly linear-leaf cambess) and insects, especially aphids.

McCown's Longspur (*Calcarius mccownii*)

STATUS A casual winter visitor. Two seen and heard along the lower Bill Williams River, 24 December 1977. Also present south of Blythe in three winters since 1981: 27 November–1 December 1981 (2), 23 December 1982 (4), and 5 February 1984 (2).

This is the rarest longspur to reach our region. It has been found somewhat regularly in the Imperial Valley, and casually elsewhere in California and southern Arizona. Most records are of birds found in large flocks of Horned Larks, or occasionally with American Pipits or other longspurs.

HABITAT This species should be looked for in the barest plowed fields, especially where flocks of Horned Larks are present.

Lapland Longspur (*Calcarius lapponicus*)

STATUS A rare and irregular transient and winter visitor from November to early March, usually found singly among flocks of Horned Larks.

This species has been recorded about 12 times in 8 different years, from 11 November (1956, at Martinez Lake) to 12 March (1977, near Poston). Most recent records are from fields near Parker and Blythe, with up to 6 in a single field south of Parker (1 spec) on 25 December 1977. This longspur winters regularly in the Imperial Valley in very small numbers and has been found at a variety of other locations throughout the desert Southwest.

HABITAT This species favors barren agricultural fields, especially those recently plowed or newly planted that also attract Horned Larks.

Chestnut-collared Longspur (*Calcarius ornatus*)

STATUS A rare or irregularly uncommon fall transient and winter resident from late October to March. Occurs primarily as singles or in small groups with larks and pipits. Max: 65; 15 October 1961, Imperial NWR.

This is the most likely longspur to occur in our area and the only species to occur in numbers. Although a majority of records are for fall (earliest, 25 September 1952 at Lake Havasu), this species has remained to winter at least in 1977–78, 1978–79, and 1983–84. In 1978, as many as 20 were west of Poston (1 spec) on 16 January, and 25 were in a field north of Ehrenberg on 29 January. The only later records are of 15, in breeding plumage and singing, north of Poston on 16 March 1979 (5 remaining until 1 April), and 6 at Cibola NWR on 17 March 1984.

Recent large concentrations include 40+ at various locations near Poston on 10 November 1978, a single flock of 40 north of Needles on 27 November 1981, and 60 near Blythe on 29 November 1981. Surprisingly, this is the rarest longspur to reach the Imperial Valley, although it is more numerous (only in fall) elsewhere in southern California. They winter commonly in southeastern Arizona.

HABITAT This longspur occurs in agricultural fields, especially dry Bermuda grass or short alfalfa. The presence of any longspurs in the LCRV is undoubtedly because of increased agricultural development in recent decades.

Bobolink (*Dolichonyx oryzivorus*)

STATUS A casual transient. Three sightings: 14 September 1954 at BWD, 29 May 1975 (male) at Katherine Landing, and 8 June 1977 (singing male) near Poston.

This species is rare but a regular late spring migrant through the desert regions of southeastern California. In fall, it is regular only along the Pacific Coast, although there are records from throughout the Southwest. This species breeds sporadically as close as eastern Arizona.

Red-winged Blackbird (*Agelaius phoeniceus*)

STATUS A common to abundant year-round resident throughout the valley with local status varying due to differential breeding, wintering, and migration of local populations. Frequently found in large flocks. Max: 17,991; 22 December 1982, Parker CBC.

This is a very numerous and conspicuous bird in agricultural portions of the valley. It is one of the few species commonly seen in those areas in summer. Red-winged Blackbirds are highly gregarious during most of the year. Breeding takes place in colonies of evenly spaced, small territories, where the males perch as high as possible to advertise their presence. Large nighttime roosts are formed in the nonbreeding season, with flocks ranging out to feed in large concentrations. Males and females often will segregate into separate flocks.

Although the resident population (*A. p. sonoriensis* Ridgway) is undoubtedly augmented by migrants from farther north, exact determination of the origins of these is difficult without additional specimens (especially fresh fall females). At least some of the wintering birds are from the northern Great Basin (*A. p. nevadensis* Grinnell). Note that the Tricolored Blackbird (*A. tricolor*) has been found within 80 km of the Colorado River, at Kelso, California, and the Salton Sea. The Tricolored Blackbird should be looked for in nonbreeding flocks of Red-winged Blackbirds.

HABITAT Marshes are used for nesting and roosting, although most feeding takes place in agricultural land. Largest flocks are often at feedlots or irrigated fields. This species also occasionally occurs in riparian woodland or near dwellings. They sometimes nest in willows or saltcedar in flooded areas or adjacent to agricultural fields. Some even nest in tall, weedy alfalfa fields.

FOOD HABITS The 161 stomach samples contained 30 insect and 19 plant taxa. Weed seeds and insect pests such as weevils, caterpillars, and various bugs formed a large portion of the overall diet, along with commercial livestock food from feedlots. Panic grass and milo were strongly represented in winter. In spring, plant material (65%) exceeded insects, but lepidopteran larvae were still commonly eaten. By summer, plant and animal materials were taken about equally, and in fall plant material (70%) again dominated.

Eastern Meadowlark (*Sturnella magna*)

STATUS Status uncertain. One specimen collected ("by accident") 17 March 1979 north of Ehrenberg, and one photographed (four others seen) west of Poston, 22 December 1982. Also, a few careful sight records such as 25 November and 20 December 1979 (2) near Parker; 28 February 1980 east of Yuma (singing); 23 December 1982 north of Ehrenberg; and 27 December 1985 at BWD.

This species could be a regular winter visitor to the valley and would represent a new species for California if found on that side of the river. The race involved is *S. m. linianae* Oberholser of Arizona, southern New Mexico, and adjacent Mexico. Aside from the single specimen, we have located this species in the LCRV using criteria developed in central Arizona where both Eastern and Western Meadowlarks winter together locally. Most important is the extent of white in the outer tail, forming distinct patterns as the birds spread their tails just before landing. Also, the overall buffier coloration (especially on flanks), different call notes, and more erratic ("Spotted Sandpiper-like") flight of the Eastern Meadowlark are helpful. This species also is found invariably in taller, denser fields (especially tall alfalfa or Bermuda grass) than Western Meadowlarks. For example, the individual at BWD was flushed from tall, dry grass along the Bill Williams River. It did not associate with the numerous Western Meadowlarks in adjacent farm fields on Planet Ranch.

Western Meadowlark (*Sturnella neglecta*)

STATUS A fairly common and local summer breeder and common winter resident primarily in agricultural portions of the valley. Max: 1,393; 22 December 1982, Parker CBC.

This is a characteristic species of open situations that has undoubtedly increased with agricultural development in the LCRV. Grinnell (1914) noted meadowlarks in only a few natural grassy areas along the river in 1910, except at Needles where this species was already associated with early attempts at irrigated agriculture. Although the LCRV resident population is augmented by migrants, a general lack of geographic variation in this species does not allow identification of source populations, except in extreme cases. At least one specimen from

Yuma, 12 October 1902, is identified as *S. n. confluenta* Rathbun from the Pacific Northwest. This race has heavier black dorsal markings, which may cause some individuals to resemble Eastern Meadowlarks in the field.

HABITAT Meadowlarks inhabit a variety of agricultural fields, especially alfalfa. Some are found in sparse riparian or desert areas when grass is available, or in marshes.

FOOD HABITS The 125 stomach samples included 28 arthropod and 7 plant taxa. The overall diet consisted mainly of beetles, earwigs, insect larvae, termites, and grass seeds. Arthropods made up 77% of the diet in fall, 66% in winter, and 95% in spring. Although frequently persecuted as a crop pest, this species' diet in agricultural situations was 72% arthropods, with many being crop pests.

Yellow-headed Blackbird (*Xanthocephalus xanthocephalus*)

STATUS A locally common summer breeder throughout the valley between March and September. Usually uncommon in winter, but large concentrations may occur in the southern parts of the valley. Max: 20,000 (all male); 1 January 1955, northeast of Yuma.

This species is similar in many regards to the Red-winged Blackbird, although its distribution in the LCRV is more concentrated and local. The two species may associate in the same flocks, except during the breeding season when pure colonies form. Yellow-headed Blackbird nesting colonies are mostly restricted to marshes, and on occasion this species will usurp a marsh from Red-winged Blackbirds. Nonbreeding concentrations are typically made up of males, and these may form as early as July. Winter numbers are highly variable, with the species often being difficult to find north of Blythe.

HABITAT The species is restricted to marshes for nesting and roosting; agricultural fields and feedlots are used for feeding year round.

BREEDING Our information on breeding biology of this species comes from a study by Voss-Roberts (1984) on the Colorado River Indian Reservation. Females began nest building by mid-April and most nests contained eggs by 22 April. Average clutch size was four, with the entire clutch hatching in 50% of all nests. Eggs began hatching on 29 April with most hatched by 8 May.

Nests were grouped and placed in the most shaded and densely vegetated cattail clumps, 10 cm above standing water. Presumably nest placement is to avoid detection by predators and also to ameliorate high afternoon temperatures and solar radiation. Predation on eggs was low (<5% of all nests), but 45% of all eggs in affected nests were lost.

Potential predators included Great-tailed Grackles which nested in the same marshes and were vigorously chased when they approached blackbird nests. Marsh Wrens were also probably predators because they nested in the same marsh and are known to puncture blackbird eggs.

FOOD HABITS In 133 stomachs there were 26 arthropod and 13 plant taxa, dominated by insect larvae, beetles, grasshoppers, and grass seeds. The general diet of this primarily agricultural species is 60% arthropods and 40% plant material. Much of the plant material was commercial seeds (wheat and milo). In spring about half of the diet was arthropods. In summer arthropods increased to 72%, then decreased to 18% in late summer and about 10% in fall.

Rusty Blackbird (*Euphagus carolinus*)

STATUS A casual transient or winter visitor; 8 records involving 14 individuals. Specimens were taken on 17 February 1950 at BWD and 6 November 1952 at Topock Marsh. Sight records are of females seen 21 November 1951 at BWD and 26 February 1959 at Imperial NWR; a male and female, 23 December 1974 near Yuma; a flock of three males and two females, 10 March 1977 at BWD; a male, 28 November 1981 at Lake Havasu City (photo), remaining through January 1982 and joined by a female, 15–28 January (photos); and a male, 25–26 November 1988 at Ehrenberg.

This species is a somewhat regular visitor in late fall and winter throughout the Southwest. In our experience, this species does not associate with other blackbirds, but occurs solitarily in wet woodland or pond edges. The five birds in March 1977 were singing and feeding among partially flooded willows along the Bill Williams River. They represent the only record of a migrant flock in the Southwest.

Brewer's Blackbird (*Euphagus cyanocephalus*)

STATUS A common to abundant winter resident between October and mid-April, often occurring in large flocks. Uncommon from September until early May. Two summer records: a female 12 June 1947 at Lake Havasu (spec) and a male seen 9 June 1983 at Lost Lake. Max: 11,717; 23 December 1977, Parker CBC.

This highly gregarious species often flocks with other blackbirds. Like the Brown-headed Cowbird, it is more closely associated with human settlement and livestock production than the preceding species. In 1910, Grinnell (1914) found it only at towns. The species has undoubtedly increased with the advent of large-scale agriculture. Flocks have been noted as early as 29 August (30 birds in 1977). Although normally gone by late April, a few have lingered as late as 18 May (1950).

At least some of the wintering birds may originate in supposedly sedentary populations along the California coast, including a specimen from Yuma, 7 November 1905, identified as *E. c. minusculus* Grinnell (Phillips et al. 1964). A small number breeds near the Salton Sea in the Coachella Valley.

HABITAT Flocks may be found in agricultural fields, pastures, feedlots, and towns. Small numbers are seen along open lakeshores and river-banks; however, this niche has largely been usurped recently by the Great-tailed Grackle.

FOOD HABITS Fourteen arthropod and 15 plant taxa were identified from 106 stomachs. As birds begin arriving in fall, a major portion of their diet is plant material (97%), with seeds of panic grass, pigweed, and Bermuda grass being dominant. During winter, plant material (86%) decreased slightly and insects (14%) increased. Insects increased dramatically (64%) in the spring diet, with leafhoppers, caterpillars, and mantids dominating.

Great-tailed Grackle (*Quiscalus mexicanus*)

STATUS A common local resident and breeder throughout the valley, most numerous in the southern half. Max: 1,094; 15 December 1984, Yuma CBC.

The species was first recorded along the Colorado River in 1964 at Imperial Dam, and was found breeding there in June 1968. It has been increasing and spreading continually ever since (Fig. 34). In 1975, small colonies existed at Blythe and Needles, and the first small flocks were at Parker and BWD in fall 1976. A colony formed at BWD the following spring. By 1980, this species had filled in many gaps in its distribution and was very common around Parker. It is fast becoming an abundant resident throughout the valley.

This species was unrecorded in Arizona before 1935. Its inexorable spread northward and westward was well documented by Phillips (1950), Phillips et al. (1964), and Monson and Phillips (1981). After its first appearance along the Colorado River, it dispersed rapidly to the Salton Sea and more slowly through the desert regions of southeastern California, eventually reaching the Pacific Coast. Its rapid success is clearly associated with the spread of human settlement and especially with agricultural and urban irrigation practices. Interestingly, two subspecies were involved in this simultaneous range expansion into the southwestern United States. It is unclear at present if populations in the LCRV represent *Q. m. nelsoni* (Ridgway), originally from Sonora, Mexico, or if they are mixed or intermediate with *Q. m. monsoni* (Phillips).

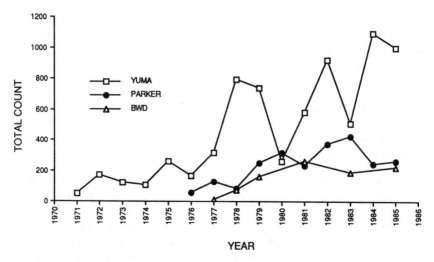

FIGURE 34. Increase in Great-tailed Grackle numbers on Christmas Bird Counts at Yuma, Parker, and Bill Williams Delta.

HABITAT Grackles are most numerous where tall trees grow in association with human residences close to water. Such situations include trailer resorts, marinas, ranch yards, and town parks. As this species increases in abundance, it is found in a wider variety of habitats. Roosts may form in marshes, and flocks feed with other blackbirds in agricultural fields and feedlots. Nesting colonies still require tall trees, however. This need may ultimately limit the breeding population in the LCRV.

FOOD HABITS Four stomach samples contained mainly grasshoppers, earwigs, beetles, and wasps. This species is thought to be an important predator on the eggs and nestlings of other birds.

Common Grackle (*Quiscalus quiscula*)

STATUS A casual visitor. One careful sighting, 19 June–9 July 1979, north of Blythe. Also a probable sighting of three was at the same location 3 September, 18 October, and 5 November 1980 (photos).

Although expanding its range steadily in the western United States, this species remains a very rare vagrant to southern California and was unverified in Arizona until 1983. The three individuals seen at the Clark Ranch in 1980 flew across the river each time into Arizona.

Bronzed Cowbird (*Molothrus aeneus*)

STATUS Recently a rare but regular summer resident and breeder locally from mid-April through July. Areas of occurrence include Davis Dam,

Parker Dam, Earp, Blythe, Ehrenberg, Laguna Dam, and Yuma. Max: 7; 26 May 1979, Clark Ranch.

This is another recent invader to the region, first recorded in the LCRV at Parker on 30 July 1950. A later sighting along the shore of Lake Havasu, 29 May 1951, was the first for California. The species appeared regularly, at least around Parker, beginning in 1952, although the present population remains very small. It is most numerous in the Yuma area, including Dome Valley.

HABITAT This cowbird is found almost exclusively around houses, especially those with irrigated lawns.

BREEDING This cowbird is a brood parasite of other birds; its most frequent host is the Hooded Oriole.

Brown-headed Cowbird (*Molothrus ater*)

STATUS A common year-round resident throughout the valley, but local distributions vary seasonally. Occurs primarily in flocks near human developments in fall and winter. Breeding birds enter riparian habitats in late March. Max: 4,007; 20 December 1978, Parker CBC.

Although not notable for its coloration or voice, this species is an important member of the local avifauna. Because of its parasitic habits and massive population increases during the past century, it is thought to have had serious, detrimental effects on the populations of several Southwestern riparian birds. In particular, the near or total extirpation of Willow Flycatcher, Bell's Vireo, and Yellow Warbler from the LCRV is most often attributed to cowbird parasitism. However, cowbirds may perhaps be considered as a symptom of the larger problem facing riparian birds—habitat loss. As riparian habitats become fragmented and engulfed by agricultural land, the potential for parasitism by the edge-loving cowbird is enhanced.

Species most adversely affected by this parasitism are habitat specialists (mainly cottonwood–willow) that breed in late spring or summer. However, other such species that are not affected by cowbirds (Yellow-billed Cuckoo, Summer Tanager) have similarly declined. Habitat generalists and those species that begin nesting in early spring have not declined significantly, in spite of heavy cowbird parasitism. In an extreme case, the Black-tailed Gnatcatcher suffers virtually complete nesting failure due to parasitism after its first, early brood. However, the gnatcatcher still attempts up to three or more broods each season. Other species have mechanisms for avoiding such losses; e.g., Crissal Thrashers eject cowbird eggs from the nests (Finch 1982).

The historic status of this species in the LCRV is not well documented. By 1910, it was already common in riparian forests in spring, although it may not have wintered locally (Grinnell 1914). Today, high overwinter survival is promoted by abundant food supplies at feedlots, cattle pastures, and suburban residences. The LCRV breeding Brown-headed Cowbirds are all presumably *M. a. obscurus* (Gmelin), the "Dwarf" Cowbird. However, *M. a. artemisiae* Grinnell may occur in winter.

HABITAT Between July and March, this species is primarily associated with human-altered habitats. There they form large flocks, usually mixed with other blackbirds, that move about the agricultural land-scape to feed at pastures, feedlots, grain fields, or in towns. Nighttime roosts may form in marshes or urban hedgerows. In March, flocks break up and birds move into riparian habitats to breed. They are numerous in all riparian habitats and usually outnumber most other locally breeding residents in early summer.

BREEDING This cowbird is a brood parasite, potentially affecting all small- or medium-sized, open-nesting passerines in the LCRV. Breeding activity begins in March, when small groups of males gather in treetops to display. Eggs are deposited from April to July; thus, only the earliest-breeding hosts are spared from parasitism throughout their breeding cycle.

FOOD HABITS The 110 stomach samples contained 13 plant and 20 arthropod taxa. Seeds were mostly of panic grass, milo, canary grass, and Bermuda grass. Caterpillars, cicadas, grasshoppers, and other insects were also eaten. Seasonal consumption of plant material was 100% in winter, 34% in spring, 49% in summer, 78% in late summer, and 95% in fall.

Orchard Oriole (*Icterus spurius*)

STATUS A casual transient and winter visitor. Six records: 15 June 1969 (singing male) and 20–21 February 1974 at Imperial Dam; 23 April 1978 at Earp; 20 May 1979 near Yuma; 1 October 1979 at Laguna Dam; and 9 November 1985 at Topock.

This oriole is a rare but regular migrant in southern California, primarily in fall. Most Arizona records are in May and June. There are additional winter records in both regions, and this species is a common winter resident farther south in western Mexico. Nearest breeding populations are in northern Sonora, Mexico.

Hooded Oriole (*Icterus cucullatus*)

STATUS A fairly common but local summer breeder from mid-March through September, with a few lingering as late as November; casual in winter.

This oriole is thought to be a relatively recent invader to the region, although it was already common by 1910, at least in the southern part of the LCRV (Grinnell 1914). Its continued spread northward is associated with residential developments, especially in regions where Northern Orioles occupy native riparian habitats.

This species has arrived as early as 6 March 1980 at Clark Ranch (a regular breeding location), and one seen 21 February 1985 at Bard may also have been an early spring arrival. An immature male and a female lingered at a residential feeder at Blythe, at least until 22 November 1976. An adult male (possibly along with an immature male) wintered successfully at feeders in Ehrenberg in 1983–84. They have been induced to remain through winter at hummingbird feeders elsewhere in southern California and Arizona. Breeding birds, at least, are *I. c. nelsoni* Ridgway, the same race that breeds throughout Arizona and California.

HABITAT The Hooded Oriole is a bird of towns and other inhabited areas, particularly those with ornamental palms and hummingbird feeders. It is less numerous in riparian woodland (mainly cottonwoods), where it is outnumbered by the Northern Oriole. This pattern is not repeated elsewhere in Arizona (on the Verde and San Pedro Rivers), where both species are equally common in cottonwoods (Higgins and Ohmart 1981; Hunter 1988). Migrants may appear in desert washes or even agricultural fields.

BREEDING The nest is a woven basket most often placed in a palm tree. It is a frequent host of the Bronzed Cowbird, which has also spread recently to the LCRV. A female Hooded Oriole was feeding a fledgling cowbird on 6 August 1976 at Parker. Another cowbird nestling was in a Hooded Oriole nest in Ehrenberg on 27 June 1979.

Northern Oriole (*Icterus galbula*)

STATUS A fairly common to common summer breeder between mid-March and late July. Most depart in early August, but lingerers and transients remain until late September or rarely later. Casual in winter with five records: 1 January–1 February 1947 at Parker (spec); males seen 24 December 1952 at Parker and 28 February 1956 at Bard; two netted in February 1966 at Yuma; and a female seen at Earp 22–24 December 1984 (at feeders). Max: 35; 9 June 1978, Parker to Poston (river census).

The Northern Orioles typically breeding in the LCRV are the Western form (Bullock's Oriole) and are most referable to the smaller *I. g. parvus* (van Rossem) of California, rather than *I. g. bullocki* (Swainson) found throughout most of Arizona. The winter specimen from Parker is referable to *I. g. parvus*. The Eastern form, *I. g. galbula* Linneaus (Baltimore Oriole), is a casual transient with records from 22 September 1956 at Imperial NWR (immature male); all summer 1977 (male mated with female Bullock's, see below) north of Blythe; 2 August 1977 (female) at Ehrenberg; and 5 April 1980 north of Ehrenberg (male).

This is one of few migratory, midsummer-breeding riparian species not to have decreased substantially in recent years. Reasons for its continued success are (1) its adaptiveness in the face of massive habitat alteration and destruction, and (2) its mechanism for avoiding parasitism by cowbirds. Both of these factors are discussed in detail below.

First individuals to arrive in spring are typically adult males, thought to be part of an early migration through the LCRV by birds breeding along the Pacific Coast. Our earliest such record is 12 March 1983 at Cibola NWR, although the 28 February record, cited above, may represent a very early migrant. Local breeders have been detected as early as 25 March (at Clark Ranch), but they are not present in numbers until early April. Breeding birds farther east do not reach Arizona until mid-April (Phillips et al. 1964). Because the LCRV breeding population is apparently of the same race (*I. g. parvus*) as that nesting in coastal California, it is difficult to separate residents from transients, even with specimens.

Northern Orioles depart the LCRV earlier in fall than most migratory breeders. They are scarce in riparian habitats after July, when presumed transients appear in nonbreeding areas such as towns and orchards. Recent records after September include an immature male, 12 October 1978, at a Blythe feeder; two on 13 October 1979 north of Poston; female-plumaged birds on 1 November 1978 at Poston and 11 November 1978 at BWD; and an immature male netted on 16 November 1971 at Yuma. All orioles in late fall should be checked carefully for the Streak-backed Oriole (*I. pustulatus*), which has been found several times in Arizona and southern California.

The male Baltimore Oriole that nested at the Clark Ranch north of Blythe was first seen 16 May 1977 and remained until 7 September. Intergrade offspring were seen at this location until at least 1980, although the fate of any later nesting attempts was not studied. This represents the only known nesting by this distinct race in California, although an identical pairing took place in 1980 along the Verde River in central Arizona. An additional intergrade male Baltimore × Bullock's Oriole was seen 27 April 1979 at Laguna Dam.

HABITAT This species is a habitat generalist in the LCRV, breeding in almost every kind of riparian vegetation, including saltcedar. Highest densities, however, are in remaining mature cottonwood–willow stands, such as at BWD, where up to 80 birds/40 ha were estimated in

1977–78. The next most important habitats are mature honey and screwbean mesquite bosques and saltcedar that has attained heights of at least 5 m. Surprisingly, they are not numerous in athel tamarisks. Younger, more open, or disturbed sites are less attractive, although a few birds may be found even in desert washes, marshes, and citrus orchards. Nesting also occurs in residential and agricultural areas supporting native cottonwoods. However, this species is replaced by the Hooded Oriole in truly urban settings. Migrants may be seen more frequently away from riparian habitats, in orchards, towns, and desert.

In their preferred cottonwood habitat at BWD, orioles concentrate their activity in the interior of the groves and avoid areas with a dense saltcedar understory. Interestingly, this species shows a tendency to become semicolonial in areas where nest sites are clumped and extended foraging trips are made outside the nest territory (Pleasants 1979). This breakdown in the usual territorial behavior may account for the much higher densities of this species compared with other canopy-nesting birds, and for the ability of orioles to exploit isolated or remnant cottonwoods, even in agricultural areas.

BREEDING The well-known nest is a pendent, woven cup suspended from a branch. Of 12 nests found at BWD in 1977 and 1978, 6 were in cottonwoods and 5 were in willows, ranging from 8 to 16 m above ground. Other nests have been found in the tops of mesquites and saltcedars and inside mistletoe clumps. Breeding activity began at BWD in mid-April, when males were aggressively setting up territories. Copulation was first observed on 26 April in 1977 and 1978. Nest building continued through the first week in May. Orioles were extremely quiet and secretive during incubation and brooding, and young were evidently silent in the nests. Family groups with recently fledged young appeared suddenly throughout the cottonwoods, beginning on 28 May, when the birds were again vocal and conspicuous. Second nestings were then begun by at least some pairs, with birds observed feeding young as late as 30 July 1978. Away from BWD, recently fledged young were noted in mesquite from 3 to 12 June, and nest-building was observed on 18 June, indicating a similar breeding chronology.

An unusual aspect of breeding, at least at BWD, was the presence of more than two birds at most nest sites. Typically, a subadult male accompanied each pair, and was frequently seen foraging side by side with the adult male. Although some subadults were observed carrying nest materials and feeding young, it is not clear whether adults were present at the same nests. This interesting situation merits further study.

Perhaps the most significant feature of the oriole's biology in the LCRV is its ability to avoid the detrimental effects of cowbird parasitism. Although apparently considered a suitable host by cowbirds, the oriole invariably ejects or damages foreign eggs in its nest (Rothstein 1977). A female Brown-headed Cowbird was seen entering an oriole nest at BWD on 8 May 1977. Another cowbird was driven from the vicinity of a nest by a subadult male on 4 May 1978. However, the outcome of these individual cases is unknown.

FOOD HABITS Orioles feed primarily by gleaning insects from live foliage in the outer portions of tall trees. In this regard, they are much more stereotyped than other medium-sized, canopy-feeding insectivores, such as Summer Tanagers and *Myiarchus* flycatchers. They often fed heavily at flowering trees and shrubs in early spring and employed several specialized behaviors such as probing clusters of dead leaves and "gaping" open curled leaves or buds.

The overall diet of this species consisted mainly of caterpillars and large insects such as cicadas and grasshoppers. In 58 stomachs, we found 20 insect and 5 plant taxa. A major proportion of the plant material was consumed in spring (22% of diet), and included seeds of Compositae and *Solanum*. Insects almost totally dominated the diet by summer (97%), and in late summer the stomachs contained 100% insects. There was a marked shift from caterpillars in spring to grasshoppers and cicadas in summer. For example, nine stomachs from BWD in midsummer contained 80% cicadas. Samples from agricultural areas contained 77% insects and 13% plant material (primarily lettuce). Birds from natural riparian habitats consumed 95% insects and 5% plant material overall.

Scott's Oriole (*Icterus parisorum*)

STATUS A casual straggler or transient. Seven records: 3 August 1954 and 10 May 1983 (singing male) at BWD; 7 May 1977 and 20 July 1978 (both singing males), and 23 August 1978 (immature male; spec) north of Ehrenberg; 21 June 1977 (singing male) north of Needles; and 4–5 September 1981 (two females) at Parker Dam.

This oriole breeds commonly in desert mountain ranges that flank the LCRV, including as near as the Dome Rock Mountains, 15 km east of Ehrenberg. These areas are undoubtedly the source of most of our records, especially those of singing males in spring and summer in desert washes draining nearby mountains.

Purple Finch (*Carpodacus purpureus*)

STATUS A rare and irregular fall and winter visitor, with records from late October (spec, 29 October) through March. A female was identified at Cibola, 6 April 1982, and a female *Carpodacus*, seen 9 April 1978 at

Ehrenberg, was probably this species. Max: 12; 22 December 1985, Parker CBC.

This species has been recorded in the LCRV about 15 times in at least 9 different years. About half of the records are of single birds, with the remainder being small groups of two to five individuals, except the larger group noted above. A group of about seven was present at Parker, 22 November 1953–14 March 1954, with one still present and singing on 4 April. This species is an irregular visitor to lowland desert localities throughout the Southwest, often corresponding with larger flights along the Pacific Coast. The LCRV birds presumably originate from these coastal populations (*C. p. californicus* Baird), although the Eastern race has been collected as close as Tucson, Arizona.

Cassin's Finch (*Carpodacus cassinii*)

STATUS A casual fall and winter visitor. Four records of 28 individuals: 1 seen and heard 29 October 1979 at Andrade, near the Mexican border; a female carefully identified 9 November 1980 north of Blythe; 6 individuals between 24 October and 7 November 1981 from Parker Dam to Cibola; and at least 20 in late December 1985 at Parker Dam and Parker (photo).

This montane finch rarely wanders to lowland areas anywhere in the Southwest. However, its recent appearance in the LCRV may, in part, represent a previously overlooked pattern of fall occurrence in the region. Larger numbers seen in 1985 corresponded with an almost unprecedented flight of this species throughout southern Arizona and adjacent California. A small flock at Parker was associated with an equal number of Purple Finches that were also present that winter.

House Finch (*Carpodacus mexicanus*)

STATUS A common to abundant year-round resident, but with large shifts in local distribution. Frequently in large flocks except during the spring breeding season. Max: 2,729; 22 December 1983, Parker CBC.

The House Finch is ubiquitous in the LCRV and may be seen any day of the year in virtually any part of the valley. Largest numbers are found in agricultural areas in fall and winter, undoubtedly including nonbreeding birds from desert areas outside the region. By late winter most birds are dispersed throughout breeding habitats and are usually in pairs.

HABITAT This species breeds in low densities (1–5 pairs/40 ha) in virtually every type of riparian woodland, as well as in desert washes and orchards. Largest resident populations, however, are probably in towns and around farmhouses. Nonbreeding flocks concentrate in grain fields, especially milo and wheat.

BREEDING Although this is a very common bird, we have little specific information on its nesting in the LCRV. Grinnell (1914) found nests with eggs as early as 18 and 26 March in 1910 and a nest with small young on 14 April. Recently we have seen fledged young on 17 April (1978). Nests have been located in a variety of desert and riparian trees, in cholla cactus, and even in cavities of giant saguaros.

FOOD HABITS The diet is almost entirely seeds, with insect remains found in only 1 of 63 stomachs. A majority of the seeds identified were Cruciferae, Portulacaceae, Amaranthaceae, and Solanaceae. Although not present in stomach samples, mistletoe berries are eaten during winter and spring.

Red Crossbill (*Loxia curvirostra*)

STATUS A casual visitor. Recorded 23 August 1953 at Parker (flock of 5, one spec), 16 November 1976 at BWD, 20 December 1976 at Parker (2), 27 November 1979 near Yuma, and 14 August 1984 north of Blythe (9 birds).

This species is a notorious and unpredictable wanderer throughout its range, with separate populations moving independently of each other. This situation brings a variety of identifiable races to the Southwest (summarized in Monson and Phillips 1981). The only Colorado River specimen, along with most others from the desert lowlands, is *L. c. sitkensis* Grinnell from the Pacific Northwest.

Pine Siskin (*Carduelis pinus*)

STATUS An irregularly uncommon fall transient in October and November, and irregular winter visitor through April; a few have lingered to early June. Max: 77; 10 April 1976, south of Yuma.

Although found every year, numbers and timing of occurrence in the LCRV are extremely variable. For example, small flocks (up to 40) were present in 1977 from early October throughout the winter and following spring (over 300 individuals reported between Topock and Imperial Dam). The following year our only records were of 4 individuals at Parker in mid-December and 1 over Ehrenberg in March. The earliest fall arrival date is 12 September (1982) at Ehrenberg (the only record that fall), and the latest date of lingering birds is 3 June (2 at Davis Dam in 1970 and 4 at BWD in 1978).

HABITAT Siskins may occur in riparian woodland, along agricultural margins, and in towns or parks.

FOOD HABITS The three stomachs examined contained filaree and inkweed seeds. When flocks are present at BWD in late winter, they feed heavily on newly opening flowers of cottonwood and willow trees.

Lesser Goldfinch (*Carduelis psaltria*)

STATUS A fairly common but local spring and summer breeder; fairly common but irregular in winter. Max: 68; 18 December 1978, Parker CBC.

This species is usually encountered singly or in very small groups (2–5) throughout the year. The only recent larger concentrations have been at BWD in late winter and early spring, when flocks move into the riparian groves to feed on the abundant cottonwood and willow flowers. Small flocks of 15–20 birds are very occasionally seen elsewhere in early spring, such as at citrus orchards near Blythe. The only sizeable breeding population at present is also at BWD. Grinnell (1914) commonly found this species in flocks along much of the Colorado River in March 1910. The extent to which populations may have declined in recent years or to which fluctuations in numbers may be influenced by influxes from outside the valley remains unknown.

HABITAT At present this species apparently breeds only in cotton-wood–willow habitats and also probably in citrus orchards and in towns. Nonbreeding birds may be found in other riparian woodland, desert washes, or agricultural areas. Grinnell (1914) reported them as common in mesquite, willow, and desert washes.

BREEDING The Lesser Goldfinch has a notoriously protracted breeding season elsewhere in southern Arizona, extending from January to November (Phillips et al. 1964). All of our breeding records, however, are for spring. Grinnell (1914) observed nest building in late March and fully grown young by mid-April. At BWD in 1978, nests were also completed in late March and fledged young were noted in late April and early May. No new nests were found after May, and small flocks formed (family groups?) from late May through July, suggesting that only one brood was raised. Although Grinnell observed nests close to the ground in arrowweed, we found most nests to be between 7 and 15 m high in willow and cottonwood trees.

FOOD HABITS Two stomachs contained grass seeds and other unidentified plant material.

Lawrence's Goldfinch (*Carduelis lawrencei*)

STATUS Highly irregular. An uncommon to fairly common transient and winter visitor in flocks from October to April during flight years, but absent in other years. Also a rare but possibly regular summer breeder locally; breeding has been confirmed at Parker (1952), north of Blythe (1978), and BWD (1978, 1979). Max: 126; 7 February 1948, Poston.

This species exhibits rather massive and regular movements eastward across southern Arizona from its California breeding range. This occurs, on average, every two to three winters. During those flights, large numbers may pass through the LCRV in fall and spring, but relatively few normally remain through winter. Individuals may appear as early as 30 August (1979), and the return flight is often noticeable from mid-February at least through April. Later spring departures are difficult to determine because of the occasional presence of breeding individuals. Probable migrants away from areas of suspected breeding include one north of Ehrenberg on 7 May 1982 and a flock of five at Imperial NWR on 2 June 1958.

It is unclear whether the limited breeding that takes place is an opportunistic extension of the eastward flight or represents an independent, though extremely limited, local population. The nestings in 1952 and 1978 followed major flight years. However, in 1979 no birds were present before 18 April, when pairs were found simultaneously at Clark Ranch north of Blythe and BWD. The latter pair was nesting. Also, a pair was seen elsewhere at BWD, 1–9 October 1978, possibly having remained after the 1978 breeding season. The few that nest elsewhere in Arizona, such as near Phoenix, also appear to "show up" in April, even after nonflight winters.

HABITAT Breeding has been noted only in cottonwood–willow habitat (BWD) and at residential areas with tall cottonwoods and other trees (Clark Ranch, Parker). Migrants and winter visitors may be found along shrubby or weedy margins in agricultural areas and in open mesquite woodland with patches of shrubs, particularly inkweed. Other migrants may be detected anywhere in the valley as they pass overhead.

BREEDING At the Clark Ranch in 1978, a pair first seen on 30 May was with a dependent juvenile on 11 June. The adult female remained there at least until 9 July. At BWD, a pair was present in late May–June 1978, but nesting was not confirmed there until 1979. On 18 April 1979 a nest was found 4 m high in a saltcedar at the edge of a large cottonwood grove. Three young were in the nest on 12 May, although the fate of this nest is unknown.

FOOD HABITS Five stomach samples contained inkweed seeds and other unidentified plant material.

American Goldfinch (*Carduelis tristis*)

STATUS An uncommon winter resident usually in small flocks between mid-October and mid-April, rarely to early or mid-May. An adult male,

4 July 1977 at BWD, was unseasonal. Max: 41; 22 December 1983, Parker CBC.

This species is the most regular of the "northern" finches to visit the LCRV, although it is never numerous. The largest single flock was of 30 birds west of Poston on 4 January 1979; 1–5 birds are more usual. All but one specimen from the area are *C. t. pallida* (Mearns), the widespread form in interior western North America. The exception is also our latest spring record; a *C. t. salicamans* (Grinnell) (from coastal California) taken 17 May 1948 at Parker. A very early group of four males was at BWD on 4 September 1981.

HABITAT This goldfinch may frequent margins of agricultural fields (especially sunflowers) and brushy riparian woodland. Individuals are often detected as they fly overhead, in any habitat.

FOOD HABITS The diet, based on two stomachs, consisted of seeds of inkweed, thistle, pigweed, and composites.

Evening Grosbeak (*Coccothraustes vespertinus*)

STATUS A rare and irregular fall visitor in October and November. Two spring records: 6 May 1902 at Yuma and 7 May 1973 at Lost Lake (two males). Max: 12; 1 October 1972, Topock.

The period of fall occurrence (about 15 records) extends only from 27 September to 17 November; these extreme dates were both of single birds north of Blythe in 1982 and 1977, respectively. There are no winter records. They are highly erratic visitors to the desert lowlands of the Southwest. Larger numbers appear during flight years in some areas, where some birds also occasionally winter.

HABITAT This species has been found in cottonwood and willow gróves (e.g., BWD), and in tall trees near human habitations.

House Sparrow (*Passer domesticus*)

STATUS A common to abundant resident and breeder in all developed parts of the valley, often in large concentrations. Max: 1,021; 22 December 1983, Parker CBC.

This species was already established and common (at least at Needles and Yuma) when ornithologists visited these areas in 1910. These weaver finches are thought to have arrived via development of the east–west railroads, some time during the previous decades (Phillips et al. 1964). Although this species is exotic to our continent, it has reached Arizona and the LCRV on its own. At present it is among the most widespread and numerous residents, nearly always in association with humans or their livestock. However, we have no specific information on the status of populations or any local ecological adaptations that may have occurred.

HABITAT This species is common in towns, other inhabited areas, and throughout agricultural lands.

FOOD HABITS Five stomach samples contained 96% plant material dominated by wheat, milo, panic grass, and pigweed.

Appendix 1
Summary of Christmas Bird Counts
Along the Lower Colorado River

	Martinez Lake–Yuma 1968, 1971–1988			Parker 1976–1985			Bill Williams 1977–1979, 1981, 1983, 1985		
Species	CBCs Present	Range Count	National High Years	CBCs Present	Range Count	National High Years	CBCs Present	Range Count	National High Years
Pacific Loon	1	0–1		1	0–1		2	0–1	
Common Loon	14	0–6		1	0–2		5	0–4	
Pied-billed Grebe	19	6–79		10	11–51		6	14–39	
Horned Grebe	6	0–7		1	0–1		3	0–3	
Eared Grebe	19	1–96		9	0–79		6	20–2,291	
Western Grebe	18	0–107					6	300–6,803	
Clark's Grebe	1	0–2					6	217–300	1983,* 1985
Blue-footed Booby	+								
American White Pelican	6	0–30							
Brown Pelican	2	0–1							
Double-crested Cormorant	19	134–2,314		10	2–104		6	1–55	
Olivaceous Cormorant	+								
American Bittern	16	0–10		6	0–8		2	0–2	
Least Bittern	13	0–9		2	0–1		1	0–1	
Great Blue Heron	19	32–306		10	23–59		6	7–35	
Great Egret	19	18–376		10	29–121		3	0–8	
Snowy Egret	19	32–274		9	0–58		2	0–3	
Cattle Egret	11	0–367		8	0–185		1	0–45	
Green-backed Heron	15	0–17		9	0–14		4	0–1	
Black-crowned Night-Heron	19	2–127		10	9–79		6	2–10	
White-faced Ibis	4	0–50		4	0–3				
Tundra Swan	6	0–3		4	0–4				

	n	Range	n	Range	n	Range
Fulvous Whistling Duck						
Greater White-fronted Goose	4	0–5	3	0–13	1	0–2
Snow Goose	15	0–28	1	0–2		
Ross' Goose	2	0–1	8	0–83		
Canada Goose	19	35–1,683	2	0–2	6	119–600
Wood Duck	2	0–1	10	18–398	2	0–14
Green-winged Teal	19	6–161	10	13–187	5	0–54
Mallard	19	11–588	8	0–657	6	31–223
Northern Pintail	19	2–2,179	3	0–2	6	6–65
Blue-winged Teal	5	0–5	5	0–16		
Cinnamon Teal	13	0–16	8	0–68	1	0–1
Northern Shoveler	18	0–383	10	2–101	4	0–20
Gadwall	17	0–53	10	22–363	6	16–199
Eurasian Wigeon	1	0–1				
American Wigeon	19	4–197	6	0–10	4	0–40
Canvasback	19	1–65	8	0–24	2	0–9
Redhead	17	0–84	10	1–122	3	0–2
Ring-necked Duck	19	11–984	3	0–3	5	0–69
Greater Scaup	1	0–1	10	8–95	4	0–23
Lesser Scaup	19	18–508	10	94–583	6	15–246
Oldsquaw			+		1	0–1
Surf Scoter					1	0–1
White-winged Scoter	1	0–1			1	0–3
Common Goldeneye	17	0–125	5	0–2	6	300–1,860
Barrow's Goldeneye					6	1–9
Bufflehead	19	23–173	10	12–464	6	11–178
Hooded Merganser	5	0–13	1	0–1	3	0–3
Common Merganser	18	0–258	9	0–245	6	8–1,863
Red-breasted Merganser	10	0–376	5	0–3	6	2–22
Ruddy Duck	19	4–173	7	0–23	4	0–40
Turkey Vulture	10	0–30	10	7–116	1	0–2
Osprey	19	2–18	1	0–1		

Species	Martinez Lake–Yuma 1968, 1971–1988			Parker 1976–1985			Bill Williams 1977–1979, 1981, 1983, 1985		
	CBCs Present	Range Count	National High Years	CBCs Present	Range Count	National High Years	CBCs Present	Range Count	National High Years
Black-shouldered Kite	1	0–2					3	0–2	
Bald Eagle	13	0–6		3	0–1		6	1–5	
Northern Harrier	19	9–57		10	32–107		6	2–9	
Sharp-shinned Hawk	17	0–8		10	9–26				
Cooper's Hawk	18	0–15		10	10–24		6	2–11	
Harris' Hawk	9	0–10		1	0–1				
Red-shouldered Hawk							1	0–1	
Swainson's Hawk	1	0–1							
Red-tailed Hawk	19	5–37		10	28–91		6	1–23	
Ferruginous Hawk	4	0–2		10	2–38	1982‡			
Rough-legged Hawk	1	0–2		4	0–2				
Golden Eagle	12	0–4		1	0–1		1	0–1	
American Kestrel	19	1–61		10	81–187		6	16–23	
Merlin	2	0–1		8	0–4		3	0–2	
Peregrine Falcon	1	0–1		1	0–2				
Prairie Falcon	7	0–1		10	1–8		6	1–2	
Gambel's Quail	19	32–677		10	290–1,206	1978, 1983	6	71–1,270	
Black Rail	12	0–7	1973, 1982, 1985						
Clapper Rail	13	0–17					1	0–2	
Virginia Rail	18	0–71	1981	9	0–7		6	13–90	
Sora	18	0–110	1981, 1983	10	2–64		6	7–81	

Species	n	Range		n	Range		n	Range
American Coot	19	332–7,010		10	356–2,736		6	35–677
Sandhill Crane	18	0–158		9	0–778		6	2–69
Killdeer				10	99–1,248			
Mountain Plover	2	0–2		3	0–82			
Black-necked Stilt	2	0–4		1	0–1			
American Avocet	10	0–24		7	0–16		1	0–20
Greater Yellowlegs	19	1–57		10	13–84			
Spotted Sandpiper						1976, 1980, 1981	6	2–10
Western Sandpiper	1	0–1		1	0–1		1	0–1
Least Sandpiper	18	0–277		10	23–552		6	4–39
Dunlin	7	0–25		3	0–1		1	0–1
Long-billed Dowitcher	17	0–83	1971	7	0–20		3	0–25
Common Snipe	19	1–55		10	3–76		6	4–28
Wilson's Phalarope	1	0–1		1	0–1		1	0–1
Bonaparte's Gull	1	0–2						
Heermann's Gull				1	0–1			
Ring-billed Gull	18	0–1,481		10	25–118		6	75–450
California Gull	5	0–3		4	0–7		5	0–6
Herring Gull	1	0–1					2	0–1
Caspian Tern	8	0–14						
Forster's Tern	4	0–62						
Rock Dove	19	1–145		10	29–163		1	0–1
White-winged Dove	2	0–2					1	0–10
Mourning Dove	19	1–504		10	322–3,118		5	0–106
Inca Dove	2	0–3		10	7–55		4	0–21
Common Ground-Dove	18	0–55	1988	10	2–34		2	0–10
Ruddy Ground-Dove	1	0–1						
Greater Roadrunner	19	1–30		10	7–72	1976–78, 1979,‡ 1980	6	1–13

Species	Martinez Lake–Yuma 1968, 1971–1988			Parker 1976–1985			Bill Williams 1977–1979, 1981, 1983, 1985		
	CBCs Present	Range Count	National High Years	CBCs Present	Range Count	National High Years	CBCs Present	Range Count	National High Years
Barn Owl	15	0–18		10	1–7		6	2–9	
Western Screech-Owl	11	0–5		7	0–6		6	2–12	
Great Horned Owl	16	0–9		10	5–13		1	0–1	
Burrowing Owl	5	0–3		10	2–33	1979			
Long-eared Owl				4	0–3				
Short-eared Owl	2	0–1		4	0–3				
Northern Saw-whet Owl							+		
Lesser Nighthawk	2	0–5	1977, 1978*	3	0–1	1977			
Common Poorwill	1	0–1	1979						
White-throated Swift	10	0–28		10	15–1,210	1979–81, 1984	5	0–1,199	1978
Anna's Hummingbird	18	0–25		10	3–9		6	6–24	
Costa's Hummingbird	13	0–7		3	0–2		3	0–7	
Rufous/Allen's Hummingbird							+		
Belted Kingfisher	18	0–63		10	6–34		6	1–12	
Lewis' Woodpecker	2	0–2		2	0–2				
Gila Woodpecker	19	1–69		10	8–40		6	11–78	
Yellow-bellied Sapsucker				2	0–1		1	0–2	
Red-naped Sapsucker	17	0–9		10	1–10		6	5–15	
Red-breasted Sapsucker	3	0–1		3	0–2		1	0–1	
Ladder-backed Woodpecker	18	0–39		10	25–90		6	11–99	

Species									
Gilded	8	0–4		8	0–9		6	1–15	
Red-shafted	19	1–78		10	7–269		6	6–151	1977,‡
Yellow-shafted				2	0–1		1	0–2	1978,‡
Greater Pewee							3	0–1	1983‡
Hammond's Flycatcher							1	0–1	
Dusky Flycatcher				3	0–2		2	0–1	1983
Gray Flycatcher	2	0–2		1	0–1		2	0–2	
Pacific-slope Flycatcher	2	0–1	1971				2	0–2	1977
Black Phoebe	19	6–105		10	61–144		6	40–119	
Eastern Phoebe	1	0–1		2	0–1		1	0–1	
Say's Phoebe	19	4–87		10	61–244	1978,‡ 1979,‡ 1980,* 1981–83	6	41–109	
Vermilion Flycatcher	16	0–13		10	1–11		5	0–6	
Ash-throated Flycatcher	6	0–2		8	0–3	1976	4	0–6	1979, 1983
Thick-billed Kingbird	1	0–1	1972*	1	0–1	1979*	1	0–1	1983*
Horned Lark	13	0–231		10	35–3,950		1	0–11	
Violet-green Swallow	1			1	0–2				
Tree Swallow	19	79–3,141		10	8–4,831		5	0–134	
Northern Rough-winged Swallow	14	0–51		10	20–171	1977–80, 1982, 1983	5	0–37	
Bank Swallow	2	0–7	1972,* 1987	1	0–1	1977			
Barn Swallow	7	0–13	1981	4	0–6	1979, 1980	1	0–1	
Scrub Jay	2	0–2	1979, 1980	5	0–4		1	0–16	
Clark's Nutcracker	1	0–1							

Species	Martinez Lake–Yuma 1968, 1971–1988			Parker 1976–1985			Bill Williams 1977–1979, 1981, 1983, 1985		
	CBCs Present	Range Count	National High Years	CBCs Present	Range Count	National High Years	CBCs Present	Range Count	National High Years
American Crow	1	0–1		2	0–1		6	7–16	
Common Raven	2	0–2		10	5–48		6	31–195	
Verdin	19	2–241	1983	10	31–333		2	0–87	
Bushtit				2	0–2				
White-breasted Nuthatch				1	0–1				
Red-breasted Nuthatch				1	0–1				
Brown Creeper	17	0–43		6	0–3		4	0–9	
Cactus Wren	19	1–18		10	17–78		6	2–74	
Rock Wren	2	0–1		10	1–19		6	12–56	
Canyon Wren	16	0–20					6	19–68	
Bewick's Wren	15	0–34		10	6–108		6	4–98	
House Wren	1	0–1		10	30–97		6	39–121	
Winter Wren				4	0–4		4	0–3	
Sedge Wren				1	0–1				
Marsh Wren	19	1–188		10	49–370	1978, 1982	6	38–430	
Golden-crowned Kinglet	1	0–1		4	0–7		3	0–6	
Ruby-crowned Kinglet	19	5–257		10	110–957		6	96–1,347	
Blue-gray Gnatcatcher	16	0–44		10	26–102				
Black-tailed Gnatcatcher	19	6–115	1974, 1975	10	43–165	1976, 1978–82	6	23–95	
Western Bluebird	3	0–11		8	0–56		2	0–17	
Mountain Bluebird	6	0–101		8	0–213		1	0–10	
Townsend's Solitaire	1	0–2		1	0–1		2	0–1	
Hermit Thrush	13	0–5		8	0–77		6	1–188	

Species	n	Range	Year	n	Range	Year	n	Range	Year
American Robin	17	0–315		10	2–389		5	0–52	
Varied Thrush	+			1	0–1		1	0–4	
Northern Mockingbird	19	1–71		10	25–165		6	3–19	
Sage Thrasher				3	0–4		1	0–1	
Brown Thrasher				1	0–1				
Bendire's Thrasher	1	0–1							
Curve-billed Thrasher	2	0–1					3	0–2	
Crissal Thrasher	19	1–26		10	23–106	1977, 1978,‡ 1979–81, 1982*	6	2–36	
Le Conte's Thrasher				3	0–2				
American Pipit	19	3–295		10	754–6,723	1978, 1979, 1982	6	12–1,107	
Sprague's Pipit				2	0–1		6	1–95	
Cedar Waxwing	4	0–22		9	0–22		6	30–157	
Phainopepla	19	6–184		10	77–798	1976,‡ 1977,* 1978–81			
Northern Shrike				1	0–1				
Loggerhead Shrike	19	10–78		10	56–194		6	28–52	
European Starling	19	12–893		10	556–6,830		6	89–523	
Bell's Vireo	1	0–1	1984	2	0–1	1978, 1981			
Solitary Vireo		0–3		6	0–3		4	0–2	
Cassin's form	5	0–1	1986					0–2	
Plumbeous form	2	0–3	1987	6	0–3		4	0–2	
Hutton's Vireo	5			1	0–1		1	0–1	
Warbling Vireo							1		1978
Tennessee Warbler	1	0–1							
Orange-crowned Warbler	18	0–104		10	48–352	1978	6	49–196	

Species	Martinez Lake–Yuma 1968, 1971–1988			Parker 1976–1985			Bill Williams 1977–1979, 1981, 1983, 1985		
	CBCs Present	Range Count	National High Years	CBCs Present	Range Count	National High Years	CBCs Present	Range Count	National High Years
Nashville Warbler				2	0–1		1	0–1	
Northern Parula				9	0–4		1	0–1	
Yellow Warbler	2	0–1		1	0–1	1978	1	0–1	
Chestnut-sided Warbler							1	0–1	1977
Magnolia Warbler									
Yellow-rumped Warbler	19	62–1,250		10	275–3,990		6	158–1,193	
Audubon's form	19	62–1,246		10	269–3,981		6	158–1,188	
Myrtle form	7	0–8		8	0–10		6	1–8	
Black-throated Gray Warbler	5	0–2		4	0–2		4	0–11	
Townsend's Warbler	1	0–1		1	0–2				
Palm Warbler	1	0–1		1	0–1				
Black-and-white Warbler	2	0–2		2	0–2		3	0–2	
American Redstart	4	0–1		5	0–1		1	0–1	
Ovenbird				1	0–1				
Northern Waterthrush				1	0–2				
Common Yellowthroat	18	0–31		10	26–117		6	1–11	
Wilson's Warbler	3	0–1		4	0–1				
Hepatic Tanager	†								
Summer Tanager				10	1–5		1	0–1	
Northern Cardinal				2	0–2		1	0–2	
Lazuli Bunting									
Green-tailed Towhee	11	0–8		10	6–24		6	2–8	
Rufous-sided Towhee	13	0–16		10	1–32		6	1–38	

Species						1982	
Chipping Sparrow	15	0–75	10	13–240		5	0–23
Brewer's Sparrow	10	0–83	10	1–934		2	0–93
Black-chinned Sparrow	2	0–3					
Vesper Sparrow	16	0–310	10	95–1,064	1978	6	2–11
Lark Sparrow	11	0–30	10	1–121		4	0–17
Black-throated Sparrow	10	0–40	6	0–18		5	0–74
Sage Sparrow	18	0–47	10	24–237	1978, 1983	5	0–12
Lark Bunting			3	0–74			
Savannah Sparrow	16	0–504	10	280–4,965	1977, 1981	6	7–516
Grasshopper Sparrow	1	0–1	2	0–1			
Fox Sparrow	1	0–1	3	0–1			
Song Sparrow	17	0–60	10	21–234		6	56–172
Lincoln's Sparrow	18	0–24	10	9–360	1979	6	13–244
Swamp Sparrow	3	0–3	6	0–8		3	0–7
White-throated Sparrow	4	0–3	3	0–3		3	0–1
Golden-crowned Sparrow			7	0–5		4	0–8
White-crowned Sparrow	19	85–991	10	2,444–19,529	1977–81	6	415–1,234
Harris' Sparrow			2	0–1			
Dark-eyed Junco	19	3–107	10	32–157		6	6–57
Oregon and Pink-sided forms	19	3–99	9	0–151		6	6–56
Slate-colored form	4	0–13	6	0–3		1	0–1
Gray-headed form	3	0–5	6	0–3		1	0–1
McCown's Longspur						1	0–2
Lapland Longspur			2	0–3			
Chestnut-collared Longspur			6	0–18			
Red-winged Blackbird	19	2–13,444	10	2,619–17,991		6	18–113
Eastern Meadowlark			2	0–5		1	0–1

Species	Martinez Lake–Yuma 1968, 1971–1988			Parker 1976–1985			Bill Williams[*] 1977–1979, 1981, 1983, 1985		
	CBCs Present	Range Count	National High Years	CBCs Present	Range Count	National High Years	CBCs Present	Range Count	National High Years
Western Meadowlark	19	13–253		10	417–1,393		6	54–310	
Yellow-headed Blackbird	14	0–3,590		8	0–1,002		1	0–7	
Rusty Blackbird	1	0–2							
Brewer's Blackbird	12	0–1,164		10	616–11,717		5	0–87	
Great-tailed Grackle	19	4–1,094		10	59–425		6	3–250	
Brown-headed Cowbird	7	0–50		10	11–4,007		2	0–21	
Northern Oriole				1	0–1				
Purple Finch				3	0–11		2	0–1	
Cassin's Finch				1	0–12				
House Finch	18	0–745		10	87–2,729		6	56–164	
Red Crossbill				1	0–2				
Pine Siskin	1	0–12		8	0–28		2	0–46	
Lesser Goldfinch	14	0–29		10	1–68		5	0–9	
Lawrence's Goldfinch				3	0–27				
American Goldfinch	3	0–4		10	2–45		5	0–53	
Evening Grosbeak				1	0–1				
House Sparrow	19	15–310		10	163–1,021		6	35–155	

*Year in which all-time national high count has been maintained, at least until 1989.
†Species recorded during count period only.
‡Year in which all-time national high count was established for species on CBCs.

Appendix 2
Hypothetical Species

Leach's Storm-Petrel (*Oceanodroma leucorhoa*)

STATUS A storm-petrel with a white rump (probably this species) was seen on Lake Havasu, 3 September 1977. Although no violent storms were associated with this sighting, several other normally pelagic species occurred at the same time.

Anhinga (*Anhinga anhinga*)

STATUS Old sight records at Laguna Dam (9 February 1913) and Yuma; not substantiated.

Black Vulture (*Coragyps atratus*)

STATUS One reported below Parker Dam, in California, on 5 September 1977. This would have been the first occurrence of this species in California if the record had been confirmed; it was rejected by the California Bird Records Committee.

White-tailed Hawk (*Buteo albicaudatus*)

STATUS One adult was repeatedly seen by the refuge manager at Imperial NWR during the winter of 1977–78.

This open-country raptor has been reported as casual in central Arizona (west to Gila Bend) and bred there historically (Phillips et al. 1964). Its present status in northwestern Mexico is poorly known and, undoubtedly, its numbers have decreased with the overgrazing of desert grasslands. Although adults are distinctive if seen well, immature White-tailed Hawks (the most likely to occur here) may be easily confused with Ferruginous, some Rough-legged, and certain races of Red-tailed Hawks.

Piping Plover (*Charadrius melodus*)

STATUS One seen briefly (bright orange legs noted) with Snowy Plovers in a plowed field at Blythe on 11 August 1979. A few winter regularly at Puerto Peñasco, Sonora, Mexico, not far to the south, making passage of a few individuals through this region likely.

Common Black-headed Gull (*Larus ridibundus*)

STATUS A worn-plumaged gull seen and photographed 14–16 June 1978 at Imperial NWR (photo) may have been this species.

Plain Titmouse (*Parus inornatus*)

STATUS One reported from a feeder in Yuma, 6 April 1988. This report, if correct, is nearly unprecedented for the deserts of southwestern Arizona and southeastern California, with only one other record from near the north end of the Salton Sea.

Sedge Wren (*Cistothorus platensis*)

STATUS One glimpsed repeatedly, but not documented, on 24 December 1981 in a harvested alfalfa field south of Parker. Also, one seen at El Gulfo, Mexico, in the Colorado River Delta, on 4 December 1983.

Philadelphia Vireo (*Vireo philadelphicus*)

STATUS One seen, but not substantiated, 16 September 1976 near Poston and another 5 September 1982 at Blythe. One described in detail from Laguna Dam 27 May 1983 is being reviewed by the California Bird Records Committee. This species occurs every year in Arizona and southern California and is to be expected as a casual transient.

Red-faced Warbler (*Cardellina rubrifrons*)

STATUS One apparently mist-netted but not documented, 19 April 1975, at Imperial NWR.

Appendix 3
Bar Graphs: Seasonal Status of Each Species

These bar graphs are a pictorial representation of the seasonal status of each species. In general, the length of each line corresponds to a species' period of occurrence in the lower Colorado River Valley. The width of the line signifies the species' abundance at that time. In many cases the graphs are an oversimplification and the text should be consulted for species of special interest. Note that a few recently added species appear in the main text but not in the bar graphs.

Along with the abundance designations are codes for the major habitats used by each species. For very rare species, the code may represent the habitat in which particular records were obtained and not necessarily all habitats of potential use. The habitat codes are as follows:

O   Open water; mainly artificial lakes and reservoirs.
R   River; flowing channel, exposed bars, and shoreline.
M   Marsh; emergent vegetation along river channel, backwaters, lakes, etc.
V   Riparian vegetation; includes all natural plant communities within the floodplain.
D   Desert; includes washes and flats immediately adjacent to river or floodplain.
A   Agricultural land; includes cultivated crops, bare fields, orchards, canals, and rural residences.
U   Urban or suburban residential communities.

The following abundance codes indicate a species' average status in its preferred habitat(s). It may be more numerous locally, or less common in other habitats. For species whose abundance varies geographically within the river valley, this is clearly stated in the text.

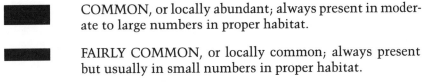

COMMON, or locally abundant; always present in moderate to large numbers in proper habitat.

FAIRLY COMMON, or locally common; always present but usually in small numbers in proper habitat.

UNCOMMON; occurs locally or patchily and in small numbers at the given season.

———————   RARE; occurs annually during the period indicated but generally in very small numbers, or extremely local and in small numbers.

....   CASUAL; records are too numerous to show individually, but occurrence is sporadic and generally unexpected.

•   INDIVIDUAL RECORD; for species recorded only a few times overall or in a given season.

•..........•   Individual record, showing a prolonged stay by the individual(s).

When a species occurrence or status varies from year to year this is indicated by a broken abundance graph. In those cases, the wider bar represents the greatest potential abundance and the narrower connecting bar the least common state. For example:

▰▰▰▰▰   Irregular abundance; bar represents minimum and maximum abundance; for example, common in some years, or fairly common in others.

- - - - - - - -   Rare and irregular; may be absent in some years, but not unexpected at that time.

*   Breeding species.

(L)   Local.

SPECIES	HABITAT	JAN	FEB	MAR	APR	MAY	JUN	JUL	AUG	SEP	OCT	NOV	DEC
Red-throated Loon	O												
Pacific Loon	O												
Common Loon	O												
Least Grebe *	M												
Pied-billed Grebe *	MRO												
Horned Grebe	O												
Red-necked Grebe	O												
Eared Grebe	OR												
Western Grebe * (L)	MO												
Clark's Grebe * (L)	MO												
Laysan Albatross	O												
Least Storm-Petrel	O												
Blue-footed Booby	O												
Brown Booby	OR												
American White Pelican	OR												
Brown Pelican	RO												
Double-crested Cormorant *	RO												
Olivaceous Cormorant	OR												
Magnificent Frigatebird	M												
American Bittern *	M												

SPECIES	HABITAT	JAN	FEB	MAR	APR	MAY	JUN	JUL	AUG	SEP	OCT	NOV	DE
Least Bittern *	M												
Great Blue Heron *	VMR												
Great Egret *	VMRA												
Snowy Egret *	VRM												
Little Blue Heron	RM												
Tricolored Heron	RM												
Reddish Egret	R												
Cattle Egret	AR												
Green-backed Heron *	VRM												
Black-crowned Night-Heron *	VMR												
Yellow-crowned Night-Heron	RM												
White Ibis	RM												
White-faced Ibis	RMA												
Roseate Spoonbill	RM												
Wood Stork	MA												
Fulvous Whistling-Duck	MA												
Tundra Swan	RM												
Greater White-fronted Goose	RA												
Snow Goose (L)	RA												
Ross' Goose	RA												

SPECIES	HABITAT	JAN	FEB	MAR	APR	MAY	JUN	JUL	AUG	SEP	OCT	NOV	DEC
ant	RA												
nada Goose	RMA												
ood Duck	RM												
reen-winged Teal	RAM												
allard	RM												
orthern Pintail	RAM												
ue-winged Teal	RM												
nnamon Teal *	RAM												
orthern Shoveler	RM												
adwall *	RM												
urasian Wigeon	R												
merican Wigeon	RM												
anvasback	RO												
edhead *	RO												
ing-necked Duck	RO												
reater Scaup	RO												
esser Scaup	RO												
ldsquaw	OR												
lack Scoter	OR												
urf Scoter	OR												

SPECIES	HABITAT	JAN	FEB	MAR	APR	MAY	JUN	JUL	AUG	SEP	OCT	NOV	DEC
White-winged Scoter	OR												
Common Goldeneye	RO												
Barrow's Goldeneye	RO												
Bufflehead	RO												
Hooded Merganser	RO												
Common Merganser	RO												
Red-breasted Merganser	RO												
Ruddy Duck *	MRO												
Turkey Vulture *	AVD												
Osprey	RM												
Black-shouldered Kite	A												
Mississippi Kite	V												
Bald Eagle	RM												
Northern Harrier	MVA												
Sharp-shinned Hawk	VA												
Cooper's Hawk *	VA												
Northern Goshawk	V												
Common Black-Hawk	V												
Harris' Hawk *	VD												
Red-shouldered Hawk *	V												

SPECIES	HABITAT	JAN	FEB	MAR	APR	MAY	JUN	JUL	AUG	SEP	OCT	NOV	DEC
Broad-winged Hawk	V												
Swainson's Hawk	AV												
Zone-tailed Hawk * (L)	V												
Red-tailed Hawk *	VAD												
Ferruginous Hawk	A												
Rough-legged Hawk	AV												
Golden Eagle	DVA												
Crested Caracara	VA												
American Kestrel *	VAD												
Merlin	VA												
Peregrine Falcon *	RMA												
Prairie Falcon *	VAD												
Ring-necked Pheasant (L)	A												
Gambel's Quail *	VDA												
Black Rail * (L)	M												
Clapper Rail *	M												
Virginia Rail *	M												
Sora	M												
Common Moorhen *	MR												
American Coot *	MOR												

SPECIES	HABITAT	JAN	FEB	MAR	APR	MAY	JUN	JUL	AUG	SEP	OCT	NOV	DEC
Sandhill Crane (L)	RAM												
Black-bellied Plover	AR												
Lesser Golden-Plover	AR												
Snowy Plover	AR												
Semipalmated Plover	AR												
Killdeer *	ARM												
Mountain Plover	A												
Black-necked Stilt *	MRA												
American Avocet	MRA												
Northern Jacana	M												
Greater Yellowlegs	RA												
Lesser Yellowlegs	RA												
Solitary Sandpiper	RA												
Willet	RMA												
Spotted Sandpiper	RM												
Upland Sandpiper	AR												
Whimbrel	RA												
Long-billed Curlew	RA												
Marbled Godwit	R												
Ruddy Turnstone	R												

SPECIES	HABITAT	JAN	FEB	MAR	APR	MAY	JUN	JUL	AUG	SEP	OCT	NOV	DEC
Black Turnstone	R					•							
Red Knot	R							•	•	•		•	
Sanderling	RA											• •	
Semipalmated Sandpiper	R								•				
Western Sandpiper	RMA										•		
Least Sandpiper	RMA					•							
Baird's Sandpiper	RMA							• •					
Pectoral Sandpiper	RMA				• •								
Dunlin	RM								•				
Stilt Sandpiper	RA						•						
Short-billed Dowitcher	RMA												
Long-billed Dowitcher	RMA					•	•	•					
Common Snipe	RMVA						•						
Wilson's Phalarope	RA					•	•						•
Red-necked Phalarope	ORA												
Red Phalarope	OR							•					
Pomarine Jaeger	O								• •				
Parasitic Jaeger	O												
Long-tailed Jaeger	O								• •				
Laughing Gull	ORA			• •			•						

SPECIES	HABITAT	JAN	FEB	MAR	APR	MAY	JUN	JUL	AUG	SEP	OCT	NOV	DEC
Franklin's Gull	ORA												
Bonaparte's Gull	OR												
Heermann's Gull	OR												
Mew Gull	OR												
Ring-billed Gull	ORA												
California Gull	ORA												
Herring Gull	ORA												
Thayer's Gull	O												
Western Gull	O												
Glaucous-winged Gull	OR												
Black-legged Kittiwake	O												
Sabine's Gull	OR												
Gull-billed Tern	R												
Caspian Tern	ROA												
Common Tern	OR												
Arctic Tern	O												
Forster's Tern	RMO												
Least Tern	R												
Black Tern	ORMA												
Black Skimmer	AR												

SPECIES	HABITAT	JAN	FEB	MAR	APR	MAY	JUN	JUL	AUG	SEP	OCT	NOV	DEC
Rock Dove *	U												
Band-tailed Pigeon	V												
White-winged Dove *	VADU												
Mourning Dove *	VAUD												
Inca Dove * (L)	U												
Common Ground-Dove *	VA												
Yellow-billed Cuckoo *	V												
Greater Roadrunner *	AVD												
Barn Owl *	AV												
Flammulated Owl	V												
Western Screech-Owl *	VD												
Great Horned Owl *	VDA												
Elf Owl * (L)	VD												
Burrowing Owl *	A												
Long-eared Owl *	V												
Short-eared Owl	AVM												
Northern Saw-whet Owl	V												
Lesser Nighthawk *	VDRA												
Common Nighthawk	VD												
Common Poorwill *	DV												

SPECIES	HABITAT	JAN	FEB	MAR	APR	MAY	JUN	JUL	AUG	SEP	OCT	NOV	DEC
Black Swift	R					•							
Chimney Swift	R					•	•						
Vaux's Swift	RV				▬█	▬			▬▬	▬			
White-throated Swift	DRA	██
Broad-billed Hummingbird	V	⋯⋯	⋯• •⋯•						•	•• ⋯	⋯ •	•⋯	
Black-chinned Hummingbird *	VUD			▬▬████████████							
Anna's Hummingbird *	UV	████████	▬▬	████████				
Costa's Hummingbird *	DV	▬█████	▬								
Calliope Hummingbird	VU				▬ ▬						
Broad-tailed Hummingbird	V				•• •						
Rufous Hummingbird	VU		▬███	▬		▬███████	▬		
Allen's Hummingbird	VU	•				•• •	• •			
Belted Kingfisher	RM	████	▬▬▬▬	████		
Lewis' Woodpecker	VA	▬▬▬▬▬▬▬ ▪			▬▬▬▬
Acorn Woodpecker	V				•		⁝	•		
Gila Woodpecker *	VDU	███
Yellow-bellied Sapsucker	V	▬▬				▬▬	
Red-naped Sapsucker	V	████▬				▬████	
Red-breasted Sapsucker	V	▬▬▬				▬▬	
Williamson's Sapsucker	V		•									⁝

SPECIES	HABITAT	JAN	FEB	MAR	APR	MAY	JUN	JUL	AUG	SEP	OCT	NOV	DEC
Ladder-backed Woodpecker *	VD												
Northern Flicker "Gilded" * (L)	VD												
Northern Flicker "Red-shafted"	VD												
"Yellow-shafted"	V												
Olive-sided Flycatcher	V												
Greater Pewee	V												
Western Wood-Pewee	V												
Willow Flycatcher *	V												
Hammond's Flycatcher	V												
Dusky Flycatcher	V												
Gray Flycatcher	V												
Pacific Slope Flycatcher	V												
Black Phoebe *	VRMA												
Eastern Phoebe	VA												
Say's Phoebe *	AVD												
Vermilion Flycatcher * (L)	AV												
Dusky-capped Flycatcher	V												
Ash-throated Flycatcher *	VD												
Brown-crested Flycatcher *	V												
Tropical Kingbird	V												

SPECIES	HABITAT	JAN	FEB	MAR	APR	MAY	JUN	JUL	AUG	SEP	OCT	NOV	DEC
Cassin's Kingbird *	AV												
Thick-billed Kingbird	V												
Western Kingbird *	AV												
Eastern Kingbird	VA												
Scissor-tailed Flycatcher	AV												
Horned Lark *	A												
Purple Martin	R												
Tree Swallow	RMA												
Violet-green Swallow *	RV												
N. Rough-winged Swallow *	RMA												
Bank Swallow	RM												
Cliff Swallow *	RMA												
Barn Swallow	RMA												
Steller's Jay	V												
Scrub Jay	V												
Pinyon Jay	V												
Clark's Nutcracker	V												
American Crow (L)	VA												
Common Raven *	VDA												
Mountain Chickadee	V												

SPECIES	HABITAT	JAN	FEB	MAR	APR	MAY	JUN	JUL	AUG	SEP	OCT	NOV	DEC
Bridled Titmouse	V												
Verdin *	VDU												
Bushtit	VD												
Red-breasted Nuthatch	V												
White-breasted Nuthatch	V												
Pygmy Nuthatch	V												
Brown Creeper (L)	V												
Cactus Wren *	VD												
Rock Wren *	D												
Canyon Wren * (L)	DV												
Bewick's Wren *	VD												
House Wren	VMA												
Winter Wren	V												
Marsh Wren *	MVA												
American Dipper	R												
Golden-crowned Kinglet	V												
Ruby-crowned Kinglet	VD												
Blue-gray Gnatcatcher	V												
Black-tailed Gnatcatcher *	VD												
Western Bluebird	VD												

SPECIES	HABITAT	JAN	FEB	MAR	APR	MAY	JUN	JUL	AUG	SEP	OCT	NOV	DEC
Mountain Bluebird	VDA												
Townsend's Solitaire	V												
Swainson's Thrush	V												
Hermit Thrush	V												
Rufous-backed Robin	VU												
American Robin *	VU												
Varied Thrush	V												
Gray Catbird	V												
Northern Mockingbird *	VAU												
Sage Thrasher	VDA												
Brown Thrasher	V												
Bendire's Thrasher	VD												
Curve-billed Thrasher	VD												
Crissal Thrasher *	VA												
Le Conte's Thrasher	D												
American Pipit	RA												
Sprague's Pipit	A												
Bohemian Waxwing	VU												
Cedar Waxwing	VU												
Phainopepla *	VD												

SPECIES	HABITAT	JAN	FEB	MAR	APR	MAY	JUN	JUL	AUG	SEP	OCT	NOV	DEC
Northern Shrike	AV												
Loggerhead Shrike *	VDA												
European Starling *	UVADM												
Bell's Vireo * (L)	V												
Gray Vireo	VU												
Solitary Vireo: *V. s. cassini*	VU												
V. s. plumbeus	VU												
Yellow-throated Vireo	V												
Hutton's Vireo	V												
Warbling Vireo	V												
Red-eyed Vireo	V												
Blue-winged Warbler	V												
Golden-winged Warbler	V												
Tennessee Warbler	V												
Orange-crowned Warbler	V												
Nashville Warbler	MA												
Virginia's Warbler	MA												
Lucy's Warbler *	RM												
Northern Parula *	RA												
Yellow Warbler *	RA												
Chestnut-sided Warbler													

SPECIES	HABITAT	JAN	FEB	MAR	APR	MAY	JUN	JUL	AUG	SEP	OCT	NOV	DEC
Magnolia Warbler	V	···•		•						••	•		•
Cape May Warbler	V									•	•		
Black-throated Blue Warbler	V									••		• ••
Yellow-rumped "Myrtle" Warbler:	VU												
"Audubon's"	VAUM												
Black-throated Gray Warbler	VD												
Townsend's Warbler	V	• •											
Hermit Warbler	V												
Black-throated Green Warbler	V										•• •		
Yellow-throated Warbler	V		•						•				
Grace's Warbler	V		•										
Prairie Warbler	V												
Palm Warbler	V									•	•		•
Bay-breasted Warbler	V						•				•		
Blackpoll Warbler	V					◄ ••	•				•		
Black-and-white Warbler	V												
American Redstart	V						•						
Prothonotary Warbler	V					•					••		
Worm-eating Warbler	V					•				•			
Ovenbird	V				•	•	•			•	•		

SPECIES	HABITAT	JAN	FEB	MAR	APR	MAY	JUN	JUL	AUG	SEP	OCT	NOV	DEC
Northern Waterthrush	MV												
Louisiana Waterthrush	V												
Kentucky Warbler	V												
MacGillivray's Warbler	V												
Common Yellowthroat *	MV												
Hooded Warbler	V												
Wilson's Warbler	V												
Painted Redstart	V												
Yellow-breasted Chat *	V												
Hepatic Tanager	V												
Summer Tanager *	V												
Scarlet Tanager	V												
Western Tanager	VD												
Northern Cardinal * (L)	V												
Pyrrhuloxia	VD												
Rose-breasted Grosbeak	VU												
Black-headed Grosbeak	V												
Blue Grosbeak *	VMA												
Lazuli Bunting	VA												
Indigo Bunting *	V												

SPECIES	HABITAT	JAN	FEB	MAR	APR	MAY	JUN	JUL	AUG	SEP	OCT	NOV	DEC
Varied Bunting	V												
Painted Bunting	V												
Dickcissel	VA												
Green-tailed Towhee	VA												
Rufous-sided Towhee	V												
Abert's Towhee *	VUA												
Cassin's Sparrow	V												
American Tree Sparrow	VM												
Chipping Sparrow	VAU												
Clay-colored Sparrow	UA												
Brewer's Sparrow	VDA												
Black-chinned Sparrow	VD												
Vesper Sparrow	A												
Lark Sparrow *	AU												
Black-throated Sparrow *	D												
Sage Sparrow	VD												
Lark Bunting	A												
Savannah Sparrow	A												
Grasshopper Sparrow	A												
Le Conte's Sparrow	A												
Sharp-tailed Sparrow	M												

SPECIES	HABITAT	JAN	FEB	MAR	APR	MAY	JUN	JUL	AUG	SEP	OCT	NOV	DEC
Fox Sparrow	VM												
Song Sparrow *	MVA												
Lincoln's Sparrow	MVA												
Swamp Sparrow	MV					•							
White-throated Sparrow	VA												
Golden-crowned Sparrow	VDA												
White-crowned Sparrow	VDAU						•						
Harris' Sparrow	A												
Dark-eyed "Oregon" Junco:	VA												
"Slate-colored"	VA												
"Pink-sided"	VA												
"Gray-headed"	VA				•								
"White-winged"	VA			•									
McCown's Longspur	A		•								•	•:	
Lapland Longspur	A												
Chestnut-collared Longspur	A			•••▸									
Bobolink	A					••			•				
Red-winged Blackbird *	MA												
Eastern Meadowlark	A			•									•
Western Meadowlark *	A												
Yellow-headed Blackbird *	MA												

SPECIES	HABITAT	JAN	FEB	MAR	APR	MAY	JUN	JUL	AUG	SEP	OCT	NOV	DEC
Rusty Blackbird	V												
Brewer's Blackbird	AUV												
Great-tailed Grackle *	MAU												
Common Grackle	AV												
Bronzed Cowbird *	U												
Brown-headed Cowbird *	VAUM												
Orchard Oriole	V												
Hooded Oriole *	VU												
Northern Oriole: "Bullock's" *	VUA												
"Baltimore" *	VUA												
Scott's Oriole	D												
Purple Finch	VU												
Cassin's Finch	V												
House Finch *	VDAU												
Red Crossbill	VU												
Pine Siskin	V												
Lesser Goldfinch *	VU												
Lawrence's Goldfinch *	V												
American Goldfinch	VA												
Evening Grosbeak	VU												
House Sparrow	AU												

Appendix 4
Common and Scientific Names of Plants and Animals Mentioned in the Text

ANIMALS

Asiatic clam (*Corbicula fluminea*)
Beaver (*Castor canadensis* Kuhle)
Cicada (*Diceroprocta apache*)
Coachwhip (*Masticophis flagellum*)
Common kingsnake (*Lampropeltis getulus*)
Cottonrat (*Sigmodon* spp.)
Coyote (*Canis latrans* Say)
Crayfish (*Procambarus clarki* (Girard))
Gopher snake (*Pituophis melanoleucus*)
Mule deer (*Odocoileus hemionus* Rafinesque)
Muskrat (*Ondatra zibethicus* Linnaeus)
Threadfin shad (*Dorosoma petenense* (Günther))

PLANTS

Arrowweed (*Tessaria sericea* (Nutt.) Shinners)
Athel tamarisk (*Tamarix aphylla* (L.) Karst.)
Bermuda grass (*Cynodon dactylon* (L.) Pers.)
Blue palo verde (*Cercidium floridum* Benth.)
California bulrush (*Scirpus californicus* (C. Meyer) Steud.)
Canary grass (*Phalaris* spp.)
Catclaw acacia (*Acacia greggii* Gray)
Cattails (*Typha latifolia* L. and *T. domingensis* Pers.)
Common reed (*Phragmites australis* (Cav.) Trin. ex. Steud.)
Coyote willow (*Salix exigua* Nutt.)
Creosote bush (*Larrea divaricata* spp. *tridentata* (Sesse & Moc. ex. D.C.) Felger & Lowe)
Duckweed (*Lemna* spp.)
Filaree (*Erodium cicutarum* (L.) L'Her.)
Fremont cottonwood (*Populus fremontii* Wats.)
Giant reed (*Arundo donax* A.)
Goodding willow (*Salix gooddingii* Ball)
Goosefoot (*Chenopodium* spp.)

Honey mesquite (*Prosopis glandulosa* var. *torreyana* (Belson) M. C. Johnst.)
Inkweed or pickleweed (*Suaeda torreyana* Wats.)
Ironwood (*Olneya tesota* Gray)
Linear-leaved cambess (*Oligomeris linifolia* (Vahl) Macbr.)
Mangrove (*Rhizophora* spp.)
Mistletoe (*Phoradendron californicum* Nutt.)
Mulefat (*Baccharis viminea* D.C.)
Oak (*Quercus* spp.)
Panic grass (*Panicum* spp.)
Pine (*Pinus* spp.)
Purslane (*Portulaca* spp.)
Quail bush (*Atriplex lentiformis* (Tour.) Wats.)
Russian thistle (*Salsola iberica* Sennen & Pau)
Sago pondweed (*Potamogeton pectinatus* L.)
Saguaro (*Carnegiea gigantea* (Engelm.) B & R)
Saltbush (*Atriplex polycarpa* (Torr.) Wats.)
Saltcedar (*Tamarix chinensis* Loureiro)
Salt-grass (*Distichlis* spp.)
Screwbean mesquite (*Prosopis pubescens* Benth.)
Sedge (*Cyperus* spp.)
Seepwillow (*Baccharis salicifolia* (R.&P.) Pers.)
Smartweed (*Polygonum* spp.)
Smotherweed (*Bassia hyssopifolia* (Pall.) Kuntze)
Thistle (*Cirsium* spp.)
Three-square bulrush (*Scirpus americanus* Pers.)
Tree tobacco (*Nicotiana glauca* Graham)
Wild grape (*Vitis* spp.)
Wolfberry (*Lycium* spp.)

Bibliography

Alcorn, S. M., S. E. McGregor, and G. Olin. 1961. Pollination of saguaro cactus by doves, nectar feeding bats, and honey bees. *Science* 133:1594–1595.

American Ornithologists' Union. 1983. *Check-list of North American Birds*, 6th ed. Allen Press, Inc., Lawrence, Kan.

———. 1985. Thirty-fifth supplement to the American Ornithologists' Union check-list of North American birds. *Auk* 102:680–686.

———. 1987. Thirty-sixth supplement to the American Ornithologists' Union check-list of North American birds. *Auk* 104:591–596.

———. 1989. Thirty-seventh supplement to the American Ornithologists' Union check-list of North American birds. *Auk* 106:532–538.

Anderson, B. W., W. C. Hunter, and R. D. Ohmart. 1989. Status changes of bird species using revegetated riparian habitats on the lower Colorado River from 1977 to 1984. Pp. 325–331 in *Proceedings of the California Riparian Systems Conference: Protection, Management, and Restoration for the 1990's*, D. Abel, tech. coord. USDA Forest Service General Technical Report PSW-110, Pacific Southwest Forest and Range Experiment Station, Berkeley, Calif.

Anderson, B. W., and R. D. Ohmart. 1977. Vegetation structure and bird use in the lower Colorado River Valley. Pp. 23–33 in *Importance, Preservation and Management of Riparian Habitat: A Symposium*, R. R. Johnson and D. A. Jones, tech. coords. USDA Forest Service General Technical Report RM-43, Rocky Mountain Forest and Range Experiment Station, Fort Collins, Colo., 217 pp.

———. 1978. Phainopepla utilization of honey mesquite forests in the Colorado River Valley. *Condor* 80:334–338.

———. 1982a. The influence of interspersion of agriculture and natural habitats on wildlife in southern California and western Arizona. Comprehensive final report submitted to U.S. Bureau of Reclamation, Lower Colorado Region, Boulder City, Nev.

———. 1982b. Revegetation for wildlife enhancement along the lower Colorado River. Final report submitted to U.S. Bureau of Reclamation, Lower Colorado Region, Boulder City, Nev.

———. 1984. A vegetation management study for the enhancement of wildlife along the lower Colorado River. Final report submitted to U.S. Bureau of Reclamation, Boulder City, Nev.

———. 1985a. Habitat use by Clapper Rails in the lower Colorado River Valley. *Condor* 87:116–136.

———. 1985b. Managing riparian vegetation and wildlife along the Colorado River: synthesis of data, predictive models and management. Pp. 123–127 in *Riparian Ecosystems and Their Management: Reconciling Conflicting Uses,*

R. R. Johnson, C. D. Ziebell, D. R. Patton, P. F. Ffolliott, and R. H. Hamre, tech. coords. USDA Forest Service General Technical Report RM-120, Rocky Mountain Forest and Range Experiment Station, Fort Collins, Colo.

———. 1988. Ecological relationships among duck species wintering along the lower Colorado River. Pp. 191–263 in *Waterfowl in Winter*, M. Weller, ed. University of Minnesota Press, Minneapolis, Minn.

Anderson, B. W., R. D. Ohmart, and S. D. Fretwell. 1982. Evidence for social regulation in some riparian bird populations. *American Naturalist* 120:340–352.

Anderson, B. W., R. D. Ohmart, and J. Rice. 1983. Avian vegetation community structure and their seasonal relationships in the lower Colorado River Valley. *Condor* 85:392–405.

Anderson, S. H., and H. H. Shugart. 1974. Habitat selection of breeding birds in an east Tennessee deciduous forest. *Ecology* 55:828–837.

Arnold, L. W. 1942. Water birds influenced by irrigation projects in the lower Colorado River Valley. *Condor* 44:183–184.

Atwood, J. L. 1988. Speciation and geographic variation in Black-tailed Gnatcatchers. *Ornithological Monographs* no. 42.

Austin, G. T. 1976. Behavioral adaptations of the Verdin to the desert. *Auk* 93:245–262.

Avise, J. C., and R. M. Zink. 1988. Molecular genetic divergence between avian sibling species: King and Clapper Rails, Long-billed and Short-billed Dowitchers, Boat-tailed and Great-tailed Grackles, and Tufted and Black-crested Titmouse. *Auk* 105:516–528.

Axelrod, D. I. 1950. Evolution of desert vegetation in western North America. *Carnegie Institute of Washington Publication* 590:215–360.

Balda, R. P. 1969. Foliage use by birds of the oak-juniper woodland and ponderosa pine forest in southeastern Arizona. *Condor* 71:399–412.

Banks, R. C., and R. G. McCaskie. 1964. Distribution and status of Wied's Crested Flycatcher in the lower Colorado River Valley. *Condor* 66:250–251.

Bartlett, J. R. 1854. *Personal Narrative of Exploration and Incidents in Texas, New Mexico, California, Sonora, and Chihuahua, Connected with the United States and Mexican Boundary Commission, During the Years 1850, '51, '52 and '53, with an Introduction by Odie B. Faulk.* D. Appleton and Co., New York, N.Y., vol. 1, 501 pp.; vol. 2, 624 pp.

Bennett, W. W., and R. D. Ohmart. 1978. Habitat requirements and population characteristics of the Clapper Rail in the Imperial Valley of California. Final report submitted to University of California, Lawrence Livermore Laboratory, Livermore, Calif.

Bent, A. C. 1963. *Life Histories of North American Marsh Birds.* Dover Publications, Inc., New York, N.Y., pp. 275–277.

———. 1964. *Life Histories of North American Birds of Prey*, pt. 1. Dover Publications, Inc., New York, N.Y., pp. 145–146.

Bolton, H. E. 1936. *Rim of Christiandom: A Biography of Eusebio Francisco Kino, Pacific Coast Pioneer.* MacMillan, New York, N.Y.

Brown, B. T. 1988. Breeding ecology of a Willow Flycatcher population in Arizona. *Western Birds* 19:25–34.

Brown, B. T., S. W. Carothers, and R. R. Johnson. 1983. Breeding range expansion of Bell's Vireo in Grand Canyon, Arizona. *Condor* 85:499–500.

Brown, B. T., and M. W. Trosset. 1989. Nesting-habitat relationships of riparian birds along the Colorado River in Grand Canyon, Arizona. *Southwestern Naturalist* 34:260–270.

Brown, D. E., ed. 1982. Biotic communities of the American Southwest—United States and Mexico. *Desert Plants* 4(1–4):1–342.

Brown, H. 1903. Arizona bird notes. *Auk* 20:43–50.

Brush, T. 1983. Cavity use by secondary cavity-nesting birds and response to manipulations. *Condor* 85:461–466.

Brush, T., B. W. Anderson, and R. D. Ohmart. 1983. Habitat selection related to resource availability among cavity-nesting birds. Pp. 88–98 in *Snag Habitat Management: Proceedings of the Symposium*, J. W. Davis, G. A. Goodwin, and R. A. Ockenfels, tech. coords. USDA Forest Service General Technical Report RM-99, Rocky Mountain Forest and Range Experiment Station, Fort Collins, Colo., 226 pp.

Butler, W. I., Jr. 1977. A White-winged Dove nesting study in three riparian communities on the lower Colorado River. Master's Thesis, unpubl., Arizona State University, Tempe, Ariz.

Buttemer, W. A., L. B. Astheimer, W. W. Weathers, and A. M. Hayworth. 1987. Energy savings attending winter-nest use by Verdins (*Auriparus flaviceps*). *Auk* 104:531–535.

Calvin, R. 1951. *Lieutenant Emory Reports: a Reprint of Lieutenant W. H. Emory's Notes of a Military Reconnaissance.* University of New Mexico Press, Albuquerque, N.M., 208 pp.

Carothers, S. W., R. R. Johnson, and S. W. Aitchison. 1974. Population structure and social organization of Southwestern riparian birds. *American Zoologist* 14:97–108.

Castetter, E. F., and W. H. Bell. 1951. *Yuman Indian Agriculture.* University of New Mexico Press, Albuquerque, N.M.

Cody, M. L. 1974. *Competition and the Structure of Bird Communities.* Princeton University Press, Princeton, N.J.

———. 1978. Habitat selection and interspecific territoriality among the syliivid warblers of England and Sweden. *Ecological Monographs* 48:351–396.

———. 1985. *Habitat Selection in Birds.* Academic Press, New York, N.Y.

Conine, K. H. 1982. Avian use of honey mesquite interior and agricultural-edge habitat in the lower Colorado River Valley. Master's Thesis, unpubl., Arizona State University, Tempe, Ariz., 58 pp.

Conine, K. H., B. W. Anderson, R. D. Ohmart, and J. F. Drake. 1978. Response of riparian species to agricultural habitat conversions. Pp. 248–262 in *Strategies for Protection and Management of Floodplain Wetlands and Other Riparian Ecosystems*, R. R. Johnson and J. F. McCormick, tech. coords. USDA Forest Service General Technical Report WO-12, U.S. Department of Agriculture, Washington, D.C., 410 pp.

Cooper, J. G. 1861. New California animals. *Proceedings of the California Academy of Sciences* 2:118–123.

———. 1869. The naturalist in California. The Colorado Valley in winter. *American Naturalist* 3:470–481.

———. 1870. *Ornithology*. Vol. 1, Land Birds. Ed. by S. F. Baird from the manuscript and notes of J. G. Cooper, published by the authority of the Legislature. University Press, Welch, Bigelow, and Co., Cambridge, Mass. Reprinted 1974, Arno Press, New York, N.Y., 592 pp.

Coues, E. 1866. List of birds of Fort Whipple, Arizona with which are incorporated all other species ascertained to inhabit the territory, with brief critical and field notes, descriptions of new species, etc. *Proceedings of the Academy of Natural Sciences* 18:39–100.

———. 1878. Birds of the Colorado Valley. A repository of scientific and popular information concerning North American ornithology. Part First. Passeres to Laniidae. USDI, U.S. Geological Survey of Territories, Miscellaneous Publications no. 11:xvi + 1–807, Government Printing Office, Washington, D.C.

Crowe, R., and S. B. Brinkerhoff. 1976. *Early Yuma: A Graphic History of Life on the American Nile*. Northland Press, Flagstaff, Ariz.

Davis, J. 1951. Distribution and variation of the Brown Towhees. *University of California Publications in Zoology* 52:1–120.

Dickey, D. R. 1923. Description of a new Clapper Rail from the Colorado River Valley. *Auk* 40:90–94.

Eddleman, W. R. 1989. Biology of the Yuma Clapper Rail in the southwestern U.S. and northwestern Mexico. Final Report, Interagency Agreement No. 4-AA-30-02060. U.S. Bureau of Reclamation, Yuma Projects Office, Yuma, Ariz., 189 pp.

Eley, T. J., Jr., and S. W. Harris. 1976. Notes: fall and winter foods of American Coots along the lower Colorado River. *California Fish and Game* 62(3):225–227.

Emory, W. H. 1848. *Notes of a Military Reconnaissance from Fort Leavenworth, in Missouri, to San Diego, in California, Including Part of the Arkansas, Del Norte, and Gila Rivers. Made in 1846–7, with the Advanced Guard of the "Army of the West."* 30th Congress, 1st Session, Exec. Doc. No. 41. Wendell and Van Benthuysen, Printers, Washington, D.C., 614 pp.

Engel-Wilson, R. W. 1981. Comparison of bird communities in salt cedar and honey mesquite along the lower Colorado, Verde, and Gila rivers. Master's Thesis, unpubl., Department of Zoology, Arizona State University, Tempe, Ariz.

Finch, D. M. 1981. Nest predation of Abert's Towhees by coachwhips and roadrunners. *Condor* 83:389.

———. 1982. Rejection of cowbird eggs by Crissal Thrashers. *Auk* 99:719–724.

———. 1983a. Seasonal variation in nest placement by Abert's Towhees. *Condor* 85:111–113.

———. 1983b. Brood parasitism of the Abert's Towhee: timing, frequency, and effects. *Condor* 85:355–359.

———. 1984. Paternal expenditure of time and energy in the Abert's Towhee. *Auk* 101:473–486.

Flores, R. E., and W. R. Eddleman. 1988. Ecology of the Black Rail in southwestern Arizona. Progress report for the period 1 September 1987–31 May 1988.

Mittry Lake, Yuma, Arizona. Report submitted to U.S. Bureau of Reclamation, Yuma Projects Office, Yuma, Ariz.

Forbes, J. D. 1965. *Warriors on the Colorado: the Yumas of the Quechan Nation and Their Neighbors.* University of Oklahoma Press, Norman, Okla., 378 pp.

Forde, C. D. 1931. Ethnography of the Yuma Indians. *University of California Publications in American Archaeology and Ethnology* 28(4), Berkeley, Calif.

Foreman, G., ed. 1941. *A Pathfinder in the Southwest. The Itinerary of Lieutenant A. W. Whipple During His Explorations for a Railway Route from Fort Smith to Los Angeles in the Years 1853 and 1854.* University of Oklahoma Press, Norman, Okla., 298 pp.

Fradkin, P. L. 1981. *A River No More: the Colorado River and the West.* University of Arizona Press, Tucson, Ariz., 360 pp.

Franzreb, K. E., and R. D. Ohmart. 1978. The effects of timber harvesting on breeding birds in a mixed-coniferous forest. *Condor* 80:431–441.

Fretwell, S. D. 1969. Dominance behavior and winter habitat distribution in juncos (*Junco hyemalis*). *Bird-Banding* 40:1–25.

———. 1972. Populations in a seasonal environment. *Monographs in Population Biology* no. 5, Princeton University Press, Princeton, N.J., 217 pp.

Gaines, D., and S. A. Laymon. 1984. Decline, status, and preservation of the Yellow-billed Cuckoo in California. *Western Birds* 15:49–80.

Garrett, K., and J. Dunn. 1981. *Birds of Southern California.* Los Angeles Audubon Society, Los Angeles, Calif.

Gatz, T. A., M. D. Jakle, and G. Monson. 1987. First nesting record and current status of the Black-shouldered Kite in Arizona. *Western Birds* 16:57–61.

Glinski, R. L. 1982. The Red-shouldered Hawk (*Buteo lineatus*) in Arizona. *American Birds* 36:801–803.

Goldstein, D. L. 1984. The thermal environment and its constraint on activity of desert quail in summer. *Auk* 101:542–550.

Grant, G. S. 1982. Avian incubation: egg temperature, nest humidity and behavioral thermoregulation in a hot environment. *Ornithological Monographs* 30:1–75.

Grinnell, J. 1914. An account of the mammals and birds of the lower Colorado Valley with especial reference to the distributional problems presented. *University of California Publications in Zoology* 12(4):51–294.

———. 1917a. Field tests of theories concerning distributional control. *American Naturalist* 51:115–128.

———. 1917b. The niche-relationships of the California Thrasher. *Auk* 34:427–433.

———. 1927. The designation of birds' ranges. *Auk* 44:322–325.

———. 1931. The type locality of the Verdin. *Condor* 30:163–168.

Grinnell, J., and A. H. Miller. 1944. The distribution of the birds of California. *Pacific Coast Avifauna* no. 27, 615 pp.

Halterman, M. D., S. A. Laymon, and M. J. Whitfield. 1989. Status and distribution of the Elf Owl in California. *Western Birds* 20:71–80.

Hanson, H. C. In press. *The White-cheeked Geese.* University of Southern Illinois Press, Carbondale, Ill.

Hastings, J. R., and R. M. Turner. 1964. *The Changing Mile: An Ecological*

Study of Vegetation Changes with Time in the Lower Mile of an Arid and Semiarid Region. University of Arizona Press, Tucson, Ariz., 317 pp.

Haughey, R. A. 1986. Diet of desert-nesting western White-winged Doves *Zenaida asiatica mearnsi.* Master's Thesis, unpubl., Arizona State University, Tempe, Ariz.

Higgins, A. E., and R. D. Ohmart. 1981. Riparian habitat analysis: Tonto National Forest. USDA Forest Service, Albuquerque, N.M.

Holmes, R. T., R .E. Bonney, Jr., and S. W. Pacala. 1979. Guild structure of the Hubbard Brook bird community: a multivariate approach. *Ecology* 60:512–520.

Hubbard, J. P. 1973. Avian evolution in aridlands of North America. *The Living Bird* 12:155–196.

Hunter, W. C. 1984. Status of nine bird species of special concern along the Colorado River. Nongame Wildlife Investigations, Wildlife Management Branch, Administrative Report 84–2, California Fish and Game Department, Sacramento, Calif., 63 pp.

———. 1988. Dynamics of bird species assemblages along a climatic gradient: a Grinnellian niche approach. Master's Thesis, unpubl., Department of Zoology, Arizona State University, Tempe, Ariz.

Hunter, W. C., B. W. Anderson, and R. D. Ohmart. 1987a. Avian community structure changes in a mature floodplain forest after extensive flooding. *Journal of Wildlife Management* 51:493–500.

———. 1987b. Status of breeding riparian-obligate birds in Southwestern riverine systems. *Western Birds* 18:10–18.

———. 1988. Use of exotic saltcedar (*Tamarix chinensis*) by birds in arid riparian systems. *Condor* 90:113–123.

Ives, J. C. 1861. *Report Upon the Colorado River of the West.* Explored in 1857 and 1858 by Lieutenant Joseph C. Ives, Corps of Topographical Engineers, under the direction of the office of explorations and surveys, A. A. Humphreys, Captain Topographical Engineers in charge. By order of the Secretary of War, 36th Congress, 1st Session, House Exec. Doc. no. 90, Government Printing Office, Washington, D.C., 353 pp.

Jaeger, E. C. 1948. Does the Poor-will "hibernate"? *Condor* 50:45–46.

———. 1949. Further observations on the hibernation of the Poor-will. *Condor* 51:105–109.

James, F. C. 1971. Ordinations of habitat relationships among breeding birds. *Wilson Bulletin* 83:215–236.

James, F. C., R. F. Johnston, N. O. Wamer, G. J. Niemi, and W. J. Boecklin. 1984. The Grinnellian niche of the Wood Thrush. *American Naturalist* 124:17–30.

Johnson, N. K., and J. A. Marten. 1988. Evolutionary genetics of flycatchers. II. Differentiation in *Empidonax difficilis* complex. *Auk* 105:177–191.

Kaufman, K. 1977. The changing seasons: An intimate look at Kathleen and other avian phenomena of autumn, 1976. *American Birds* 31:142–152.

Kelly, W. H. 1977. Cocopah ethnogeography. *Anthropological Papers of the University of Arizona* no. 29. University of Arizona, Tucson, Ariz.

Laurenzi, A. W., B. W. Anderson, and R. D. Ohmart. 1982. Wintering biology of Ruby-crowned Kinglets in the lower Colorado River Valley. *Condor* 84:385–398.

Laymon, S. A., and M. D. Halterman. 1987. Distribution and status of the Yellow-billed Cuckoo in California: 1986–1987. Draft Administration Report, Wildlife Management Division, Nongame Bird and Mammal Section, California Department of Fish and Game, Contract No. C-1845.

Lemly, A. D., and G. J. Smith. 1987. Aquatic cycling of selenium: implications for fish and wildlife. *Fish and Wildlife Service Leaflet* 12, USDI Fish and Wildlife Service, Washington, D.C., 10 pp.

Lish, J. W., and W. G. Voeiker. 1986. Field identification aspects of some Red-tailed Hawk subspecies. *American Birds* 40:197–202.

MacArthur, R. H. 1958. Population ecology of some warblers of northeastern coniferous forests. *Ecology* 39:599–619.

MacArthur, R. H., J. W. MacArthur, and J. Preer. 1962. On bird species diversity. II. Prediction of bird censuses from habitat measurements. *American Naturalist* 96:167–174.

McLean, D. D. 1969. Some additional records of birds in California. *Condor* 71:433–434.

Marshall, J. T., Jr. 1955. Hibernation in captive goatsuckers. *Condor* 57:129–134.

———. 1960. Interrelationships of Abert's and Brown Towhees. *Condor* 62:49–64.

———. 1967. Parallel variation in North and Middle American screech-owls. *Monographs of the Western Foundation of Vertebrate Zoology* 1:1–72.

Mearns, E. A. 1894. Unpublished field notes and general notes. From the files of Robert D. Ohmart, Center for Environmental Studies, Arizona State University, Tempe, Ariz.

———. 1907. Mammals of the Mexican Boundary of the United States. A descriptive catalogue of the species of mammals occurring in that region; with a general summary of the natural history, and a list of trees. Smithsonian Institution, *U.S. Natural Museum Bulletin* 56, Government Printing Office, Washington, D.C., 530 pp.

Meents, J. K., B. W. Anderson, and R. D. Ohmart. 1981. Vegetation characteristics associated with Abert's Towhee numbers in riparian habitats. *Auk* 98:818–827.

———. 1982. Vegetation relationships and food of Sage Sparrows wintering in honey mesquite habitat. *Wilson Bulletin* 94:129–138.

Miller, L. 1925. Food of the Harris Hawk. *Condor* 27:71–72.

———. 1930. Further notes on the Harris Hawk. *Condor* 32:210.

Monson, G. 1948a. Egrets nesting along Colorado River. *Auk* 65:603–607.

———. 1948b. The starling in Arizona. *Condor* 50:45.

———. 1949. Recent notes from the lower Colorado River Valley of Arizona and California. *Condor* 51:262–265.

Monson, G., and A. Phillips. 1981. *Revised Checklist of Arizona Birds*. University of Arizona Press, Tucson, Ariz., 240 pp.

Nuechterlein, G. L. 1981. Courtship behavior and reproductive isolation between Western Grebe color morphs. *Auk* 98:335–349.

Oberholser, H. C. 1975. *The Bird Life of Texas*, vol. 2. University of Texas Press, Austin, Tex.

Ohlendorf, H. M., C. M. Bunck, T. W. Aldrich, and J. F. Moore. 1986a. Relationships between selenium concentrations and avian reproduction. *Transactions of the 51st North American Natural Resources Conference*:330–342.

Ohlendorf, H. M., D. J. Hoffman, M. K. Saiki, and T. W. Aldrich. 1986b. Embryonic mortality and abnormalities of aquatic birds: apparent impacts by selenium from irrigation drainwater. *The Science of the Total Environment* 52:49–63.

Ohmart, R. D., B. W. Anderson, and W. C. Hunter.˙1985. Influence of agriculture on waterbird, wader, and shorebird use along the lower Colorado River. Pp. 117–122 in *Riparian Ecosystems and Their Management: Reconciling Conflicting Uses*, R. R. Johnson, C. D. Ziebell, D. R. Patton, P. F. Ffolliott, and R. H. Hamre, tech. coords. USDA Forest Service General Technical Report RM-120, Rocky Mountain Forest and Range Experiment Station, Fort Collins, Colo.

———. 1988. The ecology of the lower Colorado River from Davis Dam to the Mexico-United States International Boundary: a community profile. *U.S. Fish and Wildlife Service Biological Report* 85(7.19), 296 pp.

Ohmart, R. D., W. O. Deason, and C. Burke. 1977. A riparian case history: the Colorado River. Pp. 35–47 in *Importance, Preservation and Management of Riparian Habitat: A Symposium*, R. R. Johnson and D. A. Jones, tech. coords. USDA Forest Service General Technical Report RM-43, Rocky Mountain Forest and Range Experiment Station, Fort Collins, Colo., 217 pp.

Ohmart, R. D., W. O. Deason, and S. J. Freeland. 1975. Dynamics of marsh land formation and succession along the lower Colorado River and their importance and management problems as related to wildlife in the arid Southwest. *Transactions of the 40th North American Wildlife and Natural Resources Conference* 1975:240–251.

Ohmart, R. D., and P. M. Smith. 1973. *North American Clapper Rail Literature Survey*. Bureau of Reclamation, Lower Colorado Region, Boulder City, Nev.

Ohmart, R. D., and R. E. Tomlinson. 1977. Foods of western Clapper Rails. *Wilson Bulletin* 89:332–336.

Phillips, A. R. 1948. Geographic variation in *Empidonax traillii*. *Auk* 65:507–514.

———. 1950. The Great-tailed Grackles of the Southwest. *Condor* 52:78–81.

———. 1986. *Known Birds of North and Middle America*. Pt. I, Hirundinidae to Mimidae; Certhiidae. A. R. Phillips, Denver, Colo., 259 pp.

Phillips, A. R., J. Marshall, and G. Monson. 1964. *The Birds of Arizona*. University of Arizona Press, Tucson, Ariz., 212 pp.

Pleasants, B. Y. 1979. Adaptive significance of the variable dispersion pattern of breeding Northern Orioles. *Condor* 81:28–34.

Prater, A. J., J. H. Marchant, and J. Vuorinen. 1977. *Guide to the Identification and Aging of Holarctic Waders*. British Trust for Ornithology, Field Guide Seventeen, Beech Grove, Tring, Herts, 168 pp.

Radtke, D. B., W. G. Kepner, and R. Effertz. 1988. Reconnaissance investigation of water quality, bottom sediment, and biota associated with irrigation drainage in the lower Colorado River Valley, Arizona–California, and Nevada, 1986–1987. Department of the Interior, U.S. Geological Survey Water–Resources Investigation Report 88:4002.

Ratti, J. T. 1979. Reproductive separation and isolating mechanisms between sympatric dark- and light-phased Western Grebes. *Auk* 96:573–586.

Rea, A. M. 1983. *Once a River*. University of Arizona Press, Tucson, Ariz., 285 pp.

Reisner, M. 1986. *Cadillac Desert: The American West and Its Disappearing Water*. Viking Penguin, Inc., New York, N.Y.

Remsen, J. V. 1978. Bird species of special concern in California. Wildlife Management Branch, Administrative Report 78-1. California Fish and Game Department, Sacramento, Calif., 54 pp.

Repking, C. F., and R. D. Ohmart. 1977. Distribution and density of Black Rail populations along the lower Colorado River. *Condor* 79:486–489.

Rice, J., B. W. Anderson, and R. D. Ohmart. 1980. Seasonal habitat selection by birds in the lower Colorado River Valley. *Ecology* 61:1402–1411.

———. 1984. Comparison of the importance of different habitat attributes of avian community organization. *Journal of Wildlife Management* 48:895–911.

Rice, J., R. D. Ohmart, and B. W. Anderson. 1983a. Habitat selection attributes of an avian community: a discriminant analysis investigation. *Ecological Monographs* 53:263–290.

———. 1983b. Turnovers in species composition of avian communities in contiguous riparian habitats. *Ecology* 64:1444–1456.

———. 1986. Limits in a data-rich model: modeling experience with habitat management on the Colorado River. Pp. 79–86 in *Wildlife 2000: Modeling Habitat Relationships of Terrestrial Vertebrates*, J. Verner, M. L. Morrison, and C. J. Ralph, eds. An international symposium held at Stanford Sierra Camp, Fallen Leaf Lake, California, 7–11 October 1984. University of Wisconsin Press, Madison, Wis.

Robinson, T. W. 1965. Introduction, spread, and areal extent of saltcedar (*Tamarix*) in the western states. *U.S. Geological Survey Professional Paper* 491-A, Government Printing Office, Washington, D.C., 12 pp.

Root, R. B. 1967. The niche exploitation pattern of the Blue-gray Gnatcatcher. *Ecological Monographs* 37:317–350.

Root, T. 1988. Energy constraints on avian distribution and abundance. *Ecology* 69:330–339.

Rosenberg, K. V. 1980. Breeding bird community organization in a desert riparian forest. Master's Thesis, unpubl., Arizona State University, Tempe, Ariz.

Rosenberg, K. V., R. D. Ohmart, and B. W. Anderson. 1982. Community organization of riparian breeding birds: response to an annual resource peak. *Auk* 99:260–274.

Rosenberg, K. V., S. B. Terrill, and G. R. Rosenberg. 1987. Value of suburban habitats to desert riparian birds. *Wilson Bulletin* 99:642–654.

Rotenberry, J. T. 1980. Dietary relationships among shrubsteppe passerine birds: competition or opportunism in a variable environment? *Ecological Monographs* 50:93–110.

Roth, R. R. 1976. Spatial heterogeneity and bird species diversity. *Ecology* 57:773–782.

Rothstein, S. I. 1977. Cowbird parasitism and egg recognition of the Northern Oriole. *Wilson Bulletin* 89:21–32.

Russell, S. M. 1969. Regulation of egg temperatures by incubating White-winged Doves. Pp. 107–112 in *Physiological Systems in Semiarid Environments*, C. Clayton Hoff and M. L. Riedesel, eds. University of New Mexico Press, Albuquerque, N.M.

Sellers, W. D., and R. H. Hill, eds. 1974. *Arizona Climate, 1931–1972*. University of Arizona Press, Tucson, Ariz.

Serena, M. 1986. Distribution, habitat preferences, and reproductive success of Arizona Bell's Vireo (*Vireo bellii arizonae*) along the lower Colorado River in 1981. Final report to California Department of Fish and Game, Endangered, Threatened, and Rare Wildlife, Project E-W-5, Job IV-38.1.

Short, L. L., Jr. 1965. Hybridization in the flickers (*Colaptes*) of North America. *Bulletin of the American Museum of Natural History* 129:311–428.

———. 1982. Woodpeckers of the world. *Delaware Museum of Natural History Monograph Series* no. 4., Greenville, Del.

Shugart, H. H., and D. James. 1973. Ecological succession of breeding bird populations in northwestern Arkansas. *Auk* 90:62–77.

Smith, P. M. 1975. Habitat requirements and observations on the Clapper Rail. Master's Thesis, unpubl., Arizona State University, Tempe, Ariz.

Stager, K. E. 1965. An exposed nocturnal roost of migrant Vaux's Swifts. *Condor* 67:81–82.

Stegner, W. 1953. *Beyond the Hundredth Meridian*. Houghton Mifflin Co., Boston, Mass.

Stephens, F. 1903. Bird notes from eastern California to western Arizona. *Condor* 5:75–78.

Storer, R. W., and G. L. Neuchterlein. 1985. An analysis of plumage and morphological characters of two color forms of Western Grebes (*Aechmophorus*). *Auk* 102:102–119.

Strong, Jr., D. R., D. Simberloff, L. G. Abele, and A. B. Thistle, eds. 1984. *Ecological Communities: Conceptual Issues and the Evidence*. Princeton University Press, Princeton, N.J.

Swarth, H. S. 1914. A distributional list of the birds of Arizona. *Pacific Coast Avifauna* 10:1–133.

Sykes, G. 1937. The Colorado Delta. *American Geographical Society Special Publications* no. 19, W.L.G. Joerg, ed. Published jointly by Carnegie Institute, Washington, D.C., and American Geographical Society, New York, N.Y., 193 pp.

Terrill, S. B., and R. L. Crawford. 1988. Additional evidence of nocturnal migration by Yellow-rumped Warblers in winter. *Condor* 90:261–263.

Terrill, S. B., and R. D. Ohmart. 1984. Facultative extension of fall migration by Yellow-rumped Warblers. *Auk* 101:427–438.

Todd, R. L. 1987. A saltwater marsh hen in Arizona. A Federal Aid Project W-95-R Completion Report, March 1987, Arizona Game and Fish Department, Phoenix, Ariz.

Tomoff, C. S. 1974. Avian species diversity in desert scrub. *Ecology* 55:396–403.

Unitt, P. 1987. *Empidonax traillii extimus*: An endangered subspecies. *Western Birds* 18:137–192.

Van Rossem, A. J. 1942. Four new woodpeckers from the western United States and Mexico. *Condor* 44:22–26.

———. 1946. An isolated colony of the Arizona Cardinal in Arizona and California. *Condor* 48:247–248.

Voss-Roberts, K. A. 1984. Nest-site characteristics and source of egg loss in Yellow-headed Blackbird (*Xanthocephalus xanthocephalus*). Master's Thesis, unpubl., Arizona State University, Tempe, Ariz.

Walsberg, G. E. 1975. Digestive adaptations of *Phainopepla nitens* associated with the eating of mistletoe berries. *Condor* 77:169–174.

———. 1977. Ecology and energetics of contrasting social systems in *Phainopepla nitens* (Aves: Ptilogonatidae). *University of California Publications in Zoology* 108:1–63.

Walsberg, G. E., and K. A. Voss-Roberts. 1983. Incubation in desert-nesting doves: mechanisms for egg cooling. *Physiological Zoology* 56:88–93.

Whipple, A. W. 1856. Pt. I, Itinerary. Pp. 1–136 in *Report of the Explorations and Surveys, to Ascertain the Most Practicable and Economical Route for a Railroad from the Mississippi River to the Pacific Ocean, 1853–4.* 33d Congress, 2d Session, House of Representative Exec. Doc. no. 91, Vol. III, Pt. 1, 1853–4.

Whitney, B., and K. Kaufman. 1985. The *Empidonax* challenge. Pt. II, Least, Hammond's, and Dusky Flycatchers. *Birding* 17:277–287.

Wiens, J. A. 1974. Habitat heterogeneity and avian community structure in grasslands. *American Midland Naturalist* 91:195–213.

———. 1977. On competition and variable environments. *American Scientist* 65:590–597.

Wilder, H. E. 1916. Some distributional notes on California birds. *Condor* 18:127–128.

Wiley, L. 1916. Bird notes from Palo Verde, Imperial Co., California. *Condor* 18:231–232.

———. 1917. Nesting of the Harris Hawk in southeastern California. *Condor* 19:142.

Witzeman, J., J. P. Hubbard, and K. Kaufman. 1978. Southwest region. *American Birds* 32:239–243.

Younker, G. L., and C. W. Andersen. 1986. Mapping methods and vegetation changes along the lower Colorado River between Davis Dam and the border with Mexico. Final report to U.S. Bureau of Reclamation, Lower Colorado Region, Boulder City, Nev. 21 pp.

Zimmerman, D. A. 1973. Range expansion of Anna's Hummingbird. *American Birds* 27:827–835.

Zink, R. M. 1988. Evolution of Brown Towhees: allozymes, morphometrics, and species limits. *Condor* 90:72–82.

Name Index

Subject Index

Boldface numbers indicate page where species account may be found.

ABOUT THE AUTHORS

Kenneth V. Rosenberg obtained a Bachelor's degree in Wildlife Science at Cornell University, then moved to Arizona to study birds along the lower Colorado River and to complete a Master's degree in Zoology at Arizona State University under Robert D. Ohmart. After working for several years on wildlife studies for the University of California and the U.S. Forest Service in northern California, Ken went to Louisiana State University to work on his doctorate in Ornithology. He recently finished his studies of the feeding habits of tropical rainforest birds, which included work in remote areas of Peru, Bolivia, and Costa Rica.

Robert D. Ohmart is a Southwesterner by birth, training, and desire. He has been actively involved for more than 30 years in natural resources management, with a strong interest in the ecological needs of birds. He is a Professor of Zoology at Arizona State University; his research is conducted through the Center for Environmental Studies. For the past 17 years, Bob has been deeply involved in research on environmental problems that involve large river systems in the Southwest. Extreme water management activities on these rivers has dramatically altered the riparian or plant communities, which has caused major changes in bird populations and other wildlife dependent on these habitats. Thirteen years of in-depth ecological research by him and his colleagues on the abundance, distribution, food habits, and preferred habitats of the birds along the lower Colorado River form the backbone of this regional bird book.

William C. (Chuck) Hunter first became actively involved with birds and their conservation at the age of 10. His conservation ethic and knowledge of birds was cultivated during his teenage years in Jacksonville, Florida, through the National Audubon Society's Junior Audubon Program. He moved to Flagstaff, Arizona, in 1978, where he received his Bachelor of Science degree from Northern Arizona University in 1980. From 1980 to 1983, he worked as a field biologist on the Pecos River and the lower Colorado River, conducting avian inventories and other research activities addressing riparian management issues. He received his M. S. degree from Arizona State University in 1988, expanding upon his interests generated on the Colorado and Pecos rivers. During his years at Arizona State University, Chuck also worked with a number of federal and state agencies, most recently the U.S. Fish and Wildlife Service, primarily on bird and riparian management issues. He was instrumental in organizing the Arizona Riparian Council, founded in 1985, which presently serves as a facilitator of information about research needs and management strategies involving riparian systems throughout Arizona. Chuck now works for the U.S. Fish and Wildlife Service in Atlanta, Georgia.

Bertin W. Anderson came to the lower Colorado River research team from the cold Midwest in 1974 and has worked on ecological research in large river systems ever since. As an Adjunct Research Associate in the Center for Environmental Studies at Arizona State University, he has participated in and directed all of the field research on the lower Colorado River. His dedication and meticulous efforts in the revegetation of riparian communities along the river ensured their success. More recently, he has worked with The Nature Conservancy on the Kern River in California and the California Department of Fish and Game on the lower Colorado River to revegetate large acreages of habitat for wildlife.